Metallurgy Essentials

Metallurgy Essentials

Edited by **Ricky Peyret**

NY RESEARCH PRESS

New York

Published by NY Research Press,
23 West, 55th Street, Suite 816,
New York, NY 10019, USA
www.nyresearchpress.com

Metallurgy Essentials
Edited by Ricky Peyret

International Standard Book Number: 978-1-63238-321-1 (Hardback)

Printed in China.

Contents

Preface

Metallurgy is a branch of materials engineering and materials science, which is concerned with the physical and the chemical behaviour of metallic elements, their alloys and their inter-metallic compounds. Metallurgy is also referred to as the technology of metals. It is the branch of science which is used for engineering of metals and the production of metals for industrial as well as manufacturing processes. Metallurgy is a process which is categorised into two types: ferrous metallurgy which is used for iron extraction from its ores and non-ferrous metallurgy, which includes extraction of other metals. They are also referred as black metallurgy and coloured metallurgy respectively.

Metallurgy is essentially the science of refining and extracting metal from their ores, but in contemporary times, this field has dug deeper than what this definition describes. It is no longer the science to refine and extract metal from ores only, but it is also the science which explains the properties, internal structure as well as the behavior of metals. Metallurgy also describes the properties of metals, which can be altered by using different techniques.

This book contains chapters which describe different concepts of metallurgy and also its applications in various fields. I especially wish to acknowledge the contributing authors. Without them, a work of this magnitude would simply not have been feasible. I thank them for allocating much of their very valuable time to this project. Not only do I appreciate their participation, but also their adherence as a group to the time parameters set for this publication. I must also acknowledge the team at the publishing house, who has done a tremendous job. Last but not the least, I wish to thank my family and friends, who have supported me in my life through everything. I do believe that the contents in this book will be helpful to many researchers in this field around the world.

Editor

1

Fabrication of Single-Phase NiTi by Combustion Synthesis of Mechanically Activated Powders

S. Mousavi Nasab,[1] M. Aboutalebi,[1] S. H. Seyedein,[1] A. Molavi Kakhki,[2] and J. Vahdati Khaki[3]

[1] *Centre of Excellence for Advanced Materials & Processing (CEAMP), Department of Metallurgy and Materials Engineering, Iran University of Science & Technology, Tehran 16846-13114, Iran*
[2] *Department of Materials Science & Engineering, Sharif University of Technology, Tehran 11365-11155, Iran*
[3] *Department of Materials Science and Engineering, Ferdowsi University of Mashhad, Mashhad 91779-48974, Iran*

Correspondence should be addressed to S. Mousavi Nasab, saba.mousavi@gmail.com

Academic Editor: Stefano Gialanella

Single-phase NiTi was fabricated through the thermal explosion mode of combustion synthesis of mechanically activated powders. Combustion and ignition temperatures of combustion synthesis were investigated in different milling times. In this process, equiatomic powder mixtures of nickel and titanium were activated by planetary ball mill and pressed into disk-shaped pellets then heated in a tube furnace, while temperature-time profile was recorded. X-ray diffraction analysis (XRD) was performed on milled powders as well as synthesized samples. Scanning electron microscopy (SEM) was also used to study the microstructural evolution during milling. The results showed that there was a threshold milling time to obtain single-phase NiTi. It was also seen that the ignition temperature and combustion temperature were reduced significantly by increasing milling time.

1. Introduction

NiTi alloy combines the characteristics of shape memory effect and superelasticity with excellent corrosion resistance, wear resistance, mechanical properties, and good biocompatibility [1]. They have been practically used for couplings, actuators, and smart materials, as well as external and internal biomedical applications, for example, orthodontic arch wires, catheters, and orthopaedic implants, and in cardiovascular surgery, and so forth [2]. The extent of the shape memory effect and the temperature range over which it is exhibited depend strongly on the composition of the alloy, and in order to realize the maximum benefits of this effect, it is essential to have an alloy of exact stoichiometry and very good homogeneity [3].

Conventionally, NiTi intermetallics are produced by arc or induction melting followed by hot working and forming [2]. Arc melting requires multiple remelts to achieve sufficient homogeneity, while induction melting has the drawback of oxygen contamination [4]. The powder metallurgy technique has also been used for the NiTi fabrication and offers the ability to avoid the problems of casting defects due to segregation and to produce a variety of component shapes while minimizing subsequent machining operations [5]. Among the mentioned techniques, combustion synthesis (CS) has advantages of time and energy savings that make it an attractive alternative to the conventional methods for the production of various classes of materials [6]. Very homogeneous alloys of desirable stoichiometry also can be synthesized by this route, thus eliminating the subsequent thermomechanical processing for homogenization [7]. Moreover the reacted products are purer and have better mechanical properties than those prepared by conventional methods [8].

Several researches have been done on fabricating NiTi via combustion synthesis [1, 2, 7] through which the obtained products were typically a mixture of several intermetallic phases, that is, NiTi, Ti_2Ni, and Ni_3Ti. The Gibbs free energies of these phases indicate that the driving force for the formation of Ti_2Ni and Ni_3Ti is stronger than that of NiTi. Therefore, even though the formation of NiTi occurs, it is unlikely that these secondary phases of Ti_2Ni and Ni_3Ti will entirely disappear [9]. Chu et al. [10] have fabricated porous NiTi shape memory alloy with less undesired phases, using the improved self-propagating high-temperature synthesis (SHS) technique by means of providing the higher heat flux density for SHS reactions using the improved ignition technique and adopting the powder compacts with a larger volume-surface area ratio to reduce heat losses. Biswas also prepared single-phase NiTi, using the thermal explosion mode of combustion synthesis in combination with a controlled postreaction heat treatment [11].

Using mechanical activation before combustion synthesis in order to reduce secondary phases has been reported in Ti-Al [12] and Ta-Si [13] systems. However, the method has not been explored much for preparing NiTi in thermal explosion mode of combustion synthesis. In this paper premechanical activation is applied as to eliminate the Ni_3Ti phase in combustion synthesis of Ni-Ti powder mixture. Moreover the effect of milling time on combustion and ignition temperatures is investigated.

2. Experimental Procedure

The starting materials consisted of high-purity powders of nickel (99.5%, 10 μm) and titanium (98%, 150 μm) supplied by Merck Co., Germany. Ti and Ni powders were weighed, in a glove box under an argon atmosphere, to have a composition of Ni : Ti = 50 : 50 at.%, then comilled inside a planetary ball mill (PM2400) for different times (1, 3, 5, and 10 hours), using steel balls (Φ8 mm) as milling media while the ball-to-powder weight ratio and the rotational speed were 10 : 1 and 200 rpm, respectively. To prevent powders from oxidation during milling, the vials were filled with high purity argon. Ethanol was used as process control agent (PCA) to reduce agglomeration and cold welding of the particles. In addition to the milled powders, a reference mixture of equiatomic elemental powders mixed for 1 h in a Turbula apparatus was prepared. As-milled powders were pressed into disk-shaped pellets using a uniaxial load of 350 MPa. Thermal explosion of the pellets was carried out in a tube furnace, under floating argon at heating rate of 30°C/min. Thermal analysis was carried out using an S type (Pt/Pt-10Rh) thermocouple which had been embedded during compaction of the samples. The thermocouple output was collected online using a data acquisition system which was connected to a computer. In order to identify the possible reaction steps, one of the reactions was quenched on the basis of time-temperature profile results. X-ray measurements on as-milled and as-synthesized samples as well as on quenched sample were performed in a Philips PW1800 diffractometer (Cu Kα radiation: λ = 0.154 nm at 40 kV and 30 mA). The microstructural evolutions of the milled powders were analyzed by scanning electron microscopy (SEM) in a Philips MV2300 at an accelerating voltage of 15 kV.

3. Results and Discussion

3.1. Mechanical Activation. Figure 1 shows the XRD patterns of 50 Ni-50 Ti powder mixtures, after different milling times.

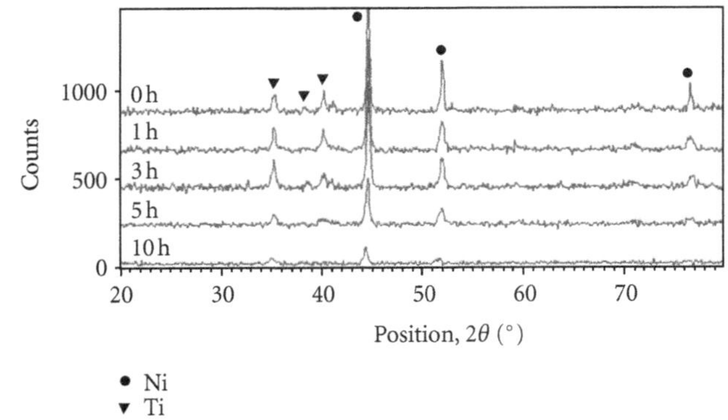

FIGURE 1: XRD patterns of 50 Ni -50 Ti powder mixtures, after different milling times.

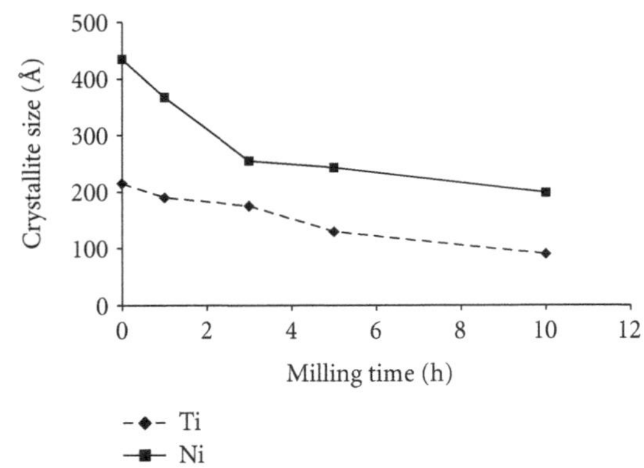

FIGURE 2: Variation of Ni and Ti crystallite size versus milling time.

As it can be seen in Figure 1, no new phase is formed during milling; that is, the alloying has not occurred. So it is to say that the selected conditions and times of the milling are appropriate in order to mechanically activate the powder mixtures.

X-ray peak broadening with increasing the milling times is also observed in Figure 1, which can be due to instrumental effects, crystallite size, and lattice strain in the material. The crystallite size can be determined by measuring the Bragg peak width at half the maximum intensity and using the Scherrer formula:

$$d = \frac{0.9\lambda}{B\cos\theta}, \quad (1)$$

where d is the crystallite size, λ is the wavelength of the X-radiation used, B is the peak width at half the maximum intensity, and θ is the Bragg angle [14]. The calculated crystallite size in different milling times for Ni and Ti is shown in Figure 2.

It can be seen that the crystallite size of Ni and Ti is decreased with increasing the milling time. Displacement of Ni X-ray peaks towards lower angles is observed after 3 h as shown in Figure 3. This indicates that lattice parameter of Ni increases with milling times due to diffusion of larger atoms of Ti into the Ni lattice resulting in an Ni(Ti) solid solution.

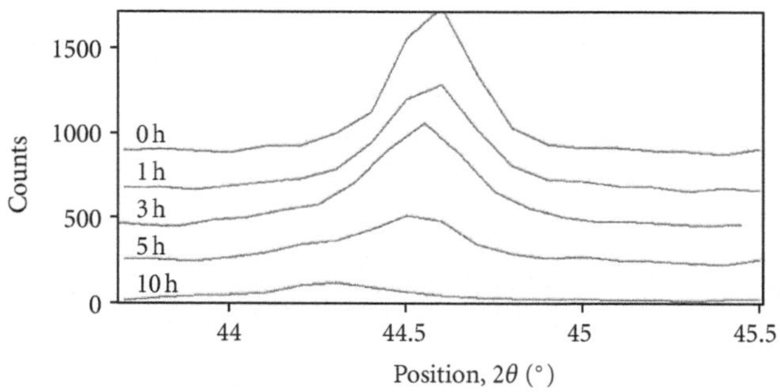

FIGURE 3: Displacement of Ni peak with increasing the milling time.

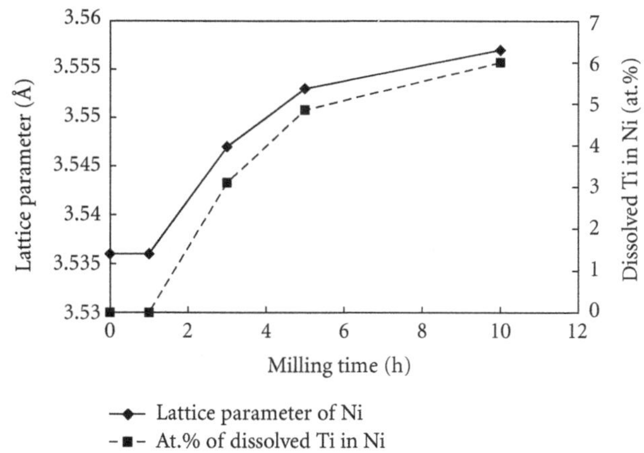

FIGURE 4: Variation of Ni lattice parameter and at.% of dissolved Ti in Ni lattice versus milling time.

According to the fact that the Ni lattice parameter increases by 0.1% per at.% of dissolved Ti [15], the dissolution of Ti in Ni at different milling times was calculated. Figure 4 shows the lattice parameter of Ni and the atomic percentage of dissolved Ti in Ni lattice versus milling time. Dissolution of Ti in Ni occurs after 3 h and increases with increasing milling time as diffusion progresses. Finally after 10 h, the Ni lattice contains 6 at.% of dissolved Ti which is in agreement with the amount of 6 at.% reported by Battezzati et al. [16]. It can be seen in Figure 4 that after milling for 5 h, the variation rate of at.% of dissolved Ti in Ni lattice decreases since it almost reaches a saturation level in the stated milling conditions.

Ni-Ti binary system is considered to be a ductile-ductile system. Benjamin and Volin [17] suggested that in the early stages of milling of ductile components, they get flattened to platelet/pancake shapes by a microforging process. These flattened particles then get cold welded together and form a composite lamellar structure of the constituent metals. With increasing milling time, the composite powder particles get work hardened, consequently getting fragmented. The processes of fracturing and welding lead to formation of micrometric agglomerates consisting of nanosized Ni and Ti convoluted layers. The progress of changes in microstructure of the Ni-Ti powder mixture during milling for 1 h and 5 h is demonstrated by the SEM images in Figures 5 and 6, respectively.

As it is obvious in Figure 5(a), the powder mixture after milling for 1 h consists of 20–40 μm particles with irregular shapes along with the almost equiaxed 3–5 μm particles. The particle with the lamellar structure which can be seen in Figure 5(b) indicates that the composite particles are formed, in accordance with the milling mechanism, after 1 h milling. Meanwhile the irregular particle in Figure 5(b) proves the presence of titanium due to its nonlamellar structure as well as its dimension. The presence of Ti particles in powder mixture after 1 h milling suggests that only a part of primary Ni and Ti particles forms composite particles with lamellar structure.

Figure 6(a) shows that formation of micrometric aggregates with a mean size of 20–40 μm is registered after 5 h milling. A more precise observation (Figure 6(b)) reveals the complicated structure of aggregates and the presence of submicron-sized particles. As pointed out earlier, formation of these micrometric aggregates is typical in the milling of ductile-ductile systems.

3.2. Combustion Synthesis. Figure 7 shows the time-temperature profiles during the heating of powder mixtures which are milled in different times. The intersection of the two slopes leads to the determination of the ignition temperature (T_{ig}) [18]. The time-temperature profiles of samples, milled for 0, 1, and 3 h, exhibit two exothermic peaks. The low temperature exotherm which is much smaller and occurred in earlier times could be related to a precombustion reaction. The high temperature exotherm which is much bigger showed the characteristic sharp rise, corresponding to the main combustion reaction.

The XRD pattern of the products obtained after the first exothermic peak of the sample milled for 0 h is shown in Figure 8.

It could be concluded that the low temperature exotherm is due to the formation of little amounts of NiTi and Ni_4Ti_3 phases. The low temperature exotherm related to a precombustion synthesis is not observed in the time-temperature profiles of the samples milled for 5 and 10 h, which is due to the significant decrease in ignition temperature of the main combustion reaction. It is worth noting that the first exothermic peak of the sample milled for 10 h corresponded to the main combustion reaction since it is sharper and more exothermic than the peaks which were due to the precombustion reactions.

Figure 9 illustrates the variation of ignition and combustion temperature versus milling time. As it can be seen in this figure, T_{ig} decreases almost 400 degrees with the increase in milling time. Such a decrease can be explained by the changes occurred in milling.

The milling process, which causes the intimate mixing of particles on a nanoscale, leads to the reduction in the crystallite size and accumulation of defects in powder particles, which introduce an additional energy to the reactant system in the form of interfacial energy and strain energy. This effectively lowers the activation barriers for reactions.

FIGURE 5: SEM images of the Ni-Ti powder mixture after milling for 1 h: (a) 1000X, (b) 15000X.

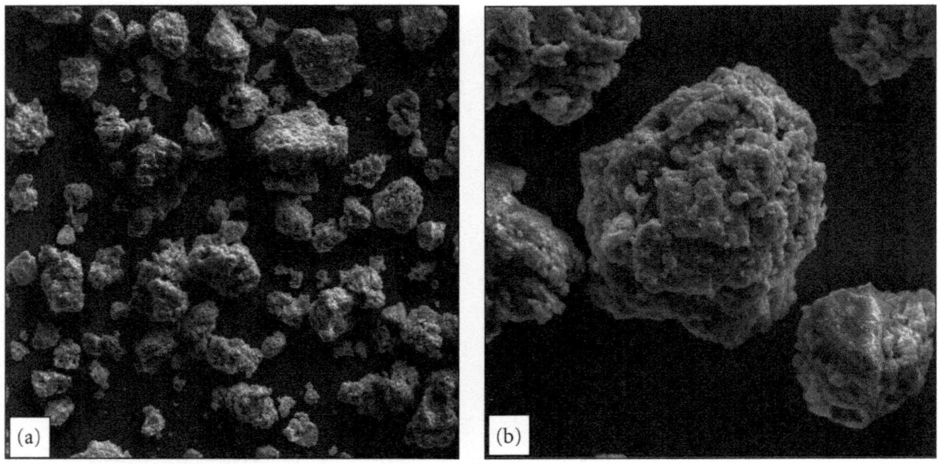

FIGURE 6: SEM images of the Ni-Ti powder mixture after milling for 5 h: (a) 1000X, (b) 5000X.

Moreover, the presence of a variety of crystal defects (i.e., dislocations, vacancies, staking faults, grain boundaries, etc.) enhances the diffusivity of elements [19]. Further, the formation of micrometric aggregates, consisting of nanosized elemental convoluted layers, decreases the diffusion distances [20].

Decrease in diffusion distances as well as activation barrier to NiTi formation reaction occurred during milling causes the ignition temperature to decrease significantly in postcombustion synthesis. According to Figure 9 it can be concluded that the above milling effects reach a maximum level after 5 h of milling in the stated conditions which results in keeping T_{ig} steady by increasing milling time from 5 to 10 h [14].

In a nonadiabatic condition of a combustion synthesis reaction, the maximum temperature of products is called combustion temperature which is equivalent to the adiabatic temperature. Adiabatic temperature (T_{ad}) is mainly dependent on reactants' temperature and reaction enthalpy (2) whereas the latter itself is a function of reactants' temperature. Decreasing in reactants' temperature from 952 K to 569 K by increasing the milling time causes the enthalpy of the reaction to change from −71.9 to −68.3 Kj/mol which is negligible. According to this slight change in enthalpy, the similar trend of ignition and combustion temperature variations shown in Figure 9 and the relation between T_{ig} and T_{ad}, that is, T_{com} presented in (2), it can be concluded that T_{com} is determined specifically by T_{ig}:

$$-\Delta H_f^0(T_0) = \int_{T_0}^{T_{ad}} C_p(\text{products})\,dT. \quad (2)$$

Figure 9 also gives the information that milling, resulting in nonequilibrium condition in powder mixtures, does not change the heat of the formation. This is because the difference between T_{ig} and T_{com} which represents the exothermic feature of the reaction remained constant in different milling times. Figure 10 indicates XRD patterns of

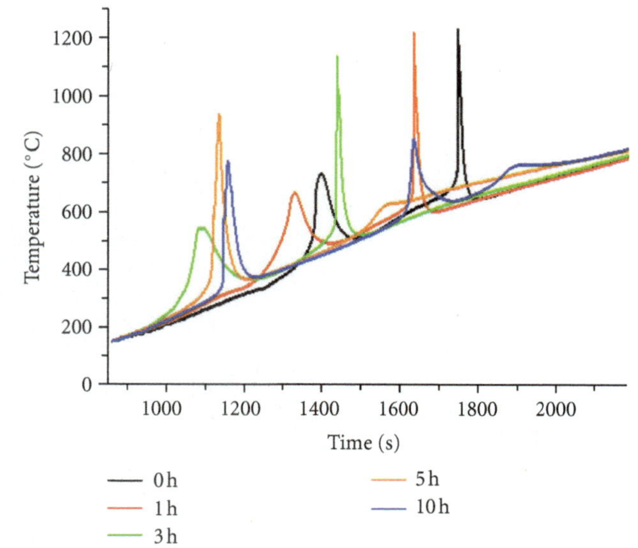

FIGURE 7: Time-temperature profiles of powder mixtures milled in different milling times.

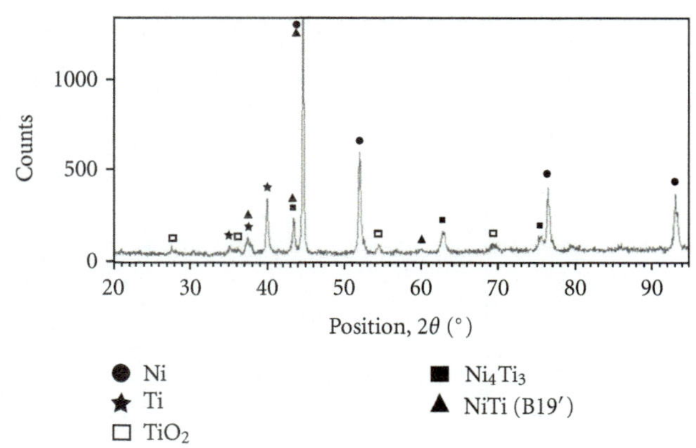

FIGURE 8: The XRD pattern of the products obtained after the first exothermic peak of the sample milled for 0 h.

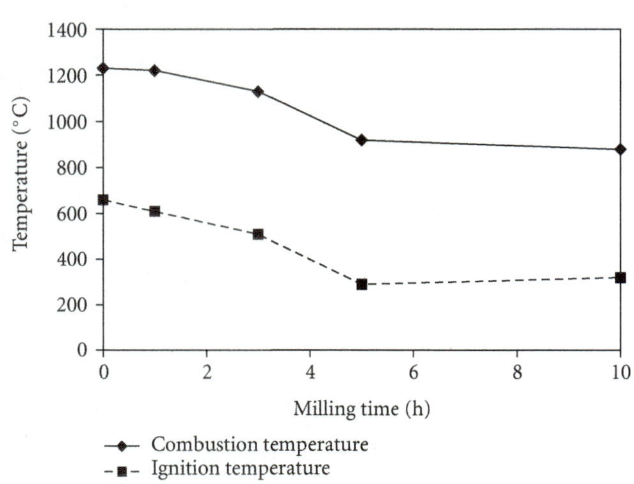

FIGURE 9: Variation of ignition and combustion temperature versus milling time.

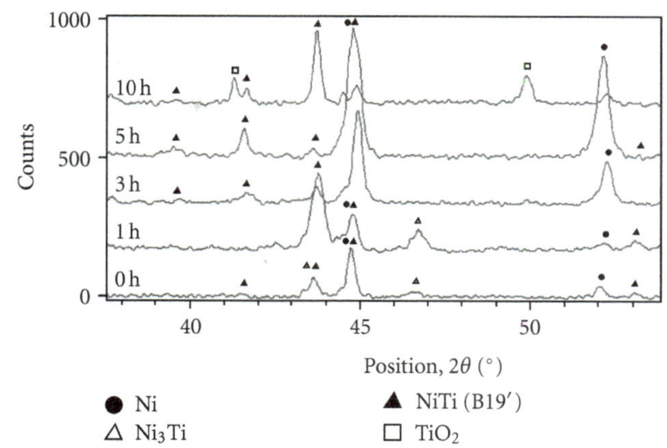

FIGURE 10: XRD patterns of powder mixtures milled in different milling times after combustion synthesis.

powder mixtures milled for different times, after combustion synthesis. As it can be seen, increasing milling time to 3 h prevents the formation of undesired Ni_3Ti phase, in the post-combustion synthesis. In order to explain this, it is to be said that milling for 3 h makes a homogenous powder mixture with less nonstoichiometric zones which reduces the possibility of Ni_3Ti formation in the postcombustion synthesis.

Moreover, as it can be seen from Figure 4, dissolution of Ti in Ni lattice occurred after 3 h milling, which creates the zones in powder mixture where the chemical composition is close to NiTi stoichiometry. The formation of such zones enhances the reactivity of Ni-Ti system which leads to increase in the probability of NiTi formation [18].

High chemical affinity of titanium with oxygen and high surface energy of the milled titanium powders which enhances the reactivity of Ti caused the formation of little amounts of TiO_2, even in a high purity argon atmosphere.

4. Conclusion

In this paper mechanically activated powders were used to fabricate NiTi intermetallic compound through combustion synthesis method. The results showed that with premechanical activation at least for 3 hours, NiTi alloy can be fabricated without undesired Ni_3Ti phase through thermal explosion mode of combustion synthesis. Moreover, decrease in diffusion distances caused by micrometric aggregation formation drastically lowers the ignition temperature and as a result lowers the combustion temperature.

References

[1] C. L. Chua, C. Y. Chung, P. H. Lin, and S. D. Wang, "Fabrication of porous NiTi shape memory alloy for hard tissue implants by combustion synthesis," *Materials Science and Engineering A*, vol. 366, no. 1, pp. 114–119, 2004.

[2] C. L. Yeh and W. Y. Sung, "Synthesis of NiTi intermetallics by self-propagating combustion," *Journal of Alloys and Compounds*, vol. 376, no. 1-2, pp. 79–88, 2004.

[3] G. K. Dey, "Micropyretic synthesis of NiTi in propagation mode," *Acta Materialia*, vol. 51, no. 9, pp. 2549–2568, 2003.

[4] H. Funakubo, *Shape Memory Alloys*, Gorden & Breach, New York, NY, USA, 1987.
[5] M. Bram, A. Ahmad-Khanlou, A. Heckmann, and B. Fuchs, "Powder metallurgical fabrication processes for NiTi shape memory alloy parts," *Materials Science and Engineering A*, vol. 337, no. 1-2, pp. 254–263, 2002.
[6] C. Zanotti, P. Giuliani, A. Terrosu, S. Gennari, and F. Maglia, "Porous Ni-Ti ignition and combustion synthesis," *Intermetallics*, vol. 15, no. 3, pp. 404–412, 2007.
[7] B. Y. Tay, C. W. Goh, Y. W. Gu, and C. S. Lim, "Porous NiTi fabricated by self-propagating high-temperature synthesis of elemental powders," *Journal of Materials Processing Technology*, vol. 202, no. 1–3, pp. 359–364, 2008.
[8] L. Qiang, Y. Jing-Yuan, M. Bai-Chun, and S. Xu-Dong, "BP neural network prediction of the mechanical properties of porous NiTi shape memory alloy prepared by thermal explosion reaction," *Materials Science and Engineering A*, vol. 419, no. 1-2, pp. 214–217, 2006.
[9] C. W. Goh, Y. W. Gu, C. S. Lim, and B. Y. Tay, "Influence of nanocrystalline Ni-Ti reaction agent on self-propagating high-temperature synthesized porous NiTi," *Intermetallics*, vol. 15, no. 4, pp. 461–467, 2007.
[10] C. L. Chu, C. Y. Chung, P. H. Lin, and S. D. Wang, "Fabrication and properties of porous NiTi shape memory alloys for heavy load-bearing medical applications," *Journal of Materials Processing Technology*, vol. 169, no. 1, pp. 103–107, 2005.
[11] A. Biswas, "Porous NiTi by thermal explosion mode of SHS: processing, mechanism and generation of single phase microstructure," *Acta Materialia*, vol. 53, no. 5, pp. 1415–1425, 2005.
[12] E. Medda, F. Delogu, and G. Cao, "Combination of mechanochemical activation and self-propagating behaviour for the synthesis of Ti aluminides," *Materials Science and Engineering A*, vol. 361, no. 1-2, pp. 23–28, 2003.
[13] F. Maglia, C. Milanese, and U. Anselmi-Tamburini, "Combustion synthesis of mechanically activated powders in the Ta-Si system," *Journal of Alloys and Compounds*, vol. 385, no. 1-2, pp. 269–275, 2004.
[14] C. Suryanarayana, "Mechanical alloying and milling," *Progress in Materials Science*, vol. 46, no. 1-2, pp. 1–184, 2001.
[15] W. Gu, C. W. Goh, L. S. Goi, C. S. Lim, and A. E. W. Jarfors, "Solid state synthesis of nanocrystalline and/or amorphous 50Ni-50Ti alloy," *Materials Science and Engineering A*, vol. 392, no. 1-2, pp. 222–228, 2005.
[16] L. Battezzati, G. Cocco, L. Schiffini, and S. Enzo, "Thermal properties of mechanically alloyed Ni50Ti50 powders," *Materials Science and Engineering C*, vol. 97, pp. 121–124, 1988.
[17] J. S. Benjamin and T. E. Volin, "Mechanism of mechanical alloying," *Metallurgical and Materials Transactions*, vol. 5, no. 8, pp. 1929–1934, 1974.
[18] C. Gras, E. Gaffet, F. Bernard, and J. C. Niepce, "Enhancement of self-sustaining reaction by mechanical activation: case of an Fe-Si system," *Materials Science and Engineering A*, vol. 264, no. 1-2, pp. 94–107, 1999.
[19] P. Mossino, "Some aspects in self-propagating high-temperature synthesis," *Ceramics International*, vol. 30, no. 3, pp. 311–332, 2004.
[20] F. Charlot, E. Gaffet, B. Zeghmati, F. Bernard, and J. C. Niepce, "Mechanically activated synthesis studied by X-ray diffraction in the Fe-Al system ," *Materials Science and Engineering A*, vol. 262, no. 1-2, pp. 279–288, 1999.

2

Precipitation of σ Phase Using General Diffusion Equation with Comparison to Vitek Diffusion Model in Dissimilar Stainless Steels

Chih-Chun Hsieh and Weite Wu

Department of Materials Science and Engineering, National Chung Hsing University, 250 Kuo-Kuang Road, Taichung 402, Taiwan

Correspondence should be addressed to Weite Wu, jeromehsieh@gmail.com

Academic Editor: Chih-Ming Chen

This study performs a precipitation examination of the σ phase using the general diffusion equation with comparison to the Vitek model in dissimilar stainless steels during multipass welding. Experimental results demonstrate that the diffusivities (D^{δ}_{Cr}, D^{δ}_{Ni}, and D^{δ}_{Si}) of Cr, Ni, and Si are higher in δ-ferrite than (D^{γ}_{Cr}, D^{γ}_{Ni}, and D^{γ}_{Si}) in the γ phase, and that they facilitate the precipitation of the σ phase in the third pass fusion zone. The Vitek diffusion equation can be modified as follows: $(\text{Diffusivity})_{\text{General}}$ = Modified Constant $\times$ $(\text{Diffusivity})_{\text{Vitek}}$.

1. Introduction

ASSs (austenitic stainless steels) are used widely in high-temperature conditions such as energy conversion systems. For instance, to ensure economy in central power systems, sections of boilers subjected to lower temperatures are constructed from FSS (ferritic stainless steel) for economic reasons [1]. However, ASS and FSS composites have acid, alkali, and high-temperature resistances, and so have many applications in a wide variety of industries, such as chemical, oil refining, artificial fiber, food, and medicine [2].

When stainless steels are heated to temperatures between 700°C and 1000°C for prolonged periods of time, several harmful second phases, such as σ, χ, and π, can precipitate [3–6]. While the σ phase was first observed in the Fe-Cr system, it has also been observed in Fe-Mo, Fe-V, and Cr-Mo-Ni alloy systems [7]. The crystallographic lattice of the σ phase is a tetragonal structure with 30 atoms per unit cell. The σ phase is the most important of these intermetallic phases due to its impact on the mechanical properties, corrosion resistance or weldability of stainless steels, as well as other properties [8–11].

Previous studies on the properties of stainless steels containing the σ phase have focused on microstructural observation and the mechanical property in the fusion zone (FZ) of similar stainless steels [12–18]. Very few researchers have discussed the precipitation of the σ phase through the diffusion theory in the fusion zones (FZ) of dissimilar stainless steels. The precipitation behavior of the σ phase in fusion zones during the welding process is unclear for ferritic and austenitic stainless steels. Hsieh and Wu [19] first reported the precipitation of the σ phase using the Vitek diffusion model in the multipass fusion zone of dissimilar stainless steels.

Autogenous GTAW (gas tungsten arc welding) was used to manufacture dissimilar stainless steels for this present study. The precipitation behavior of the σ phase in the multipass fusion zone of dissimilar stainless steels taking place during the multipass GTAW process has been discussed via a calculation of the general diffusion equation with comparison to the Vitek diffusion model. The objectives were to delineate the precipitation mechanism of the σ phase in dissimilar stainless steels at various welding passes, as well as to contribute to the understanding of the precipitation behavior of the σ phase during the multipass GTAW process in dissimilar stainless steels.

2. Experimental Procedures

Two types of stainless steels, namely, 304 and 430, were employed in the experiment. Both are representatives of

TABLE 1: Chemical composition of the raw samples.

Material	Element (wt%)						
	Cr	Ni	Si	Mn	C	Mo	Fe
304 S.S.	19.2	10.9	1.0	2.0	0.08	1.22	Bal.
430 S.S.	18.3	—	0.75	1.0	0.12	1.18	Bal.

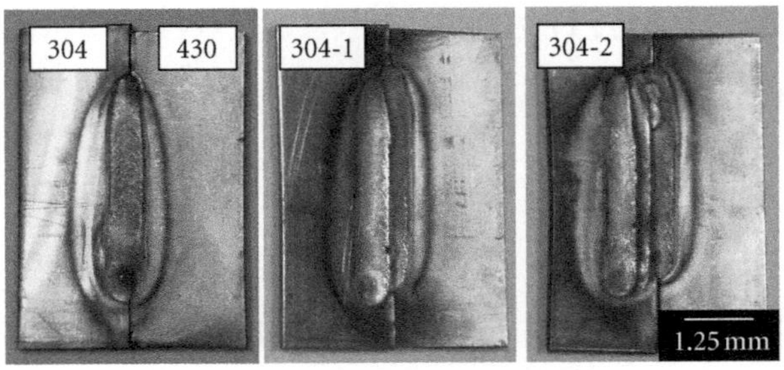

(a)

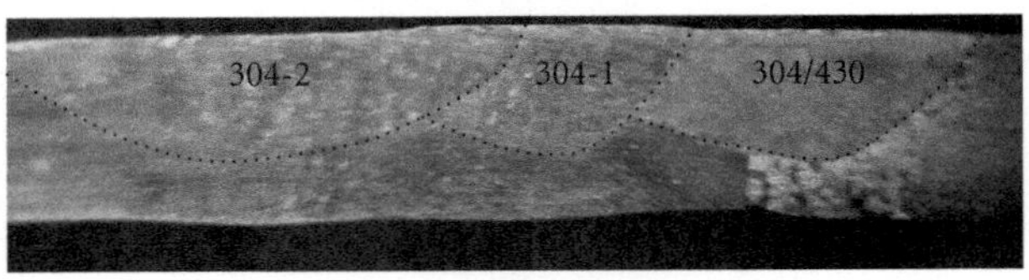

(b)

FIGURE 1: Manufacturing process of the dissimilar stainless steels. (a) Top view; (b) cross-section.

fully austenitic, that is, containing a small number of ferrite phases and fully ferritic microstructures, respectively. Their chemical compositions are listed in Table 1. The dimensions of the raw materials were 70 mm × 25 mm × 3 mm.

The composites of the dissimilar stainless steels were created using an automatic GTAW process. The appearance of the weld metal from the dissimilar stainless steels is shown in Figure 1. The multipass welding (304/430, 304-1, and 304-2) was performed without filler at a welding current of 100 A and an arc voltage of 11 V with a travel speed of 120 mm/min. A series of symbols for 304/430, 304-1, and 304-2 were defined as the first, second, and third pass welds, respectively.

Thin foils for observation with a transmission electron microscope (JEOL JEM-2010, HRTEM) were prepared from the third pass fusion zone. Specimens were thinned to 50 μm by abrasion on SiC papers, and electropolished (Struers Tenupol-5, Twin Jet Electro-Polisher) with a twin-jet unit at 35 V in a mixture of 25% glycerol, 5% perchloric acid, and 75% ethanol at a temperature of $-10°$C. The method of selected area diffraction (SAD) was employed to identify the δ-ferrite, σ phase, and γ phase. Their chemical compositions were identified using an energy dispersive spectrometer (Oxford Link ISIS, EDS).

The peak temperatures and the endothermic and exothermic peaks were examined using a differential scanning calorimeter (NETZSCH DSC 404 C) in order to discuss the precipitation of $\delta \rightarrow \sigma + \gamma_2$. The specimens were heated from 600°C to 1300°C at a heating rate of 20°C·sec^{-1}. Subsequently, the diffusion parameters were calculated using the general diffusion equation to explain the phase transformation of $\delta \rightarrow \sigma$. These diffusion parameters included diffusivity (D) and the atomic jump frequencies of the solute atoms (Γ).

TABLE 2: TEM-EDS analysis of the third pass fusion zone (304-2).

Welding region	Phase	Element (wt%)			
		Fe	Cr	Ni	Si
304-2	δ	60.26	30.11	1.08	5.10
	σ	55.63	35.58	2.04	3.32

3. Results and Discussion

3.1. TEM Identification of the δ-Ferrite and σ Phase. Figures 2(a)–2(d) showed the TEM diffraction pattern of the δ-ferrite and σ phase in the third pass fusion zone (304-2). Lots of grains were found in the third pass fusion zone, as shown in Figure 2(a). Two phases can be observed clearly from a dark field image of Figure 2(b) and are identified as the δ-ferrite and σ phase using an identification of selected area diffraction, as represented in Figures 2(c) and 2(d). From the above described results, the σ phase can precipitate in the δ-ferrite after the third pass welding.

Table 2 showed the TEM-EDS result that the purpose is to analyze the compositions of δ-ferrite and σ phase in the third pass fusion zone (304-2). The σ phase had a higher Cr and a lower Si contents than δ-ferrite. This is because that the precipitation of σ phase in a high Cr region of δ-ferrite can consume Si content. Hence, the Cr content had a main role in forming the σ phase, and Si had an enhanced effect on the precipitation of σ phase.

3.2. Diffusion Calculation. The formation of a σ phase includes temperature and elemental diffusion. Hence, this section will determine the precipitation temperature of a σ phase using a differential scanning calorimeter (DSC) in an AISI 304 multipass welding system, as shown in Figure 3. The diagram shows two exothermic and two endothermic peaks with peak temperatures of 861°C, 934°C, 678°C, and 893°C, respectively. This is for the purpose of plotting the composition of the σ phase in this study (35 wt% Cr, 4 wt% Ni) to the Fe-Cr-Ni phase diagram shown in Figure 4. Here, the solidification order was $L \rightarrow \delta + L \rightarrow \delta \rightarrow \delta + \gamma \rightarrow \gamma + \delta + \sigma$. The above-mentioned results are marked out in Figure 4. It was determined that the σ phase precipitates from 400°C to 934°C.

3.3. Calculation of the General Diffusion Equation. In the previous published paper [19], the diffusivities D^{δ}_{Cr}, D^{δ}_{Ni}, D^{γ}_{Ni}, and D^{γ}_{Cr} were calculated at various peak temperatures (678°C, 861°C, 893°C, and 934°C), and the $\delta \rightarrow \sigma + \gamma_2$ eutectoid decomposition was discussed in accordance with the Vitek diffusion equation. In this section, the general diffusion equation calculation will be compared with the Vitek general equation. Its primary purpose is to examine the differences between the general diffusion and the Vitek diffusion equation. The general diffusion equation is shown

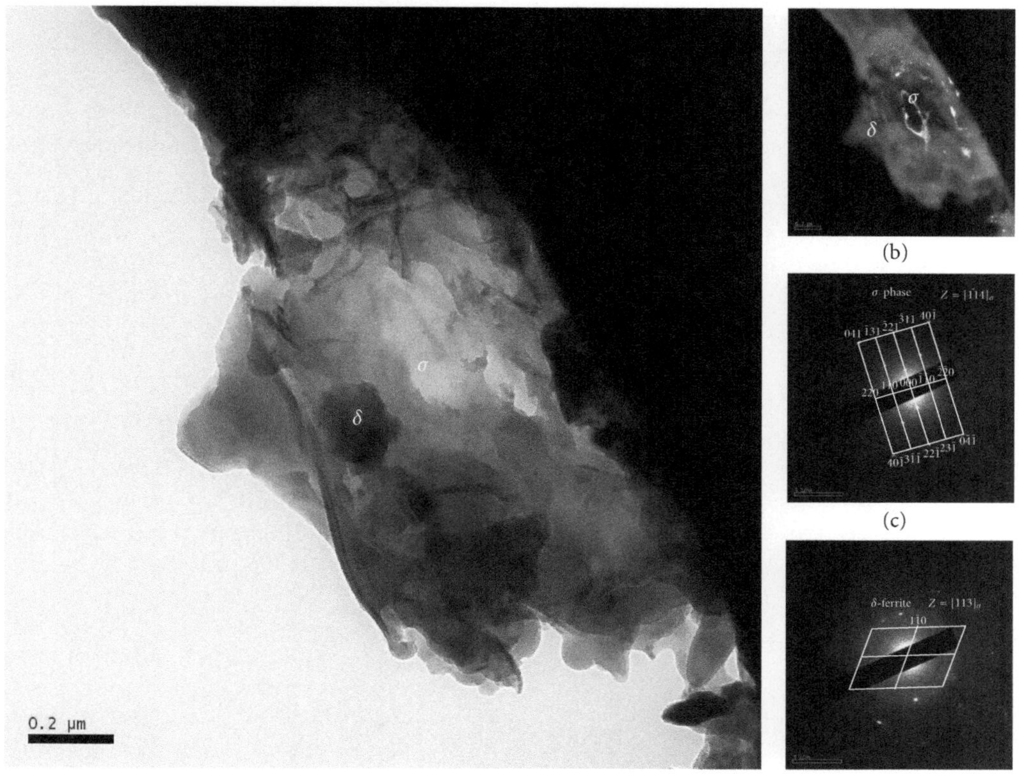

FIGURE 2: TEM phase identification of the third pass fusion zone. (a) Blight filed image, (b) dark field image, and (c, d) SAD pattern of the δ-ferrite and σ phase.

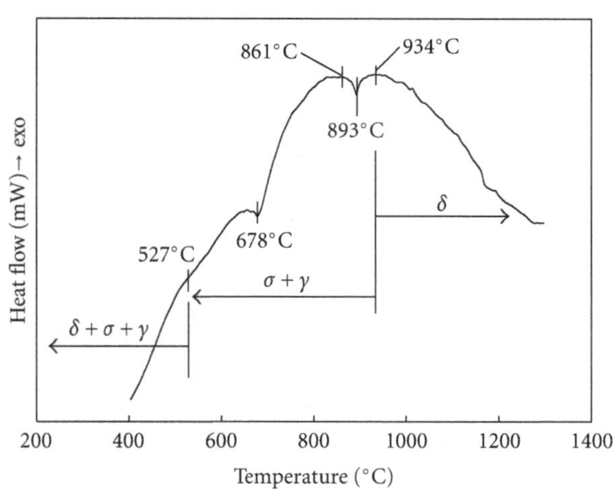

FIGURE 3: DSC thermal analysis of the third pass fusion zone.

in (1). Table 3 illustrates the diffusivities of various elements in stainless steel [20]. The diffusivities of Cr, Ni, and Si in the δ, σ, and γ phases at various peak temperatures for the $\delta \rightarrow \sigma + \gamma_2$ eutectoid decomposition were estimated through the use of the following equation and Table 3:

$$D = D_0 \cdot \exp\left(\frac{-Q}{RT}\right), \quad (1)$$

where D is the diffusion coefficient, $\mathrm{cm^2 \cdot sec^{-1}}$; D_0 is the approximate constant; Q is the diffusion activation energy, $\mathrm{KJ \cdot mole^{-1}}$; R is the gas constant (= 8.314 $\mathrm{J \cdot (mole \cdot K)^{-1}}$); T is the thermodynamic temperature, K.

Two temperatures (800°C and 1100°C) were chosen from Table 3 and substituted into (1). Subsequently, (2) and (3) were obtained. Equations (4) and (5) were derived by simplifying (2) and (3). Creating one equation from (4) and (5), (4) was divided by (5). After simplification, (6) and (7) were acquired:

$$D^{\delta}_{\mathrm{Cr}(800^\circ\mathrm{C})} = 10^{-11} = D_0 \cdot \exp\left[\frac{-Q}{8.314(800+273)}\right], \quad (2)$$

$$D^{\delta}_{\mathrm{Cr}(1100^\circ\mathrm{C})} = 10^{-9} = D_0 \cdot \exp\left[\frac{-Q}{8.314(1100+273)}\right], \quad (3)$$

$$10^{-11} = D_0 \cdot \exp(-1.12 \times 10^{-4} Q), \quad (4)$$

$$10^{-9} = D_0 \cdot \exp(-8.76 \times 10^{-5} Q), \quad (5)$$

$$10^{-2} = \exp(-1.12 \times 10^{-4} Q + 8.76 \times 10^{-5} Q), \quad (6)$$

$$10^{-2} = \exp(-2.44 \times 10^{-5} Q). \quad (7)$$

Take the natural logarithm from both sides of (7). After determining the natural logarithm, (8) and (9) were

Table 3: Diffusion coefficients of elements of the δ and γ phases in stainless steel.

Element	Diffusion coefficient (cm^2/sec)									
	$\delta(\alpha)$-iron					γ-iron				
	Temperature (°C)					Temperature (°C)				
	20	400	800	1100	1400	20	400	800	1100	1400
$\delta(\gamma)$-Fe	10^{-46}	10^{-19}	10^{-12}	10^{-9}	10^{-7}	10^{-53}	10^{-22}	10^{-14}	10^{-11}	10^{-9}
Al			10^{-11}	10^{-9}	10^{-7}	10^{-36}	10^{-15}	10^{-9}	10^{-8}	
B						10^{-20}	10^{-9}	10^{-6}	10^{-5}	
C	10^{-17}	10^{-8}	10^{-5}	10^{-4}		10^{-27}	10^{-13}	10^{-8}	10^{-6}	
Cr	10^{-42}	10^{-18}	10^{-11}	10^{-9}	10^{-7}	10^{-37}	10^{-18}	10^{-13}	10^{-11}	10^{-9}
Co	10^{-44}	10^{-19}	10^{-12}	10^{-10}		10^{-68}	10^{-26}	10^{-16}	10^{-12}	
Cu			10^{-11}	10^{-9}		10^{-49}	10^{-20}	10^{-12}	10^{-10}	
H	10^{-5}	10^{-3}	10^{-3}	10^{-3}		10^{-10}	10^{-5}	10^{-4}	10^{-3}	
Mn						10^{-53}	10^{-22}	10^{-14}	10^{-11}	10^{-9}
Mo	10^{-46}	10^{-18}	10^{-11}	10^{-9}	10^{-7}	10^{-49}	10^{-21}	10^{-13}	10^{-11}	10^{-9}
N	10^{-17}	10^{-8}	10^{-6}	10^{-5}		10^{-31}	10^{-13}	10^{-8}	10^{-7}	
Nb								10^{-13}	10^{-11}	10^{-9}
Ni	10^{-45}	10^{-18}	10^{-11}	10^{-9}		10^{-54}	10^{-23}	10^{-15}	10^{-11}	10^{-9}
P	10^{-34}	10^{-12}	10^{-8}	10^{-7}	10^{-6}			10^{-14}	10^{-9}	10^{-8}
S	10^{-34}	10^{-12}	10^{-8}	10^{-7}	10^{-6}			10^{-14}	10^{-9}	10^{-8}
Si	10^{-50}	10^{-21}	10^{-13}	10^{-11}	10^{-7}	10^{-48}	10^{-20}	10^{-13}	10^{-12}	10^{-9}
Ti	10^{-47}	10^{-19}	10^{-12}	10^{-9}	10^{-7}	10^{-49}	10^{-21}	10^{-13}	10^{-11}	10^{-9}
V			10^{-12}	10^{-9}	10^{-7}			10^{-14}	10^{-11}	10^{-9}
W	10^{-60}	10^{-26}	10^{-17}	10^{-10}		10^{-68}	10^{-30}	10^{-20}	10^{-16}	

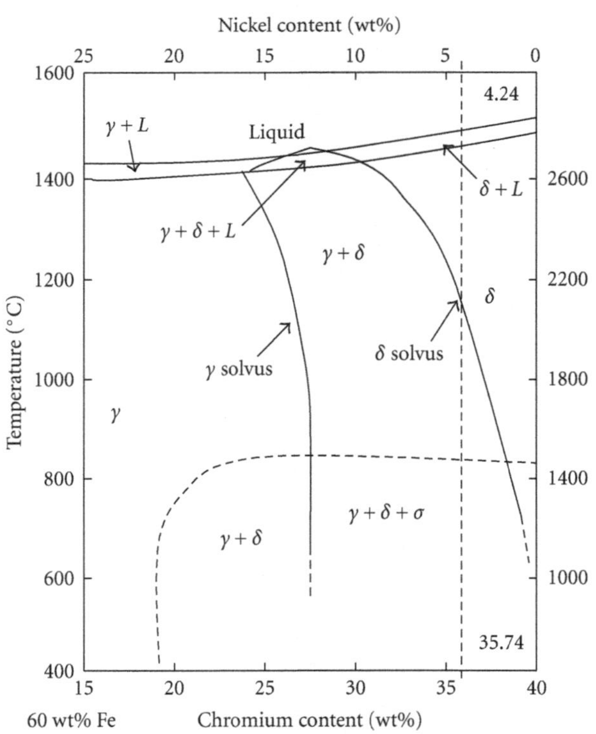

Figure 4: The Fe-Cr-Ni phase diagram (60% Cr).

obtained. By calculating (9), the diffusion activation energy was calculated:

$$\ln(10^{-2}) = \ln[\exp(-2.44 \times 10^{-5} Q)], \quad (8)$$

$$-4.605 = -2.44 \times 10^{-5} Q, \quad (9)$$

then $Q = 188729\,\text{KJ} \cdot \text{mole}^{-1}$.

The $Q = 188729\,\text{KJ} \cdot \text{mole}^{-1}$ and 678°C were substituted into (3) and (1), respectively. Hence, (10) and (11) were obtained.

One equation was created from (10) and (11) and sorted simultaneously, as shown in (12)–(14). Finally, the diffusivity of Cr in δ-ferrite at 678°C ($D^{\delta}_{\text{Cr}(678°\text{C})}$) can be calculated.

In accordance with the above calculated method, a series of diffusivities (D^{δ}_{Cr}, D^{δ}_{Ni}, D^{γ}_{Ni}, D^{γ}_{Cr}, D^{δ}_{Si}, and D^{γ}_{Si}) in the δ and γ phases at different peak temperatures (861°C, 893°C, and 934°C). The calculated results are listed in Table 4. $D^{\delta}_{\text{Cr}} = D^{\delta}_{\text{Ni}}$, indicates that it held the same diffusivity ($6.58 \times 10^{-13}\,\text{cm}^2 \cdot \text{sec}^{-1}$) of Cr and Ni in δ-ferrite during the $\delta \rightarrow \sigma + \gamma_2$ eutectoid decomposition. Therefore, the Cr and Ni were easier to diffuse in the δ-ferrite than in the γ-phase. This can be attributed to higher (D^{δ}_{Cr} and D^{δ}_{Ni}) and lower diffusivities (D^{γ}_{Ni}). Consequently, the σ phase and secondary austenite (γ_2) were precipitated in the δ-ferrite particles in

TABLE 4: Diffusivities of the elements of Cr, Ni, and Si in δ-ferrite and γ-austenite at various peak temperatures in accordance with the general diffusion equation ($D_{general}$).

Temperature (°C)	Diffusivity (cm^2/sec)					
	D^{δ}_{Cr}	D^{γ}_{Ni}	D^{δ}_{Ni}	D^{γ}_{Cr}	D^{δ}_{Si}	D^{γ}_{Si}
678	6.58×10^{-13}	4.38×10^{-18}	6.58×10^{-13}	6.58×10^{-15}	6.39×10^{-15}	2.56×10^{-14}
861	3.09×10^{-11}	9.26×10^{-15}	3.09×10^{-11}	3.09×10^{-13}	3.13×10^{-13}	1.75×10^{-13}
893	5.36×10^{-11}	3.33×10^{-14}	5.36×10^{-11}	5.37×10^{-13}	5.46×10^{-13}	2.32×10^{-13}
934	1.04×10^{-10}	1.08×10^{-13}	1.04×10^{-10}	1.04×10^{-12}	1.08×10^{-12}	3.23×10^{-13}

TABLE 5: Jump frequencies of the solute atoms of Cr, Ni, and Si in δ and γ at various peak temperatures.

Temperature (°C)	Jump frequencies of solute atoms (sec^{-1})					
	Γ^{δ}_{Cr}	Γ^{γ}_{Ni}	Γ^{δ}_{Ni}	Γ^{γ}_{Cr}	Γ^{δ}_{Si}	Γ^{γ}_{Si}
678	6370	0.04	6370	61	62	238
861	299000	90	299000	2877	3040	1629
893	518000	310	518000	5001	5307	2165
934	1012317	1004	1012317	9695	10400	3003

the third pass fusion zone. However, there is no prediction for the diffusivity of Si in the δ-ferrite and γ phase (D^{δ}_{Si} and D^{γ}_{Si}) in regard to the Vitek diffusion equation. The empirical formula for Si in the δ and γ phases was derived from the general diffusion equation, as illustrated in (15) and (16). Table 4 shows the $D^{\delta}_{Si} > D^{\gamma}_{Si}$ indicating that Si had a higher diffusion rate in δ-ferrite than in the γ-phase. However, the Si can promote $\delta \rightarrow \sigma + \gamma_2$ eutectoid decomposition from the above-described result. This is the primary reason that precipitation of the σ phase occurs through a Fe-Cr-Si-rich intermetallic compound in the third pass fusion zone of this study:

$$10^{-11} = D_0 \cdot \exp\left[\frac{-188729}{8.314(800+273)}\right], \quad (10)$$

$$D^{\delta}_{Cr(678^{\circ}C)} = D_0 \cdot \exp\left[\frac{-188729}{8.314(677.9+273)}\right], \quad (11)$$

$$\frac{10^{-11}}{D^{\delta}_{Cr(678^{\circ}C)}} = \exp(-21.15+23.87), \quad (12)$$

$$D^{\delta}_{Cr(678^{\circ}C)} = \frac{10^{-11}}{\exp(-21.15+23.87)}, \quad (13)$$

$$D^{\delta}_{Cr(678^{\circ}C)} = \frac{10^{-11}}{15.18}, \quad (14)$$

then $D^{\delta}_{Cr(678^{\circ}C)} = 6.58 \times 10^{-13}\,cm^2 \cdot sec^{-1}$,

$$D^{\delta}_{Si} = \exp\left(-\frac{22700}{T}\right), \quad (15)$$

$$D^{\gamma}_{Si} = \exp\left(-\frac{11337}{T}\right). \quad (16)$$

3.4. The Jump Frequencies of the Cr, Ni, and Si Atoms in δ and γ Phases. Atom jump frequency (Γ) can be calculated using the above-mentioned calculation of diffusivities. This section will explain the mechanisms of eutectoid decomposition from an atom jump perspective. Atom jump frequency can be calculated as shown in the following equation:

$$\Gamma = \frac{6D}{\lambda^2}, \quad (17)$$

where Γ is the jump frequency of solute atom (sec^{-1}), D is the diffusion coefficient ($cm^2 \cdot sec^{-1}$), and λ represents space between the two atoms (nm).

Because the atom spacing of the δ-ferrite and γ phase is $\sqrt{3}a/2$ and $\sqrt{2}a/2$, respectively, it is understood that the lattice constants of the δ-ferrite and γ phases are about 2.87 nm and 0.359 nm. Then, the values of diffusivities from Table 5 were substituted into (17), and a series of the jump frequencies were obtained, as shown in Table 4. This Table shows $\Gamma^{\delta}_{Cr} = \Gamma^{\delta}_{Ni}$ in the third pass fusion zone. This means that the solute atoms of Cr and Ni had a higher atom jump frequency with a temperature range of 678°C, 861°C, 893°C, and 934°C in the eutectoid decomposition of $\delta \rightarrow \sigma + \gamma_2$, as shown in Figure 5. This tendency is in agreement with previous results ($D^{\delta}_{Cr} = D^{\delta}_{Ni}$). Especially notable at 934°C, the jump frequencies of Cr and Ni solute atoms were about 10^7 times per second in δ-ferrite in the third pass fusion zone. Conversely, the jump frequency of the Ni solute atom in the γ phase was very small (just 1004 times) at 934°C, as shown in Figure 6. Hence, the σ phase and secondary austenite (γ_2) formed easily in δ-ferrite. Furthermore, the jump frequency of the Si solute atom in δ-ferrite (Γ^{δ}_{Si}) was higher than that of Cr and Ni in δ-ferrite (Γ^{δ}_{Cr} and Γ^{δ}_{Ni}). As shown in Figure 7, the diffusivity of the Si solute atom increased with an increase in the jump frequency of the Si solute at various peak temperatures. Hence, the Si can be understood to accelerate the eutectoid decomposition of $\delta \rightarrow \sigma + \gamma_2$.

3.5. Modified Vitek Diffusion Equation. As previous Hsieh and Wu's study in Table 6 [19], the difference in diffusivities between the D^{δ}_{Cr} and D^{δ}_{Ni} was very small according to the Vitek model calculation. This is despite the fact that

Table 6: Diffusivities of the Cr and Ni in δ-ferrite and γ-austenite at various peak temperatures in accordance with the Vitek model.

Peak temperature (°C)	Diffusivity ($cm^2 \cdot sec^{-1}$)			
	D^{δ}_{Cr}	D^{γ}_{Ni}	D^{δ}_{Ni}	D^{γ}_{Cr}
678	1.31×10^{-13}	1.11×10^{-16}	1.00×10^{-13}	3.78×10^{-16}
861	1.74×10^{-11}	4.73×10^{-14}	1.34×10^{-11}	1.60×10^{-13}
893	3.52×10^{-11}	1.13×10^{-13}	2.71×10^{-11}	3.84×10^{-13}
934	8.14×10^{-11}	3.18×10^{-13}	6.26×10^{-11}	1.08×10^{-12}

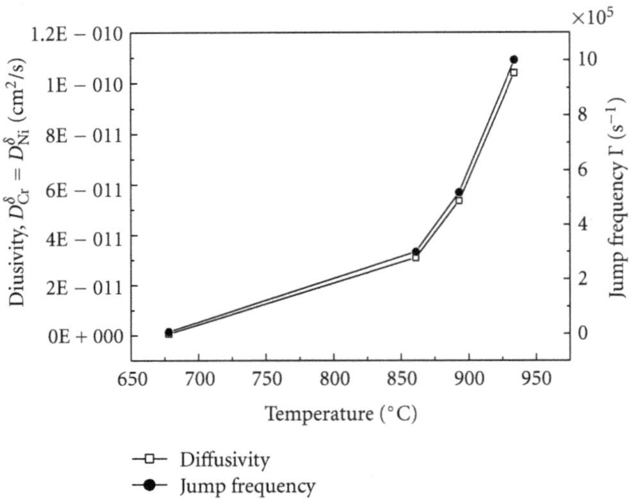

Figure 5: The diffusivities of D^{δ}_{Cr} and D^{δ}_{Ni} and the jump frequencies at various peak temperatures.

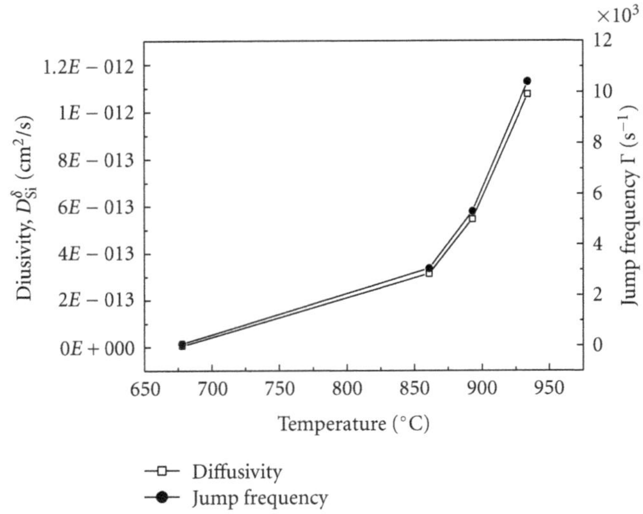

Figure 7: The diffusivity of D^{δ}_{Si} and the jump frequencies at various peak temperatures.

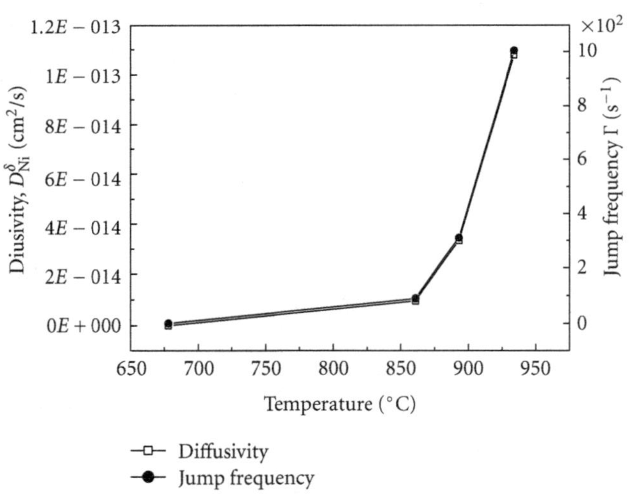

Figure 6: The diffusivity of D^{γ}_{Ni} and the jump frequencies at various peak temperatures.

Table 7: Modified diffusion equation at various peak temperatures.

Temperature (°C)	Coefficient			
	C^{δ}_{Cr}	C^{γ}_{Ni}	C^{δ}_{Ni}	C^{γ}_{Cr}
678	5.02	0.33	6.58	17.4
861	1.77	0.19	2.30	1.93
893	1.52	0.29	1.98	1.39
934	1.27	0.34	1.66	0.96

* $(D)_{General} = (C) \times (D)_{Vitek}$.

the order of the D^{δ}_{Cr} is equivalent to the D^{δ}_{Ni}. Generally speaking, the general diffusion equation is more precise than other diffusion empirical equations. For this study, the Vitek diffusion equation was modified and its accuracy improved, as indicated in (18). The modifications are listed in Table 7 where it is shown that the modified coefficient had a greater at 678°C than other peak temperatures. In total, the difference between the general diffusion equation and the Vitek diffusion equation did not exceed 10^2. The Vitek model showed a quite accuracy for predicting the diffusion coefficient; however, the modified Vitek model can increase its precision:

$$(D)_{General} = (C) \times (D)_{Vitek}, \qquad (18)$$

where $(D)_{General}$ is a diffusivity of general diffusion equation calculation ($cm^2 \cdot sec^{-1}$), (C) is a modified coefficient, and $(D)_{Vitek}$ is a diffusivity of Vitek diffusion model calculation ($cm^2 \cdot sec^{-1}$).

4. Conclusions

In this study, a new method for the examination of the $\delta \rightarrow \sigma + \gamma_2$ phase transformation within multipass fusion zones was investigated. The new method employed the general diffusion equation with comparison to the Vitek diffusion model during multipass dissimilar stainless steel welding. The significant results were as follows.

(1) The precipitation of the σ phase in the δ-ferrite was observed using TEM-SAD and TEM-EDS after the multipass welding.

(2) The diffusion coefficients of Si in δ-ferrite and γ-austenite were predicted using the general diffusion equation with comparison to the Vitek diffusion model.

(3) The Vitek diffusion equation was modified as in the following equation: $(D)_{\text{General}} = (C) \times (D)_{\text{Vitek}}$ in the multipass fusion zone.

Acknowledgment

The authors are obligated to thank the National Science Council of Taiwan, for their financial support under Projects no. NSC 101-2811-E-005-001 and NSC 101-2623-E-005-002-ET.

References

[1] S. Missori and C. Koerber, "Laser beam welding of austenitic-rerritic transition joints," *Welding Journal*, vol. 76, no. 12, pp. 125S–134S, 1997.

[2] B. Sun, Y. J. Li, Q. Chi, and X. S. Zhang, "Analysis of the microstructure in the weld zone of $0Cr_{18}Mo_2$ and $1Cr_{18}Ni_9Ti$ dissimilar stainless steel," *Chemical Engineering and Machinery*, vol. 31, no. 1, pp. 17–23, 2004.

[3] P. Michael, S. Oliver, and G. Thomas, "Effect of intermetallic precipitations on the properties of duplex stainless steel," *Materials Characterization*, vol. 58, pp. 65–71, 2007.

[4] R. F. Steigerwald, "Effects of metallic second phases in stainless steels," *Corrosion*, vol. 33, no. 9, pp. 338–343, 1977.

[5] H. Kiesheyer and H. Brandis, "Investigation of phase equilibria in the ternary system Fe-Cr-Mo in the solid state," *Zeitschrift fuer Metallkunde*, vol. 67, no. 4, pp. 258–263, 1976.

[6] C. J. Park, M. K. Ahn, and H. S. Kwon, "Influences of Mo substitution by W on the precipitation kinetics of secondary phases and the associated localized corrosion and embrittlement in 29% Cr ferritic stainless steels," *Materials Science and Engineering A*, vol. 418, pp. 211–217, 2006.

[7] M. Raghavan, G. A. Mueller, G. A. Vaughn, and S. Floreen, "Determination of isothermal sections of nickel rich portion of Ni-Cr-Mo system by analytical electron microscopy," *Metallurgical Transactions A*, vol. 15, no. 5, pp. 783–792, 1984.

[8] L. Karlsson, L. Ryen, and S. Pak, "Precipitation of intermetallic phases in 22% Cr duplex stainless weld metals," *Welding Journal*, vol. 74, pp. 28S–40S, 1995.

[9] T. H. Chen, K. L. Weng, and J. R. Yang, "The effect of high-temperature exposure on the microstructural stability and toughness property in a 2205 duplex stainless steel," *Materials Science and Engineering A*, vol. 338, no. 1-2, pp. 259–270, 2002.

[10] J. O. Nilsson, P. Kangas, and T. Karlsson, "Mechanical properties, microstructural stability and kinetics of σ-phase formation in 29Cr-6Ni-2Mo-0.38N superduplex stainless steel," *Metallurgical and Materials Transactions A*, vol. 31, pp. 35–45, 2000.

[11] L. Jinchun, W. Tuyan, and R. Yves:, "δ phase precipitation and its effect on the mechanical properties of a super duplex stainless steel," *Materials Science and Engineering A*, vol. 174, pp. 149–156, 1994.

[12] A. Turnbull, P. E. Francis, M. P. Ryan, L. P. Orkney, A. J. Griffiths, and B. Hawkins, "A novel approach to characterizing the corrosion resistance of super duplex stainless steel welds," *Corrosion*, vol. 58, no. 12, pp. 1039–1048, 2002.

[13] D. Y. Lin, G. L. Liu, T. C. Chang, and H. C. Hsieh, "Microstructure development in 24Cr–14Ni–2Mn stainless steel after aging under various nitrogen/air ratios," *Journal of Alloys and Compounds*, vol. 377, pp. 150–154, 2004.

[14] C. C. Hsieh, D. Y. Lin, and W. Wu, "Precipitation behavior of σ phase in 19Cr–9Ni–2Mn and 18Cr–0.75Si stainless steels hot-rolled at 800°C with various reduction ratios," *Materials Science and Engineering A*, vol. 467, pp. 181–189, 2007.

[15] C. C. Hsieh, T. C. Chang, D. Y. Lin, and M. C. Chen, "Dispersion strengthening behavior of σ phase in 304 modified stainless steels during 1073 K hot rolling," *Metals and Materials International*, vol. 13, pp. 359–363, 2007.

[16] C. C. Hsieh, D. Y. Lin, and W. Wu, "Precipitation behavior of σ phase in fusion zone of dissimilar stainless steel welds during multi-pass GTAW process," *Metals and Materials International*, vol. 13, pp. 411–416, 2007.

[17] C. C. Hsieh, D. Y. Lin, and T. C. Chang, "Microstructural evolution during the $\delta/\sigma/\gamma$ phase transformation of the SUS 309LSi stainless steel after aging under various nitrogen atmospheric ratios," *Materials Science and Engineering A*, vol. 475, no. 1-2, pp. 128–135, 2008.

[18] C. C. Hsieh, X. Guo, and W. Wu, "Dendrite evolution of delta (δ) ferrite and precipitation behavior of sigma (σ) phase during multipass dissimilar stainless steels welding," *Metals and Materials International*, vol. 16, pp. 349–356, 2010.

[19] C. C. Hsieh and W. Wu, "Discussing the precipitation behavior of σ phase using diffusion equation and thermodynamic simulation in dissimilar stainless steels," *Journal of Alloys and Compounds*, vol. 506, pp. 820–825, 2010.

[20] E. Folkhard, G. Rabensteiner, E. Perteneder, H. Schabereiter, and J. Tösch, *Welding Metallurgy of Stainless Steels*, Springer, New York, NY, USA, 2004.

3

Validation of FeSiMg Alloy Production Model for the Experimental Process

Saeed Ghali, Mamdouh Eissa, and Hoda El-Faramawy

Steel Technology Department, Central Metallurgical Research & Development Institute (CMRDI), Helwan 11421, Egypt

Correspondence should be addressed to Saeed Ghali, a3708052@yahoo.com

Academic Editor: Brij Kumar Dhindaw

This paper investigates the effect of limestone, bauxite, fluorspar, and quartzite additions on the magnesium recovery in FeSiMg production from dolomite. Also, it illustrates the validation of the previous designed model. According to the model, magnesium content in the product alloy is calculated by the equation $[\mathrm{Mg}] = (\mathrm{MgO}_0)[\mathrm{Si}_0][e^{Kt[(\mathrm{MgO}_0)-[\mathrm{Si}_0]]} - 1]/((\mathrm{MgO}_0)[e^{Kt[(\mathrm{MgO}_0)-[\mathrm{Si}_0]]}] - [\mathrm{Si}_0])$, where [Mg] is the concentration of magnesium metal in ferrosilicon magnesium alloy in mol/L, $[\mathrm{Si}_0]$ and (MgO_0) are the initial concentrations of silicon and magnesium oxide in charge in mol/L, t is the time in seconds, and K is the reaction rate constant (3.26588×10^{-7} $\mathrm{LSec}^{-1}\,\mathrm{mol}^{-1}$). The results of the production process are compared with the model results. The deviation between the actual and predicted magnesium content decreases as fluorspar, limestone, and quartzite increase up to 12.8wt.%, 8wt.%, and 8wt.%, respectively, with increase in the amount of the additives, the magnesium content in the produced alloy becomes far from the predicted values. It was found also that the addition of bauxite increases the gap between the actual and predicted values of magnesium content. It was found that the deviation of the actual magnesium content from the predicted depends mainly on the viscosity of the slag.

1. Introduction

Ferrosilicon magnesium alloys are used to produce all types of ductile cast iron. It is added to the molten iron for producing a structure containing graphite in nodular or spheroidal form [1]. This form minimizes the embrittlement effect of graphite on the metal matrix with the result of producing cast iron with better machinability, toughness, and tensile properties. Ferrosilicon magnesium consists mainly of magnesium silicides combined with the silicides of iron, calcium and rare earth metals. The silicothermic process is one of the most attractive techniques for producing FeSiMg alloy. This is attributed to much lower working temperature compared with carbothermic process. Also, it is of much lower cost compared with the alumino-thermic process. The oldest trials to reduce magnesium oxide with silicon were carried out by Kubaschewski and Evans [2] and Misra et al. [3]. Their results [2, 3] showed a low recovery of magnesium under this working conditions. They attributed this phenomenon to the formation of magnesium orthosilicate. Misra et al. [4] have also found that silicothermic reduction of pure magnesium oxide results in a poor magnesium recovery, due to the fact that a part of the magnesium oxide combines with the silica. Recovery of magnesium can be improved by carrying out the reduction of magnesium oxide in the presence of a metallic oxide that can form silicates more stable than magnesium orthosilicate. According to the thermodynamic data, lime can be considered as a suitable additive for improving the reduction process of magnesium oxide due to the fact that its silicate is more easily formed than magnesium silicate [3, 5, 6]. Thus, better recovery is expected when a mixture of lime and magnesite was used. Calcinated dolomite seems to be a suitable cheap raw material for the production of ferrosilicon magnesium alloy. Using silicothermic reduction process to produce magnesium-bearing alloys, which can be carried out in submerged arc furnace.

At temperature range 1650−1750°C [7–9], silicon will react with the constituents of charge that have lower affinity to oxygen than silicon to form SiO_2. Then, SiO_2 will react with CaO and MgO to form low-temperature molten silicates. The product of calcium silicate and magnesium silicate will react with molten silicon metal according to the reactions

$$2[\mathrm{CaO} \cdot \mathrm{SiO}_{2(\mathrm{L})}] + 5\mathrm{Si}_{(\mathrm{L})} \longrightarrow 2\mathrm{CaSi}_2 + 3\mathrm{SiO}_2 \qquad \Delta G = -8515.447 + 1.5411T, \tag{1}$$

$$2[\mathrm{MgO} \cdot \mathrm{SiO}_{2(\mathrm{L})}] + 2\mathrm{Si}_{(\mathrm{L})} \longrightarrow \mathrm{Mg}_2\mathrm{Si} + 3\mathrm{SiO}_2 \qquad \Delta G = -6711.941 + 1.1478T. \tag{2}$$

The above reactions followed by the fact that calcium oxide has greater affinity to silica, and, therefore, it form more stable suitable silicate than magnesium oxide [10, 11]:

$$2[\mathrm{MgO} \cdot \mathrm{CaO} \cdot \mathrm{SiO}_{2(\mathrm{L})}] + 2\mathrm{Si}_{(\mathrm{L})} \longrightarrow \mathrm{Mg}_2\mathrm{Si} + \mathrm{SiO}_2 + 2[\mathrm{CaO} \cdot \mathrm{SiO}_2] \qquad \Delta G = -6148.5398 + 1.031T. \tag{3}$$

2. Aim and Scope

This paper aims at investigating the influence of bauxite, quartzite, limestone, and fluorspar on the extent of deviation of magnesium content from the predicted values by using the suggested model by Ghali [12]. This model used the following equation to calculate the magnesium content in the produced FeSiMg alloy:

$$[\mathrm{Mg}] = \frac{(\mathrm{MgO}_0)[\mathrm{Si}_0]\left[e^{Kt[(\mathrm{MgO}_0)-[\mathrm{Si}_0]]} - 1\right]}{(\mathrm{MgO}_0)\left[e^{Kt[(\mathrm{MgO}_0)-[\mathrm{Si}_0]]}\right] - [\mathrm{Si}_0]}, \tag{4}$$

where [Mg] is the concentration of magnesium metal in ferrosilicon magnesium alloy in mol/L, $[\mathrm{Si}_0]$ and (MgO_0) are the initial contents of silicon and magnesium oxide in charge, respectively, in mol/L, t the time in second, and K the reaction rate constant 3.26588×10^{-7} L Sec^{-1} mol^{-1}.

3. Experimental

The FeSiMg production data [13], which are produced by the experimental melting in a submerged arc furnace, are given in Tables 1 and 2. The molar concentrations of SiO_2, Al_2O_3, CaF_2, and CaO were calculated.

The data [13] from the previous study was considered as the main source for the evaluation of the new model. The different production parameters were considered and correlated to predict the different relations and dependences in the product process.

The amount and effect of different additives on the recovery of magnesium and composition of the product alloy were considered. Magnesium content is measured by using wet analysis method [14].

4. Results and Discussion

The data used in this investigation were obtained from El-Faramawy et al. [13] and Ghali [12]. The effect of fluxes (limestone, fluorspar, quartzite, and bauxite) on magnesium content and recovery in FeSiMg production process was investigated. The differences between the predicted values, by applying Ghali model [12], and the actual magnesium content and recovery in the FeSiMg alloy produced by metallothermic process using silicon as a reducing material [13] were considered. The results of four series of bench scale experiments having constant amount of dolomite with various weight percentages of limestone, quartzite, fluorspar, and bauxite, respectively (Table 1) were selected as the base for theoretical calculation.

Figure 1 shows the influence of weight ratio limestone/dolomite on the difference between the predicted and actual magnesium content. It indicates that the difference decreases by increasing limestone/dolomite mass content ratio up to 8 wt.%. Further increase of limestone wt.% results increases the difference between predicted and actual magnesium mass content. This figure also shows that at optimum limestone mass content the predicted values of magnesium mass content are in good agreement with the actual values. This could be attributed to two opposing effects. The first one is the effect of increasing CaO content in SiO_2-rich slag leading to the formation of $2CaO{\cdot}SiO_2$, $3CaO{\cdot}SiO_2$, and $CaO{\cdot}SiO_2$ [6, 15–19]. According to the thermodynamic data free energy of formation of pure crystalline silicate from their component oxides as given in (5) and (6) [9], these compounds are formed first and are very stable leading to free MgO to be reduced

$$\mathrm{CaO} + \mathrm{SiO}_2 = \mathrm{CaO} \cdot \mathrm{SiO}_2, \qquad \Delta G\,(\mathrm{KJ/mol}) = -2368.75 + 0.392T, \tag{5}$$

$$\mathrm{MgO} + \mathrm{SiO}_2 = \mathrm{MgO} \cdot \mathrm{SiO}_2, \qquad \Delta G\,(\mathrm{KJ/mol}) = -2312.78 + 0.308T. \tag{6}$$

The other one is the negative effect of increasing the content of these high melting compounds $3CaO{\cdot}SiO_2$, $2CaO{\cdot}SiO_2$, and $CaO{\cdot}SiO_2$ with respective melting temperatures of 2070°C, 2130°C, and 1564°C, resulting in increasing the slag viscosity. Furthermore, the presence of CO_2 gas, from the decomposition of limestone, leads to more oxidation of magnesium. The outcome of these two opposite effects leads to an optimum value. In other words, the increase of the CaO content in the charge is accompanied by an increase in the activity of MgO in the slag. This is due to liberation of MgO from $MgO{\cdot}SiO_2$ and formation of $CaO{\cdot}SiO_2$ as shown in (5) and (6). This leads to increasing magnesium recovery and hence the gap between the predicted and actual magnesium content decreases. Further increase of limestone content is accompanied by an increase in slag viscosity. This is due to the formation of $2CaO{\cdot}SiO_2$, and $3CaO{\cdot}SiO_2$ which have high melting points 2070°C and 2130°C, respectively. This deteriorates the diffusion of reactants to the reaction zone. Thus increasing the difference between the predicted and actual magnesium content in the product alloy.

The effect of fluorspar content in the charge on the deviation of the actual magnesium content, given in Table 1 [13], from the predicted values using Ghali model [12] is illustrated in Figure 2. From this figure it is clear that the difference between the predicted magnesium contents

Table 1: Materials balance of the experimental heats.

Heat no.	Input, gm								Output		Predicted Mg mass content in %
	Dolomite	FeSi	Fluorspar	Limestone	Al	Quartzite	Bauxite	$CeSo_4$	Metal mass	Mg mass content in %,	
1	1250	750	100	50				25	725	2.25	4.38
2	1250	750	100	70				25	590	2.86	5.30
3	1250	750	100	100				25	550	4.1	5.54
4	1250	750	100	130				25	430	2.0	6.90
5	1250	750	100	150				25	441.5	1.3	6.61
6	1250	600	160		50	50		25	480	1.7	5.48
7	1250	600	160		50	75		25	558	1.8	4.60
8	1250	600	160		50	100		25	622	3.2	4.04
9	1250	600	160		50	150		25	813	1.6	2.96
10	1250	600	40		50			25	750	1.76	4.11
11	1250	600	80		50			25	489	3.5	6.06
12	1250	600	120		50			25	550	3.5	5.19
13	1250	600	160		50			25	575	4.24	4.78
14	1250	600	200		50			25	595	3.25	4.45
15	1250	600	400		50			25	550	3.25	4.05
16	1250	950	160				50	25	983	2.45	3.19
17	1250	950	160				100	25	828	2.84	3.67
18	1250	950	160				150	25	731	3.05	3.99
19	1250	950	160				200	25	805	2.52	3.50
20	1250	950	160				250	25	750	2.63	3.62

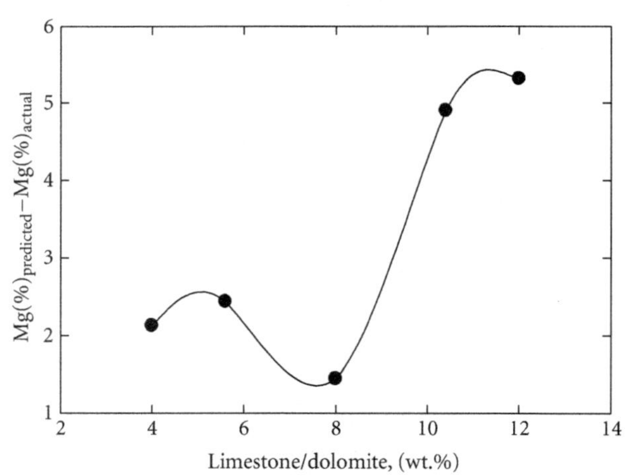

Figure 1: Effect of limestone/dolomite weight% on the gap between the predicted and actual magnesium content.

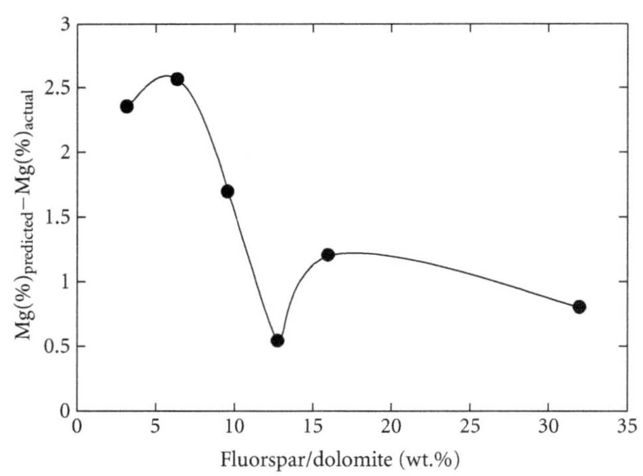

Figure 2: Influence of fluorspar/dolomite weight% on the difference between the predicted and actual magnesium content.

and actual values decreases by increasing fluorspar/dolomite weight ratios in the charge up to 12.8 weight content in %. Further increase in the fluorspar percentage leads to a large gap between predicted and actual values of magnesium content.

Figure 3 shows the effect of quartzite/dolomite weight ratios on the deviation of the actual magnesium content from the predicted values. It is clear that the gap between the actual and predicted magnesium content decreases by increasing quartzite/dolomite weight ratio up to 8 weight content in %. Further increase in quartzite/dolomite weight ratio leads to increasing the difference between the actual and predicted magnesium content.

On the other hand, increasing the bauxite content in the charge is accompanied by higher-degree deviation of the actual magnesium content from the predicted values as given in Figure 4.

Table 2: Chemical composition of the charge constituents.

Constituents	Chemical composition, mass content in % Calcinated dolomite	Fluorspar	Rare earth metals	Limestone	Quartzite	Bauxite	FeSi	Al
SiO_2	1.4	12.6		3.88	95	6.43		
Fe_2O_3	1.45	0.35		0.5	0.2			
CaO	62.4		1	51.78				
MgO	33.6		1	0.6	0.3			
L.O.I. at 1000°C	0.43			41.3				
Al_2O_3	1	2.4		0.8	2.5	85		
Na_2O			1	1.64				
K_2O				0.35				
CaF_2		82						
$CaCO_3$		2.2						
P_2O_3			0.01					
CeO_2/ReO			45					
Fe			0.005	0.34	0.14		23.3	
Pb			0.001					
P_2O_5		0.001						
SO_3			0.03					
FeO						1.8		
C							0.09	
S							0.003	
P							0.031	
Al							1.41	99
Ca							0.31	
Si							74.8	

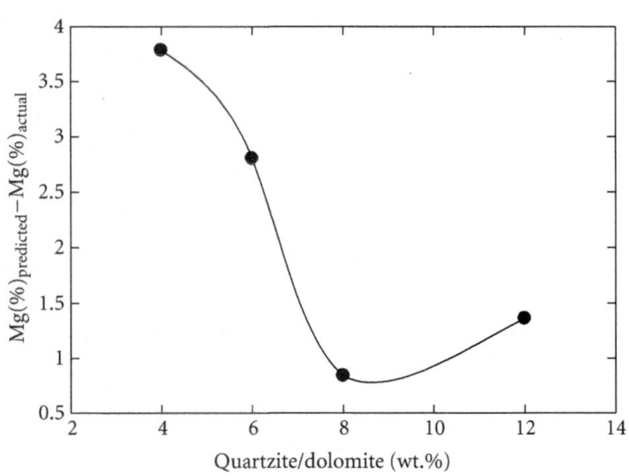

Figure 3: Effect of quartzite/dolomite weight% on the difference between the predicted and actual magnesium content.

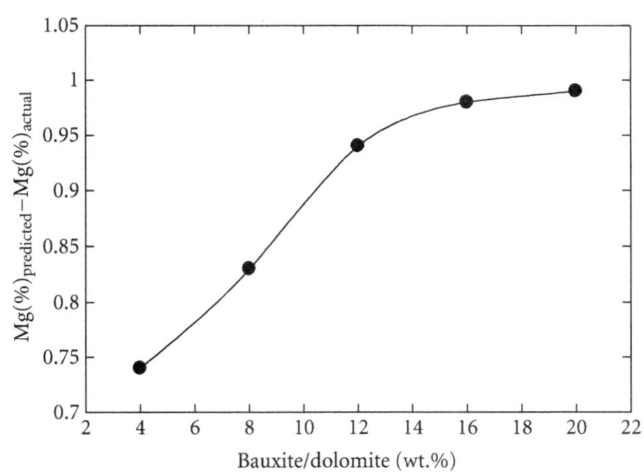

Figure 4: Effect of bauxite/dolomite weight% on the difference between the predicted and actual magnesium content.

Reduction of dolomite by ferrosilicon takes place according to the following sequence: (1) the diffusion of MgO and CaO through the slag layer to the slag/metal interface, (2) chemical reaction between the reactants at the slag/metal interface, (3) the transformation of the metal droplets through slag/metal interface to the molten metal, (4) diffusion of non-metallic products of reaction through slag/metal interface to the slag layer. The diffusion rate of either the reactants or the products through the slag layer is controlled by the physical properties [20] of the molten slag. So, it is expected that the magnesium content will effectively be influenced by slag viscosity. Increasing viscosity of slag obstructs the diffusion of reactants and products to and from the reaction zone. Therefore, lower degree of deviation of actual magnesium content from the predicted one—as given in Figure 2—can be a result of increasing the fluorspar content. This phenomenon could be attributed to an increase of the slag fluidity due to high fluorspar content, which results in a higher rate of metallic droplets diffusion. On the other hand, addition of more fluorides to silicate slag results in evolution of silicon tetra fluoride (SiF_4) vapour [21]. Also, there is a diffusion constant of silicon between metal and slag at a given temperature.

Therefore, as the slag is saturated with SiO_2, the silicon content in the alloy increases also leading to lower deviation in magnesium content between the actual and predicted values as illustrated in Figure 3. With further addition of quartzite, the excess SiO_2 tends to form a less stable compound such as $Ca_3Mg(SiO_2)$ [22, 23]. This compound is dissociated to Ca_2SiO_4 with a high melting point leading to more viscous slag.

The obtained results also showed the negative effect of bauxite addition on the deviation of the actual magnesium content from the predicted values, which could be attributed to the higher viscosity of high alumina slag [24–28].

Figure 4 shows the effect of bauxite (alumina content) on the magnesium content and the difference between the predicted and actual magnesium content. It is clear that the difference between the predicted Mg mass content in % and the actual Mg mass content in % increases as the alumina increase. This behavior can be attributed to the effect of Al_2O_3 on the viscosity of slag. As Al_2O_3 content increases, the slag viscosity becomes higher. The high viscosity has negative significant effect on the diffusion rate. The diffusion rates of either the reactants or products through the slag layer decrease due to the effect of alumina. This leads to decrease in magnesium recovery and hence the gap between the predicted and actual magnesium content increases [13].

5. Conclusions

The following conclusions can be made.

(i) The deviation between the actual and predicted magnesium content decreases as the magnesium content in the produced alloy increases.

(ii) The highest magnesium content can be obtained using weight% of limestone/dolomite, quartzite/dolomite and fluorspar/dolomite 8, 8, and 12.8, respectively. This is in good agreement with the theoretically predicted values.

(iii) The addition of bauxite decreases the magnesium recovery due to high viscosity.

(iv) The physical properties, mainly the viscosity of the molten charge, play a significant role in the reduction process during the production of ferrosilicon magnesium alloy from dolomite.

Acknowledgments

M. Eissa is the head of the Steel Technology Department, Central Metallurgical Research & Development Institute (CMRDI), Egypt. S. Ghali is a researcher at the Steel Technology Department, Central Metallurgical Research & Development Institute (CMRDI), Egypt. H. El-Faramawy is a professor at the Steel Technology Department, Central Metallurgical Research & Development Institute (CMRDI), Egypt.

References

[1] C. Labrecque and M. Gagne, "Review ductile iron: fifty years of continuous development," *Canadian Metalurgical Quarterty*, vol. 37, no. 5, pp. 343–378, 1998.

[2] O. Kubaschewski and E. L. Evans, *Metallugical thermochemistry*, Pergamon Press, London, UK, 1958.

[3] R. N. Misra, V. S. Sampath, and P. P. Bhatnagar, "Studies on silico-thermic reduction of dolomite for the production of magnesium," *Trans Indian Institute of Metals*, vol. 17, pp. 145–154, 1961.

[4] R. N. Misra, V. S. Sampath, and P. P. Bhatnagar, *Light Metal Industry in India*, National Metallurgical Lab, Jamshedpur, India, 1961.

[5] 3761247,25 U.S., J. M. Avery, September,(1973).

[6] J. M. Toguri and L. M. Pidgeon, *Canadian Journal of Chemistry*, vol. 39, pp. 540–547, 1961.

[7] M. B. Hillert, B. Sundman, X. Wang, and T. Barry, "A reevaluation op the rankinite phase in the CaO-SiO_2 system," *Calphad*, vol. 15, no. 1, pp. 53–58, 1991.

[8] S. Kambayashi and S. E. Kato, "A thermodynamic study of (magnesium oxide + silicon dioxide) by mass spectrometry at 1973 K," *The Journal of Chemical Thermodynamics*, vol. 16, no. 3, pp. 241–248, 1984.

[9] F.-Z. Ji, D. Sichen, and S. Seetharaman, "Viscosities of multicomponent silicate melts at high temperatures," *International Journal of Thermophysics*, vol. 20, no. 1, pp. 309–323, 1999.

[10] H. C. Lin and W. R. Foster, "Stability Relations of Bredigite (5CaO· MgO· 3SiO_2)," *Journal of the American Ceramic Society*, vol. 58, no. 1-2, p. 73, 1975.

[11] C. M. Schlaudt and D. M. Roy, "The Join Ca_2SiO_4-$CaMgSiO_4$," *Journal of the American Ceramic Society*, vol. 49, no. 8, pp. 430–432, 1966.

[12] S. N. Ghali, "A model for prediction of Mg content in FeSiMg alloy produced by reduction of magnesium oxide," in *Proceedings of the Minerals to Materials Conference (M2M '08)*, p. 32, Cairo, Egypt, December 2008.

[13] H. El-Faramawy, M. Eissa, T. Mattar, A. Fathy, and S. Ghali, "Ferrosilicon magnesium production," *Scandinavian Journal of Metallurgy*, vol. 32, no. 1, pp. 37–46, 2003.

[14] A. Vogel, *A Textbook of Quantitative Inorganic Analysis Including Elemelllary Instrumental Analysis*, Woolwich Polytechnic, London, UK, 1992.

[15] L. M. Pidgeon, *Transactions of the Metallurgical Society of AIME*, vol. 227, pp. 821–834, 1963.

[16] K. Tomasek, I. Imris, and O. Seseiec, "Magnesium production possibilities from slovak dolomite and magnesite raw materials and wastes," in *Proceedings of the Light Metals Conference*, pp. 213–220, Montreal, QC, Canada, 1996.

[17] B. Ellingstaeter and T. Rosenqvist, *Journal of Metals*, vol. 6, p. 111, 1956.

[18] L. M. Pidgeon and J. A. King, "The vapour pressure of magnesium in the thermal reduction of MgO by ferrosilicon," *Discussions of the Faraday Society*, vol. 4, pp. 197–206, 1948.

[19] J. R. Wynnyckyj and L. M. Pidgeon, "Equilibria in the silicothermic reduction of calcined dolomite," *Metallurgical Transactions*, vol. 2, no. 4, pp. 979–986, 1971.

[20] I. M. Morsi, K. A. El Barawy, M. B. Morsi, and S. R. Abdel-Gawad, "Silicothermic reduction of dolomite ore under inert atmosphere," *Canadian Metallurgical Quarterly*, vol. 41, no. 1, pp. 15–28, 2002.

[21] K. C. Mills and B. J. Keene, "Physicochemical properties of molten CaF_2 -based slags," *International metals reviews*, vol. 26, no. 1, pp. 21–69, 1981.

[22] R. B. Winkler, *Vacuum Metallurgy*, Elsevier, London, UK, 1971.

[23] W. T. Hughes, C. E. Ransky, and E. F. Emley, "Reaction kinetics in the production of magnesium by Dolomite-fer-rosilicon (Pidgeon) process," in *Proceedings of the Advanced Extreme Metropolitan Proceedings, Symposium*, pp. 429–454, London, UK, 1968.

[24] E. A. H. Berg, *Huttemmann Monatsh*, vol. 106, p. 397, 1959.

[25] R. N. Singh, *Steel India*, no. 7, p. 73, 1984.

[26] V. N. Potamin, S. V. Shavrim, and M.Z. Panfilov, *Russ Metall*, p. 63, 1976.

[27] C. W. Dannatt and H. J. T. Ellingham, "II. Roasting and reduction processes: Roasting and reduction processes—a general survey," *Discussions of the Faraday Society*, vol. 4, pp. 126–139, 1948.

[28] J. S. Machin and T. B. Yee, "Viscosity Studies of System CaO-MgO-Al_2O_3-SiO_3: IV, 60 and 65% SiO_2 ," *Journal of the American Ceramic Society*, vol. 37, no. 4, pp. 177–186, 1954.

Mathematical Models for the Novel Cathode Convexes in a Reduction Cell

Z. He, T. Xia, W. Xiong, Q. Shen, and B. Li

Key Laboratory for Ferrous Metallurgy and Resources Utilization of Ministry of Education, Wuhan University of Science and Technology, Wuhan 430081, China

Correspondence should be addressed to Z. He; hezhu@wust.edu.cn

Academic Editor: Seetharaman Sridhar

A less magnitude of liquid aluminum deformation is required to shorten the anode-cathode distance so as to lower the electric energy consumption of the aluminum reduction cell. A mathematical model aimed to describe the electrolyte/aluminum two-phase flow in reduction cells, based on the computational fluid dynamics method, was developed to study the impacts of the cathode convexes on the electrolyte/aluminum interface deformation. The results showed that the magnitude of the two-phase interface deformation was reduced for about 17.2% with the novel cathode convexes; while at the same time, the washout of the melt on the ledge was also enhanced.

1. Introduction

Aluminum and its aluminum material is the second large metal materials in the world whose consumption is only next to the steel. Aluminum metal was produced by the electrolysis of aluminum oxide dissolved in an aluminum-electrolyte fluoride mixture, which requires large amounts of electrical energy. Therefore, there is an urgent demand to lower energy consumption in the process of aluminum production. To decrease the energy consumption, the anode-cathode-distance (ACD) in reduction cells needs to be cut down [1]. However, due to the electrolyte/aluminum interface instability, the ACD cannot be too small to avoid reduction cell voltage fluctuation which may result in cell failure [2].

The instability between the layer of molten aluminum and a slightly lighter electrolyte plays a defining role in the energy saving of aluminum reduction cells. Consequently, many researchers have focused on uncovering the mechanism of this instability. By using particle image velocimetry, Cooksey and Yang [3] experimentally investigated the onflow caused by anode gas and its bubble behavior in a full-scale, three-anode liquid model aluminum reduction cell. Pedchenko et al. [4] proposed a new solution for experimental modeling of the interfacial instability in which a tiny electrode stick was immerged into liquid metal of In-Ga-Sn on substitute of molten aluminum overlaid by electrolyte. Due to the difficulty to experimentally study the complex fluid flow phenomenon including molten aluminum, bath, and bubble agitation in aluminum reduction cells, most researchers focused on numerically modeling the magnetic fluid flow in reduction cells. Based on the nonlinear shallow water model, Kadkhodabeigi [5] built up a two dimensional model to study the electrolyte/aluminum interface instability and this model is superior to the linear model because linear model cannot determine the flow mode of the electrolyte and aluminum. Bojarevics and Pericleous [6] presented a nonlinear cell stability analysis for the 500 kA cell based on the shallow water model and showed the cathode surface unevenness and the presence of the anode bottom channels would sharply affect the cell stability. Li et al. [7] developed an improved finite-element model by building the contact mechanisms between steel bars and cathode blocks to make sure the voltage drops are properly distributed in the metal and the cathode. Based on the commercial package Ansys Li and Zhou [8] investigated the effects of various busbar configuration parameters on the stability of the reduction cell and proposed the improved busbar configuration. Bojarevics and Pericleous [9] proposed a simple shallow water model which allowed account of the channels and the interface stability is tested and compared for the channel effects.

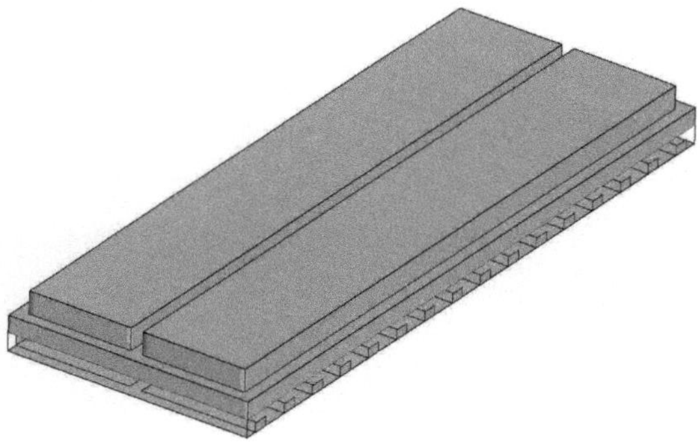

FIGURE 1: Geometry of the reduction cell.

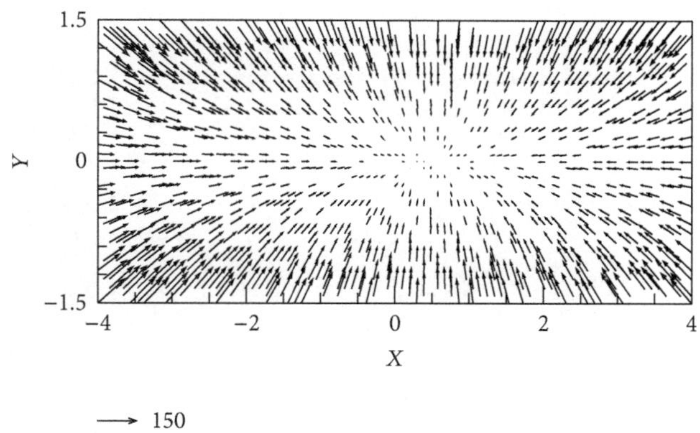

FIGURE 2: Distribution of the horizontal electric-magnetic force in the middle plane of the liquid aluminum.

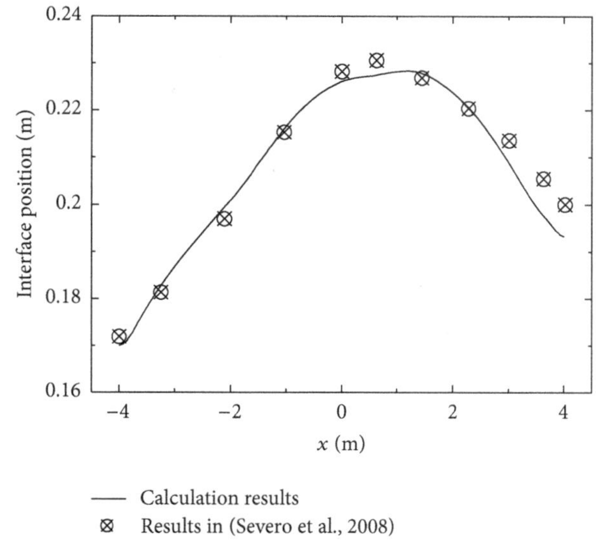

FIGURE 3: Height of the electrolyte/aluminum interface at the middle line.

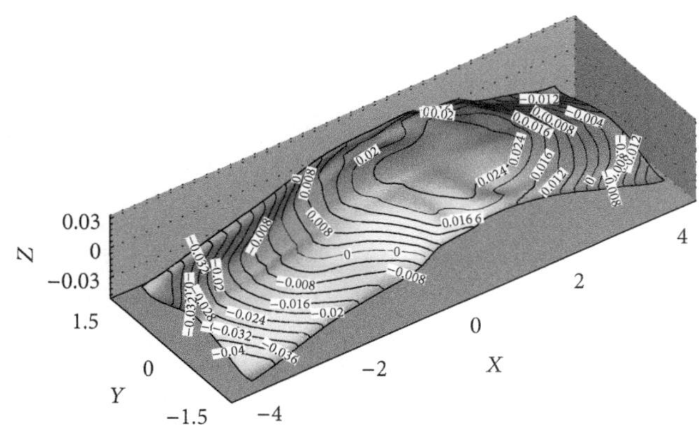

FIGURE 4: Shape of the electrolyte/aluminum interface.

In their following studies, Bojarevics and Pericleous [10] extended their model to time dependent case and account for the electrolyte channels. Liu et al. [11] developed a current efficiency predictive model based on multiphase multicomponent flow, which provides a new approach for predicting current efficiency of aluminum reduction cells.

What have been discussed above showed that though many works have been done to investigate the electrolyte/aluminum interface instability, neither reduction cells with inert anode [12] nor drained cathode cells [13] can decrease the energy consumption significantly in industrial scale experiment. Recently a new kind of aluminum reduction cell with novel cathodes has been tested industrially and the results showed that its energy consumption decreased for more than 1200 kWh/t-Al when compared with other reduction cells [14]. In the present work, the effects of novel cathode convexes on the interface instability were investigated to provide a theoretical foundation for improving the structure of the novel cathode and energy saving effects.

2. Numerical Models

Molten aluminum and electrolyte are treated as incompressible viscous fluid and electromagnetic force is the driving factor for melt to flow. Three dimensional Reynolds Average Navier-Stokes Equation was generally used to solve melt flow. The k- ε turbulence model coupled with standard wall functions was adopted to account for the turbulence effects and second order discretization on governing equations was employed to improve the accuracy of the solution. Volume of Fluid (VOF) method was used to compute the aluminum/electrolyte two-phase flow field.

The control equations of the aluminum/electrolyte flow field include the following.

Continuity equation:

$$\frac{\partial \overline{u_j}}{\partial x_j} = 0. \tag{1}$$

Momentum equation:

$$\rho \frac{\partial \overline{u_i}}{\partial t} + \rho \overline{u_j} \frac{\partial \overline{u_i}}{\partial x_j} = -\frac{\partial \overline{p}}{\partial x_i} + \frac{\partial}{\partial x}\left(\mu_{\text{eff}}\left(\frac{\partial \overline{u_i}}{\partial x_j} + \frac{\partial \overline{u_j}}{\partial x_x}\right)\right) + F, \tag{2}$$

where ρ is the density, $\overline{u_i}$ is the time averaged velocity component, $\overline{p}$ is the time averaged pressure, μ_{eff} is the effective

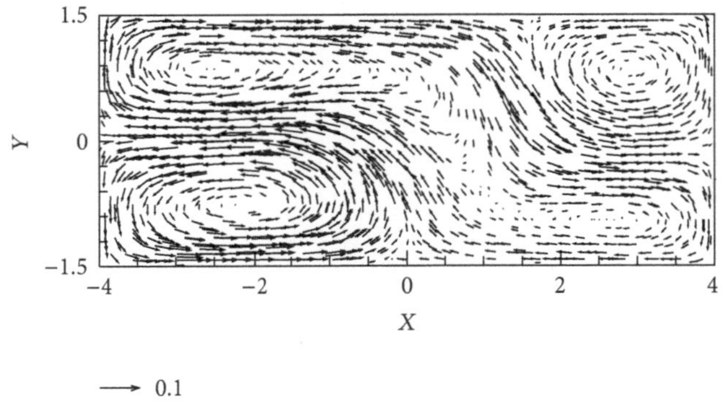

FIGURE 5: Distribution of the horizontal velocity vectors in the middle plane of the liquid aluminum.

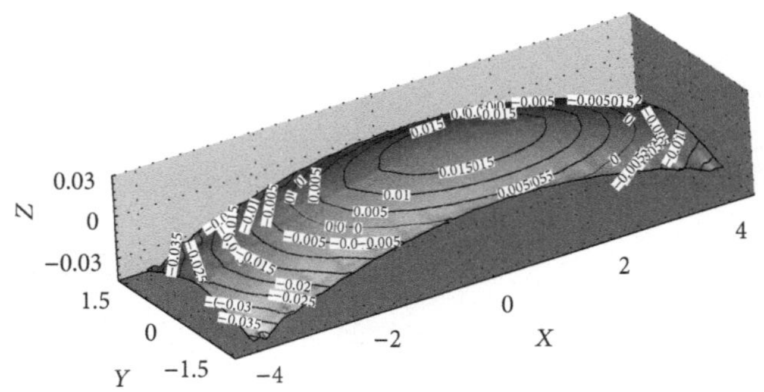

FIGURE 6: Shape of the electrolyte/aluminum interface with cathode convexes.

viscosity, and F is the volumetric force, which includes electromagnetic force $F_{em} = J \times B$ and gravity F_g, respectively.

k-equation:

$$\frac{\partial(\rho k u_i)}{\partial x_i} = \frac{\partial}{\partial x_j}\left[\left[\mu + \frac{\mu_t}{\sigma_k}\right]\frac{\partial k}{\partial x_j}\right] + G_k - \rho\varepsilon. \quad (3)$$

ε-equation:

$$\frac{\partial(\rho \varepsilon u_i)}{\partial x_i} = \frac{\partial}{\partial x_j}\left[\left[\mu + \frac{\mu_t}{\sigma_k}\right]\frac{\partial \varepsilon}{\partial x_j}\right] + \frac{C_{1\varepsilon}\varepsilon}{k}G_k - C_{2\varepsilon}\rho\frac{\varepsilon^2}{k}, \quad (4)$$

where G_k represents the generation of turbulence kinetic energy due to the mean velocity gradients,

$$G = \mu_t \frac{\partial u_j}{\partial x_i}\left(\frac{\partial u_i}{\partial x_j} + \frac{\partial u_j}{\partial x_i}\right),$$
$$\mu_{\text{eff}} = \mu + \mu_t. \quad (5)$$

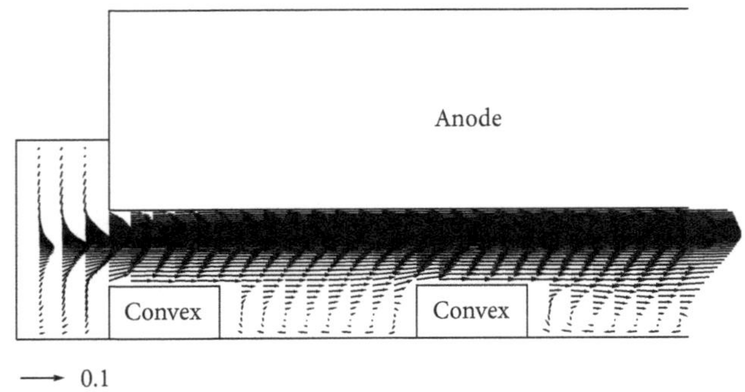

FIGURE 7: Velocity vector distribution at the cross plane ZX, with $Y = -1$ (with convexes).

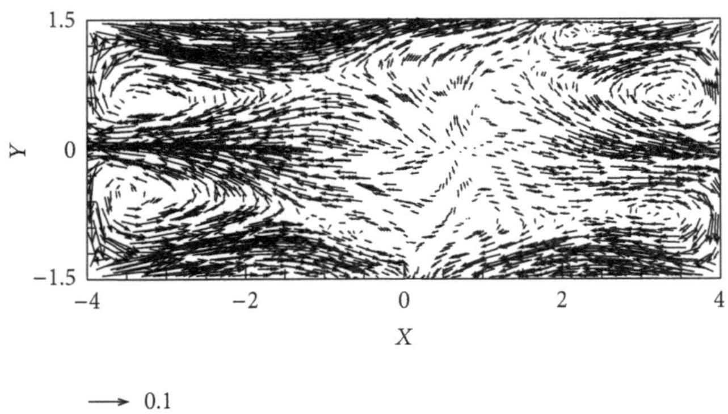

FIGURE 8: Distribution of the horizontal velocity vectors in the middle plane of the liquid aluminum with cathode convexes.

μ_t is turbulent viscosity:

$$\mu_t = \rho C_\mu \frac{k^2}{\varepsilon}. \quad (6)$$

The typical constants of the k-ε model of Launder and Spalding [15] are: $c_{1\varepsilon} = 1.44$, $c_{2\varepsilon} = 1.92$, $c_\mu = 0.09$, $\sigma_k = 1.00$, $\sigma_\varepsilon = 1.30$.

Volume fraction equation:

$$\frac{\partial \alpha_q}{\partial t} + u_i \frac{\partial \alpha_q}{\partial x_i} = 0,$$
$$\sum_{q=1}^{n} \alpha_q = 1, \quad (7)$$

where α_q is the volume fraction of the qth phase, $q = 1, 2$ represent aluminum and electrolyte, respectively.

The computational model in the present study is the second benchmark case in [16]; its geometry is shown in Figure 1. The computational parameters are listed in Table 1. In this benchmark case, magnetic field and current density distributions are simple analytical expressions, but their shape and magnitude are typical for real cells. No interanode gaps in the longitudinal direction are used.

Firstly, Laplace equation $\Delta V = 0$ was solved to get the current density distribution in the reduction cell, where V

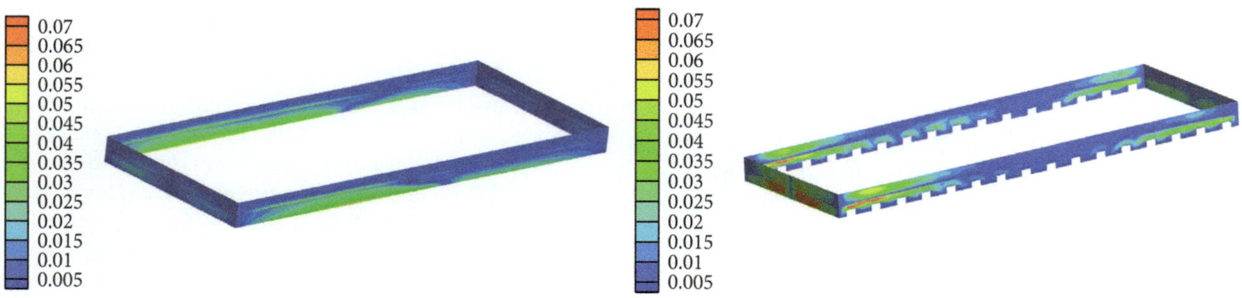

(a) Without cathode convexes

(b) With cathode convexes

FIGURE 9: Wall shear stress distributions.

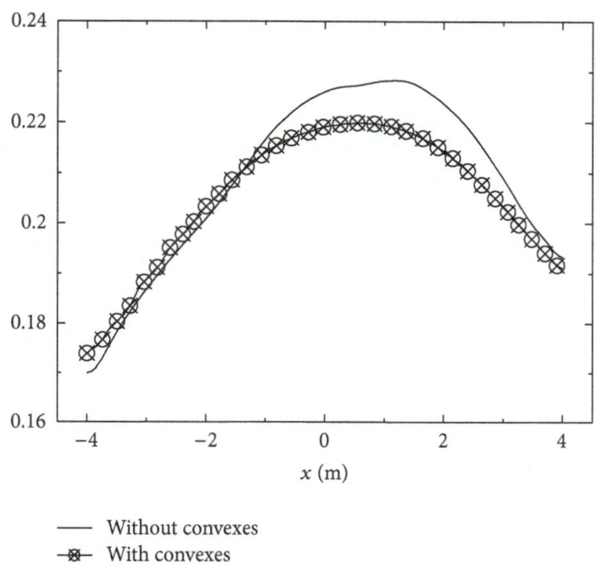

FIGURE 10: Height of the electrolyte/aluminum interface at the middle line.

TABLE 1: Computational parameters.

Aluminum height	Electrolyte height	ACD	Aluminum density	Electrolyte density
0.2 m	0.18 m	0.045 m	2270 kg/m^3	2070 kg/m^3

is the electrical potential. At the top of the cathode block, the current density is specified as the boundary condition: $J_z = -5625 - 2500y^2$ (A/m^2). At the top of anode blocks, zero electrical potential is specified. The electromagnetic force density fields, which drive the molten aluminum and electrolyte flow, are obtained as vector product of electric current density and magnetic field. The three components of the magnetic field inside the bath and the metal can be gotten in [16]. Electromagnetic force distributions in the middle plane of the molten aluminum are shown in Figure 2.

3. Results and Discussion

To validate the present model, the calculated results are compared with those presented in [16], as shown in Figure 3. The interface location is calculated where the volume fraction of the metal and of the bath is 0.5. The results in Figure 3 shows that the predictions of the present model agree well with those in [16] with a deviation not exceeding 5%, which laid the basis of further investigation. The aluminum/electrolyte interface shape is shown in Figure 4. Under the influence of the electromagnetic force which is the results of the horizontal magnetic field and the vertical electric current, molten aluminum was upheaved in the middle part. The peak uplift amplitude is about 2.4 cm and the max downwelling amplitude is about 4 cm, which approximately equal to the fluctuation magnitude in real cells. Current efficiency can be further improved by adequate fine adjustment in real reduction cells.

Figure 5 shows the velocity vector distributions in the middle plane of the molten aluminum. Four large vortices are formed in flow field and the size of the left vortex is larger than the right one. The velocity magnitude of the molten aluminum around the left vortex is larger, which makes the washout on the side ledge more severe.

Interface shape of the reduction cell with cathode convexes is shown in Figure 6. Similar to the case without convexes, molten aluminum was upheaved in the middle part, but the upheaval magnitude was weaken evidently when cathode convexes were used in the reduction cell. The peak uplift amplitude is about 1.5 cm and the max downwelling amplitude is about 3.8 cm. The peak uplift and downwelling amplitude are reduced by 37.5% and 5%, respectively, and the fluctuation magnitude is reduced by 17.2%.

With the cathode convexes, the magnitude of the electromagnetic force distributions in the middle plane of the molten aluminum is decreased for about 10%, but the distribution is very similar to what is shown in Figure 2 (not shown here). The usage of the cathode convexes brings about the secondary flow of the molten aluminum between the cathode convexes as shown in Figure 7. This secondary flow could also be the reason for weaken interfacial fluctuation magnitude. As a result, the usage of cathode convex makes it possible to further decrease ACD and lower the energy consumption. Figure 8 shows the velocity vector distributions in the middle plane of the molten aluminum when cathode convexes are used. The flow mode of the molten aluminum has been changed greatly when compared with that shown in Figure 5. The washout of the molten aluminum on the side ledge is more serious than that in reduction cells without cathode convexes, which can be also observed in Figure 9. The wall shear stress on the side ledge is obviously larger when cathode convexes are used as shown in Figure 9.

Figure 10 compared the interface profile on the longitudinal axis located at the center of the cell with and without convex. It clearly showed the inhibition effects of the cathode convexes on the molten aluminum fluctuation.

4. Conclusions

A steady three-dimensional model of the electrolyte/aluminum two-phase flow in reduction cells was developed in the present study, in which time dependent interface motion is not considered. Based on this model electrolyte/aluminum interface fluctuations in the reduction cell with novel cathode convexes were investigated. The results showed that the model is able to represent the flow mode of the electrolyte/aluminum two-phase flow in reduction cells adequately. The model predicted that the magnitude of the two-phase interface deformation was reduced for about 17.2% with novel cathode convexes; while at the same time, the washout of the melt on the ledge was also enhanced.

Acknowledgments

This work is supported by the National Natural Science Foundation of China (no. 50934005) and the Open Research Fund of Key Laboratory for Ferrous Metallurgy and Resources Utilization of Ministry of Education, Wuhan University of Science and Technology.

References

[1] Y. Tian, "Discussion on Best Polar Distance of Industrial Aluminum Cell," *Non-Ferrous Mining and Metallurgy*, vol. 25, pp. 23–25, 2007.

[2] O. Zikanov, A. Thess, P. A. Davidson, and D. P. Ziegler, "A new approach to numerical simulation of melt flows and interface instability in hall-héroult cells," *Metallurgical and Materials Transactions B*, vol. 31, no. 6, pp. 1541–1550, 2000.

[3] M. A. Cooksey and W. Yang, "PIV measurements on physical models of aluminum reduction cells," *in Light Metals*, pp. 359–365, 2005.

[4] A. Pedchenko, S. Molokov, J. Priede, A. Lukyanov, and P. J. Thomas, "Experimental model of the interfacial instability in aluminium reduction cells," *Europhysics Letters*, vol. 88, no. 2, Article ID 24001, 2009.

[5] M. Kadkhodabeigi, "Two dimensional model of melt flows and interface instability in aluminum reduction cells," in *Light Metals*, vol. 4, pp. 443–448, Wiley, New York, NY, USA, 2008.

[6] V. Bojarevics and K. Pericleous, "Shallow water model for aluminium electrolysis cells with variable top and bottom," in *Light Metals*, vol. 2, pp. 403–408, 2008.

[7] J. Li, W. Liu, Y. Lai, and Y. Liu, "An improved finite-element model for electromagnetic analysis in aluminum cells," *JOM*, vol. 60, no. 2, pp. 58–61, 2008.

[8] M. Li and J. Zhou, "Numerical simulation of busbar configuration in large aluminum electrolysis cell," *Journal of Central South University of Technology*, vol. 15, no. 2, pp. 271–275, 2008.

[9] V. Bojarevics and K. Pericleous, "Solutions for the metal-bath interface in aluminium electrolysis cells," in *Light Metals*, pp. 569–574, TMS, 2009.

[10] V. Bojarevics and K. Pericleous, "Time dependent MHD models for aluminium reduction cells," in *Jim Evans Honorary Symposium: Proceedings of the Symposium Sponsored by the Light Metals Division of The Minerals, Metals and Materials Society (TMS)*, pp. 199–206, 2010.

[11] Z. Liu, W. Li, Q. Zhao, J. Zhou, and Y. Wang, "Current efficiency predictive model and its calibration and validation," in *Light Metals*, pp. 935–938, 2012.

[12] J. Bruggeman, "Technical contributions of late Warren Haupin—a review," in *Light Metals*, pp. 495–500, 2008.

[13] J. P. Peng, N. X. Feng, Y. L. Jiang, Y. W. Wang, and J. You, "Test of drained aluminum electrolysis cell with TiB_2/G graphitized cathode at high current density," *Chinese Journal of Nonferrous Metals*, vol. 18, no. 4, pp. 738–744, 2008 (Chinese).

[14] N. Feng, "Low energy consumption aluminum reduction cell with novel cathodes," China Patent, ZL200710010523.4, 2007.

[15] B. E. Launder and D. B. Spalding, "The numerical computation of turbulent flows," *Computer Methods in Applied Mechanics and Engineering*, vol. 3, no. 2, pp. 269–289, 1974.

[16] D. S. Severo, V. Gusberti, A. F. Schneider, E. C. V. Pinto, and V. Potocnik, "Comparison of various methods for modeling the metal-bath interface," in *Light Metals*, pp. 413–418, 2008.

5

Inhibition Effect of Substituted Thiadiazoles on Corrosion Activity of N80 Steel in HCl Solution

M. Yadav, Sumit Kumar, and Debasis Behera

Department of Applied Chemistry, Indian School of Mines, Dhanbad 826004, India

Correspondence should be addressed to M. Yadav; yadav_drmahendra@yahoo.co.in

Academic Editor: Elena V. Pereloma

The inhibition effect of some prepared compounds, namely, thiadiazole derivatives, on N80 steel corrosion in 15% HCl solutions has been studied by using the weight loss, electrochemical polarization, and electrochemical impedance spectroscopy techniques. It was found that the inhibition efficiency of the thiadiazole derivatives, namely, 2-amino-5-(4-methoxyphenyl)-1,3,4-thiazole (AMPT), 2-amino-5-phenyl-1,3,4-thiazole (APT), and 2-amino-5-(4-chlorophenyl)-1,3,4-thiazole (ACPT), increases with the increase in concentration. Inhibition efficiency follows the order AMPT > APT > ACPT. The effect of temperature on the corrosion was investigated by the weight loss method, and some thermodynamic parameters were calculated. The inhibitive action may be attributed to the adsorption of inhibitor molecules on the active sites of the metal surface following Langmuir adsorption isotherm. Polarization measurements indicated that thiadiazole derivatives act as mixed-type corrosion inhibitor. The adsorption of thiadiazole derivatives on N80 surface exposed to inhibitor-containing solutions was confirmed using SEM and FT-IR spectra.

1. Introduction

N80 carbon steel has been generally used as the main construction material for down hole tubular, flow lines, and transmission pipelines in petroleum industry. In most industrial processes, acidic solutions are commonly used for pickling, industrial acid cleaning, acid descaling, oil well acidifying, and so forth [1–5]. It is commonly noticed that 15–28% of hydrochloric acid is used for the acidization of petroleum oil wells [6, 7]. Because of their aggressiveness, iron and its alloys get corrodes during these acidic applications particularly with the use of hydrochloric acid and sulphuric acid, which results in terrible waste of both resources and money [8]. Addition of inhibitor remains the necessary procedure to secure the metal against acid attack. Therefore, corrosion inhibitors for HCl solutions have attracted more attention. Most of the well-known corrosion inhibitors are organic compounds containing polar groups having nitrogen, sulphur, and/or oxygen atoms and heterocyclic compounds with polar functional groups and conjugated double bonds [9, 10]. These compounds can adsorb on the metal surface and partially blocking the active sites on the surface, thereby reducing the corrosion rate. Among different nitrogen- and sulphur-containing compounds, thiosemicarbazide has been reported to be a potential inhibitor for different metals [11, 12]. Most of the investigation is related to the application of common inhibitors like various derivatives of the compound of aminothiazoles [13], benzotriazoles [14–16], thioimidazole [17], mercapto-5-triazole [18], and thiadiazole [19, 20] which have been used as potential inhibitors for this system.

Keeping in the view of the above excellent behavior of the organic compounds containing nitrogen and sulphur as corrosion inhibitors, some compounds of substituted thiadiazoles, namely, 2-amino-5-(4-methoxyphenyl)-1, 3,4-thiadiazole (AMPT), 2-amino-5-phenyl-1,3,4-thiadiazole (APT), and 2-amino-5-(4-chlorophenyl)-1,3,4-thiadiazole (ACPT), have been synthesized, and their corrosion inhibition property was studied by the weight loss technique and electrochemical techniques.

2. Experimental Section

2.1. Materials

2.1.1. Composition of N80 Sample. The working electrode and specimens for weight loss experiments were prepared from oil well N80 steel having the percentage composition (wt.%) as shown in Table 1. The specimens were mechanically abraded with 320, 400, 600, 800, 1000, and 1200 emery paper, washed in acetone and double-distilled water, then dried, and put into the cell.

2.1.2. Solutions. The aggressive solutions were made up of 37% HCl (Rankem). The concentration of the used inhibitors ranged from 20 ppm to 200 ppm in 15% HCl. All solutions were prepared in double-distilled water.

2.1.3. Synthesis of Corrosion Inhibitor. The inhibitors AMPT, APT, and ACPT were synthesized in the laboratory by a reported method [21] as shown in Scheme 1. An equimolar mixture of substituted benzoic acid (0.05 mol), thiosemicarbazide (0.05 mol), and phosphorous oxychloride (15 mL) was refluxed gently for half an hour. After cooling, water was added (45 mL) and the mixture was refluxed for 4 h and filtered. The solution was neutralized with saturated solution of potassium hydroxide. The precipitate was filtered and recrystallized from ethanol. Their purity was confirmed by TLC. The structure of inhibitors is shown in Table 2.

Table 1: Composition of specimens.

C	Mn	Si	P	S	Cr	Fe
0.31	0.92	0.19	0.01	0.008	0.20	Remainder

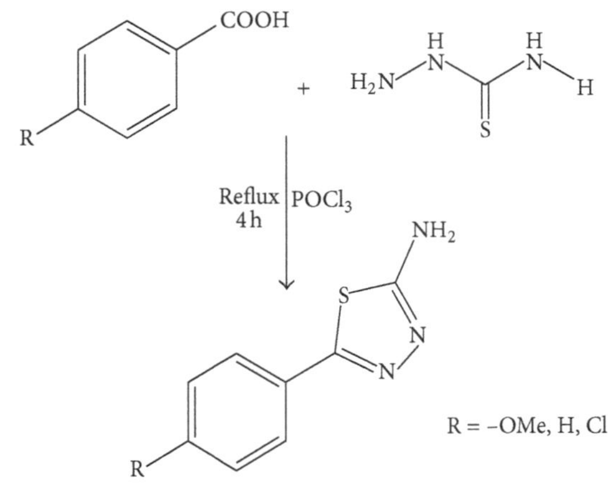

Scheme 1

2.2. Methods

2.2.1. Weight Loss Method. The N80 specimens were cut in the dimension of 3 cm × 3 cm × 0.1 cm with a small hole of 2 mm at the upper edge of specimen for weight loss studies, and the size of the electrodes was 1 cm × 1 cm with 4 cm long tag for electrochemical studies with an exposed area of 1 cm^2 towards aggressive solution. The specimens were mechanically abraded successively with 1/0, 2/0, 3/0, and 4/0 grade emery papers. The surface was thoroughly washed with soap, running tap water, and distilled water and finally was degreased with acetone. The samples were dried and stored in vacuum desiccators before immersion in the test solution. Weight loss measurements were performed at 303 K for 6 h by immersing the N80 steel coupons into 250 mL of 15% hydrochloric acid in the absence and the presence of various amounts of inhibitors. After the elapsed time, the specimens were taken out, washed, dried, and weighed accurately. The experimental studies were performed at temperature range of 303 K to 333 K. The inhibition efficiency (IE%) and surface coverage (θ) were determined by using following equation [22]:

$$\mathrm{CR}\,(\mathrm{mmpy}) = \frac{87.6\Delta W}{DAT}, \tag{1}$$

where ΔW = weight loss (mg), D = density of specimen (g/cm^3), A = area of specimen (cm^2), and T = exposure time (hours). The degree of surface coverage (θ) by inhibitors can be expressed in terms of corrosion rate as

$$\theta = \frac{\mathrm{CR}_0 - \mathrm{CR}_{\mathrm{inh}}}{\mathrm{CR}_0},$$
$$\mathrm{IE\%} = \frac{\mathrm{CR}_0 - \mathrm{CR}_{\mathrm{inh}}}{\mathrm{CR}_0} \times 100, \tag{2}$$

where CR_0 = corrosion rate in the absence of inhibitor and $\mathrm{CR}_{\mathrm{inh}}$ = corrosion rate in the presence of inhibitor.

2.2.2. Electrochemical Method. The electrochemical experiments were carried out in a three-necked glass assembly containing 250 mL of the electrolyte with different concentrations of inhibitors (from 20 ppm to 200 ppm by weight) dissolved in it. The electrochemical studies were carried out with N80 steel strips having an exposed area of 1 cm^2. The rest of the samples were isolated from the working environment by a cold mounting process, for which epoxy resin had been used. A conventional three-electrode cell consisting of N80 steel as a working electrode, platinum as counter electrode, and a saturated calomel electrode as reference electrode was used. Polarisation studies were carried out using an AMEL Potentiostat 2053, Italy, connected to a personal computer. Prior to electrochemical measurements, the working electrode was first immersed into the test solution for 30 min to establish a steady-state open circuit potential. After the establishment of the open circuit potential, dynamic polarization curves were obtained with a scan rate of 1 mV/s in the potential range from −470 to −390 mV with scan rate of 10 m Vs^{-1}. All potentials were measured against SCE. All experiments were performed at 30 ± 0.2°C in an electronically controlled air thermostat. The inhibition efficiency (IE %) was calculated using the following equation:

$$\%\ \mathrm{IE} = \left[\frac{(I_0 - I_{\mathrm{inh}})}{I_0}\right] \times 100, \tag{3}$$

where I_0 = corrosion current in the absence of inhibitor and I_{ihn} = corrosion current in the presence of inhibitor.

AC-impedance studies were carried out in three-electrode cell assembly containing N80 steel as the working electrode, platinum as counter electrode, and saturated calomel as reference electrode. The impedance measurements

TABLE 2: Name and structural formulae of compounds thiadiazole derivatives.

Compound	Structure	Name and abbreviation
1	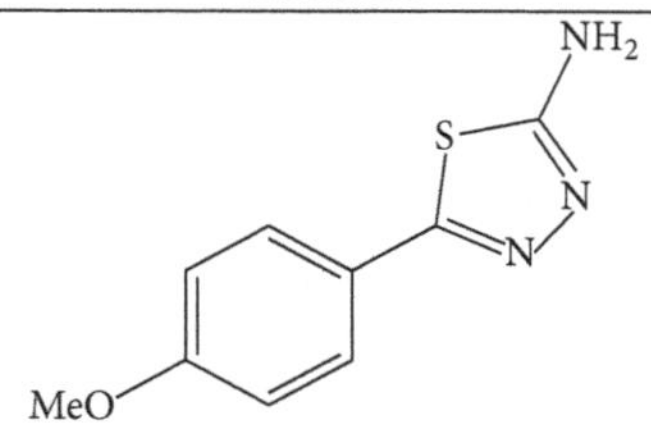	2-Amino-5-(4-methoxyphenyl)-1,3,4-thiadiazole (AMPT)
2	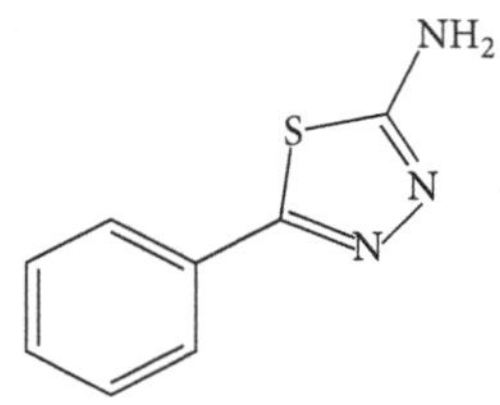	2-Amino-5-phenyl-1,3,4-thiadiazole (APT)
3	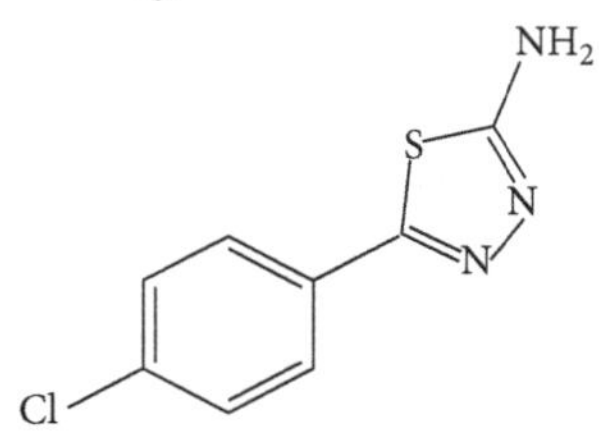	2-Amino-5-(4-chlorophenyl)-1,3,4-thiadiazole (ACPT)

were carried out at an open circuit potential (OCP), after 30 min immersion of the N80 steel electrode in the corrosive medium. The impedance data were acquired in the frequency range of 100 KHz–50 mHz with an AC voltage amplitude of 10 mV. The charge transfer resistance (R_{ct}) and double-layer capacitance (C_{dl}) were determined from Nyquist plots. The inhibition efficiencies were calculated from charge transfer resistance values by using the following formula:

$$\%\ \text{IE} = \left[\frac{R_{ct(inh)} - R_{ct}}{R_{ct(inh)}}\right] \times 100, \quad (4)$$

where R_{ct} = charge transfer resistance in absence of inhibitor; $R_{ct(Inh)}$ = charge transfer resistance in presence of inhibitor.

2.2.3. Scanning Electron Microscopic (SEM) Analysis. The N80 specimens of size 1.0 cm × 1.0 cm × 0.06 cm were abraded with a series of emery papers (grades 320, 500, 800, and 1200) and then washed with distilled water and acetone. After immersion in 15% HCl in the absence and the presence of optimum concentration of inhibitor (AMPT) at 30°C for 6 h, the specimen was cleaned with distilled water, dried with a cold air blaster, and then the SEM images were recorded using JEOL JSM-6380 LA analytical scanning electron microscope.

2.2.4. FT-IR Spectroscopy. For the FTIR studies the N80 steel coupons were immersed in 15% HCl containing optimum concentration of inhibitors (AMPT, APT, and ACPT) for 24 h. The test coupons were then removed from the solution, washed thoroughly with distilled water and dried. The film formed on the N80 surface in the presence of the inhibitor was collected by nonmetallic scrapper from the surface of the steel for spectral analysis with Fourier transform infrared spectroscopy (FT-IR). FT-IR spectra were recorded between 4000 cm^{-1} and 400 cm^{-1}, and the scraped films were recorded using a Perkin Elmer Model 2000, by KBr pellet method.

TABLE 3: The variation of percentage inhibition efficiency obtained from weight loss studies.

Concentration (ppm)	(IE percent) inhibition efficiency		
	AMPT	APT	ACPT
20	63.27	60.32	57.32
50	67.72	64.51	60.65
100	74.84	70.45	67.47
150	80.27	77.58	73.51
200	82.32	79.03	75.71

3. Results and Discussion

3.1. Weight Loss Measurement. Corrosion inhibition efficiency (IE%) offered by inhibitors (AMPT, APT, and ACPT) has been evaluated by weight loss technique after 6 h of immersion at 303 K which is given in Table 3. From this table it is apparent that inhibition efficiency (IE%) was increased with the increase in the concentration of the inhibitor [23]. The maximum of 82.3% inhibition efficiency was obtained at the 200 ppm inhibitor (AMPT) concentration. No considerable effect on the inhibition efficiency (IE%) was observed by a further increase in the inhibitor concentration in the acid solution. From this table, the order of the inhibition efficiency follows the order AMPT > APT > ACPT. Corrosion inhibition studies were also carried out at different temperatures.

TABLE 4: Corrosion parameters, namely, corrosion rate (CR), surface coverage (θ), and inhibition efficiency IE (%) of N80 steel in 15% HCl in the presence and the absence of the inhibitor at different temperatures, obtained from weight loss measurements.

Inhibitor	Conc (ppm)	303 K			313 K			323 K			333 K		
		CR (mmpy)	θ	IE%	CR (mmpy)	θ	IE%	CR (mmpy)	θ	IE%	CR (mmpy)	θ	IE%
Blank	—	2.672	—	—	9.084	—	—	17.943	—	—	60.348	—	—
AMPT	20	0.983	0.632	63.2	2.906	0.680	68.0	4.719	0.737	73.7	14.00	0.768	76.8
	50	0.863	0.677	67.7	2.416	0.734	73.4	3.893	0.783	78.3	10.00	0.833	83.3
	100	0.673	0.748	74.8	1.607	0.823	82.3	2.440	0.864	86.4	6.155	0.898	89.8
	150	0.529	0.802	80.2	1.317	0.855	85.5	1.920	0.893	89.3	5.129	0.915	91.5
	200	0.446	0.823	82.3	1.090	0.880	88.0	1.596	0.911	91.1	4.586	0.924	92.4
APT	20	1.060	0.603	60.3	3.148	0.653	65.3	5.678	0.683	68.3	16.66	0.724	72.4
	50	0.948	0.645	64.5	2.831	0.688	68.8	4.670	0.739	73.9	12.69	0.789	78.9
	100	0.790	0.704	70.4	2.128	0.766	76.6	3.583	0.800	80.0	9.975	0.839	83.9
	150	0.600	0.775	77.5	1.769	0.805	80.5	3.019	0.832	83.2	8.189	0.864	86.4
	200	0.552	0.793	79.3	1.349	0.851	85.1	2.486	0.861	86.1	6.914	0.894	89.4
ACPT	20	1.140	0.573	57.3	3.455	0.619	61.9	6.263	0.651	65.1	19.60	0.675	67.5
	50	1.053	0.606	60.6	3.332	0.633	63.3	5.363	0.701	70.1	14.83	0.754	75.4
	100	0.869	0.674	67.4	2.503	0.725	72.5	4.419	0.754	75.4	12.40	0.794	79.4
	150	0.708	0.735	73.5	2.082	0.770	77.0	3.676	0.795	79.5	10.72	0.822	82.2
	200	0.649	0.757	75.7	1.637	0.819	81.9	3.136	0.825	82.5	8.762	0.855	85.5

Corrosion parameters, namely, corrosion rate (CR), surface coverage (θ), and inhibition efficiency (IE %) of N80 steel in 15% HCl in the presence and absence of inhibitor at different temperature, obtained from weight loss measurements, are shown in Table 4. As shown from this table, by increasing the concentration of the inhibitor, corrosion rate decreases; as a result, the inhibition efficiency was increased with rise in temperature [24]. For all concentrations of AMPT, the systems at 333 K gave the highest efficiencies. The maximum efficiency of 92% was obtained for 200 ppm AMPT at 333 K. The variation in corrosion rate with different concentration of thiadiazole derivatives in 15% HCl at different temperature is shown in Figure 1. It is evident from Figure 1 that the inhibitor (AMPT, APT, and ACPT) has shown remarkable corrosion inhibition at a higher temperature. The variation of inhibition efficiency with the temperature is shown in Figure 2. From this figure, order of the inhibition efficiency of thiadiazole derivatives is as follows: AMPT > APT > ACPT. High efficiency of these compounds was attributed to the presence of extensively delocalized π electrons of the phenyl rings, planarity, and the presence of lone pair of electrons on N and S atoms which favored a greater adsorption of inhibitors on the metal surface [25].

3.1.1. Thermodynamic and Activation Parameters. Thermodynamic and activation parameters play important role in understanding the inhibitive mechanism. The weight loss measurements were done in the temperature range of 303–333 K in the absence and the presence of different concentrations of inhibitors (AMPT, APT, and ACPT) in 15% HCl for N80 steel. The apparent activation energy (E_a) for dissolution of N80 steel in 15% HCl can be expressed using the Arrhenius equation,

$$\log \text{CR} = \frac{-E_a}{2.303RT} + \log A, \quad (5)$$

TABLE 5: Activation parameter for N80 in 15% HCl solution in the absence and the presence of the inhibitor obtained from weight loss measurements.

Inhibitor	Concentration (ppm)	E_a (kJ mol^{-1})	ΔH^* (kJ mol^{-1})	ΔS^* (Jmol^{-1} K^{-1})
Blank	—	85.06	84.14	40.99
AMPT	20	72.93	66.09	−26.78
	50	66.67	64.75	−32.29
	100	60.32	62.80	−40.62
	150	62.32	59.13	−54.68
	200	63.78	49.40	−88.48
APT	20	75.40	71.61	−8.63
	50	70.50	66.87	−24.38
	100	70.15	65.50	−30.71
	150	72.25	67.63	−25.73
	200	69.41	66.00	−32.34
ACPT	20	78.53	71.73	−6.60
	50	72.64	68.45	−18.37
	100	72.66	68.72	−19.13
	150	74.13	69.77	−17.30
	200	71.15	69.44	−20.50

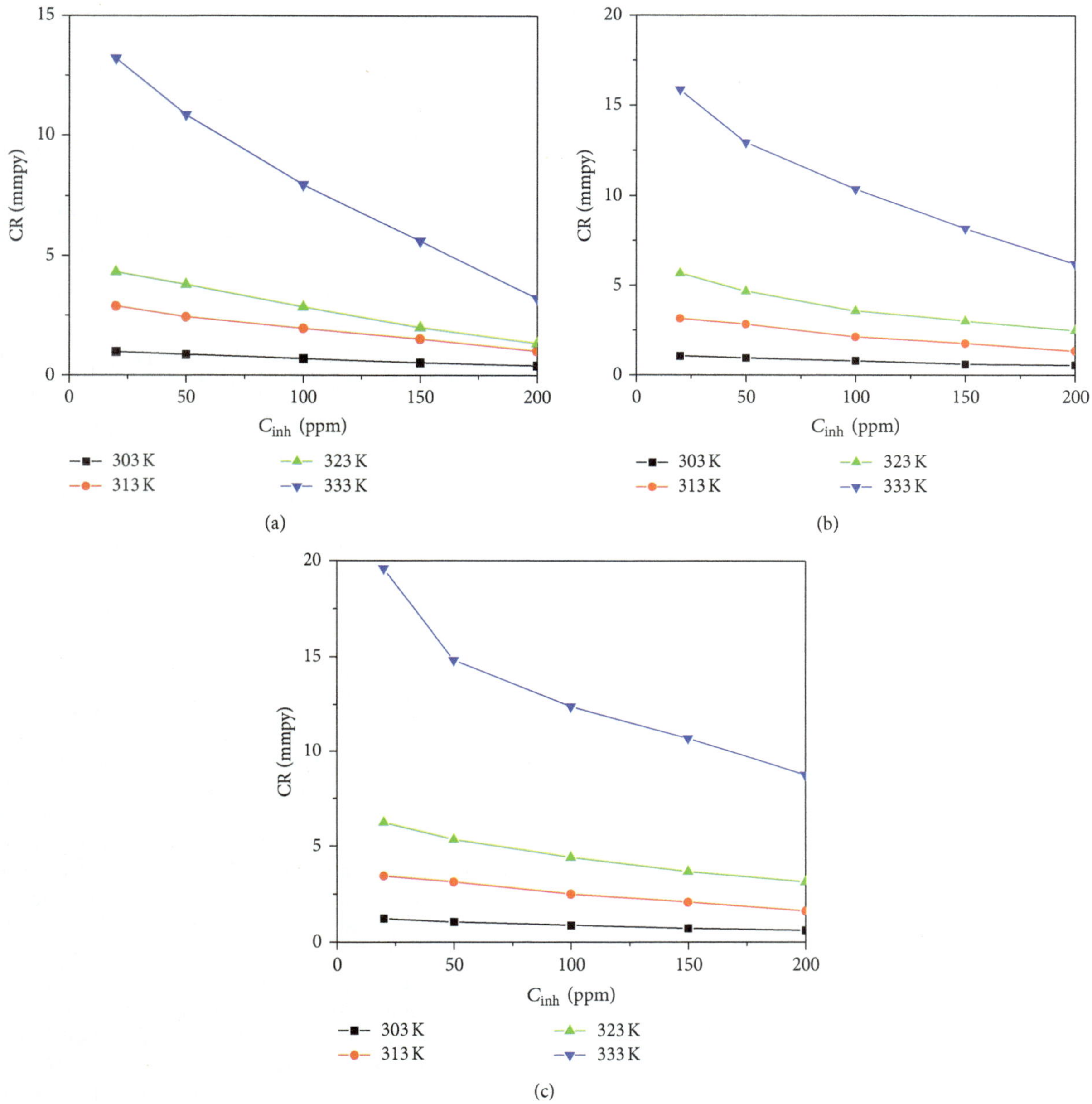

FIGURE 1: A plot of corrosion rate versus inhibitor's concentrations at different temperatures. (a) AMPT, (b) APT, and (c) ACPT.

where CR is the corrosion rate, E_a is the apparent activation energy, R is the molar gas constant ($8.314\,\mathrm{J\,K^{-1}\,mol^{-1}}$), T is the absolute temperature, and A is the Arrhenius pre-exponential factor. Figure 3 represents the Arrhenius plot of log CR against $1/T$ for the corrosion of N80 steel in 15% HCl solution with or without the presence of AMPT, APT, and ACPT at concentrations ranging from 20 ppm to 200 ppm. From Figure 3, the slope ($-E_a/R$) of each individual line was determined, and the activation energy has been calculated by using (E_a = (slope) × 2.303 × R) and the values of E_a were summarized in Table 5. It is evident from Table 5 that the decrease in activation energy E_a indicates the retardation in corrosion rate which could have occurred because of the adsorption of the inhibitors at the surface of the metal [26].

The value of enthalpy of activation of the ΔH^* and the entropy of activation ΔS^* can be calculated by using the following formula:

$$\mathrm{CR} = \frac{RT}{Nh}\exp\left(\frac{\Delta S^*}{R}\right)\exp\left(-\frac{\Delta H^*}{RT}\right), \qquad (6)$$

where CR is the corrosion rate, H^* is the enthalpy of activation, S^* is the entropy of activation, h is Planck's constant, and N is the Avogadro number, respectively.

A plot of log (CR/T) against $1/T$ (Figure 4) gave straight lines with slope of ($-\Delta H^*/2.303R$) and intercept of (log $R/Nh + \Delta S^*/2.303R$) from which the activation thermodynamic parameters (ΔH^* and ΔS^*) are estimated and listed in Table 5. The results which are presented in Table 5 show that

Table 6: Electrochemical parameter and percentage inhibition efficiency obtained from polarisation studies for N80 steel in 15% HCl solution in the presence or the absence of the inhibitors at 303 K.

Inhibitor	Concentration (ppm)	E_{corr} (mV_{SCE})	b_a ($mV\ dec^{-1}$)	b_c ($mV\ dec^{-1}$)	I_{corr} ($\mu A\ cm^{-2}$)	% IE
Blank	—	−427	95	135	573	
AMPT	20	−434	55	104	178	68.87
	50	−437	64	82	160	72.02
	100	−438	77	93	134	76.54
	150	−445	157	142	95	83.37
	200	−447	111	128	80	86.51
APT	20	−428	74	147	203	64.52
	50	−432	61	73	182	68.21
	100	−433	53	84	154	73.21
	150	−436	92	98	106	81.58
	200	−438	108	44	103	82.03
ACPT	20	−430	68	102	229	60.02
	50	−431	75	91	205	64.05
	100	−432	88	79	177	69.07
	150	−434	97	108	133	76.81
	200	−436	118	156	114	80.07

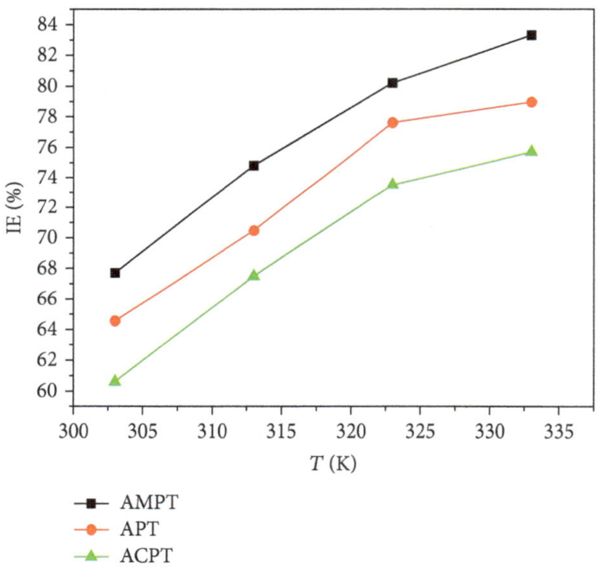

Figure 2: Variation of inhibition efficiency of the studied compounds with the temperature containing optimum concentration of inhibitor.

the enthalpy of activation is all positive. The positive sign of the enthalpy reflects the endothermic nature of the N80 steel dissolution process [27]. The negative value of ΔS^* (Table 5) for all the three inhibitors indicates that activated complex in the rate determining step represents an association rather than a dissociation step, meaning that a decrease in disorder takes place during the course of transition from the reactant to the activated complex, and the positive value of ΔS^* refers to an increase in the disorder [28].

3.2. Potentiodynamic Polarization Measurements. The anodic and cathodic polarisation curves for the corrosion of N80 steel in 15% HCl in the presence and the absence of varying concentrations of inhibitors (AMPT, APT, and ACPT) at 303 K are shown in Figure 6. The linear Tafel segments of the cathodic and anodic curves were extrapolated to the point of intersection to obtain the corrosion potential (E_{corr}) and corrosion current density (i_{corr}). The electrochemical parameters such as corrosion potential (E_{corr}), corrosion current density (i_{corr}), anodic (b_a) and cathodic (b_c) Tafel slope, and percentage inhibition efficiency (IE%) determined from polarisation curve are summarized in Table 6. The data in Table 6 clearly show that the current density decreases with the presence of thiadiazole inhibitors (AMPT, APT, and ACPT); this indicated that inhibitors adsorbed on the metal surface, and hence the inhibition efficiency increases with the increase in the inhibitor concentration up to an optimum value. Thereafter, the increase in the inhibitor concentration resulted in negligible increase in inhibition efficiency and the presence of inhibitor caused small changes in (E_{corr}) value. This implies that the inhibitor acts as a mixed-type inhibitor, affecting both anodic and cathodic reactions [29]. According to Li et al. [30], if the displacement in E_{corr} is more than ±85 mV/SCE relating to E (corrosion potential of the blank), the inhibitor can be considered as a cathodic or anodic type. If the change in E_{corr} is less than 85 mV, the corrosion inhibitor may be regarded as a mixed type. But the maximum displacement in the present case is less than 20 mV/SCE, which indicates that AMPT, APT, and ACPT act as mixed-type inhibitors. It is concluded that the inhibitor's molecules retard the corrosion process without changing the mechanism of corrosion process in the medium of investigation. However, the shift of E_{corr} values towards positive direction on increasing the concentration

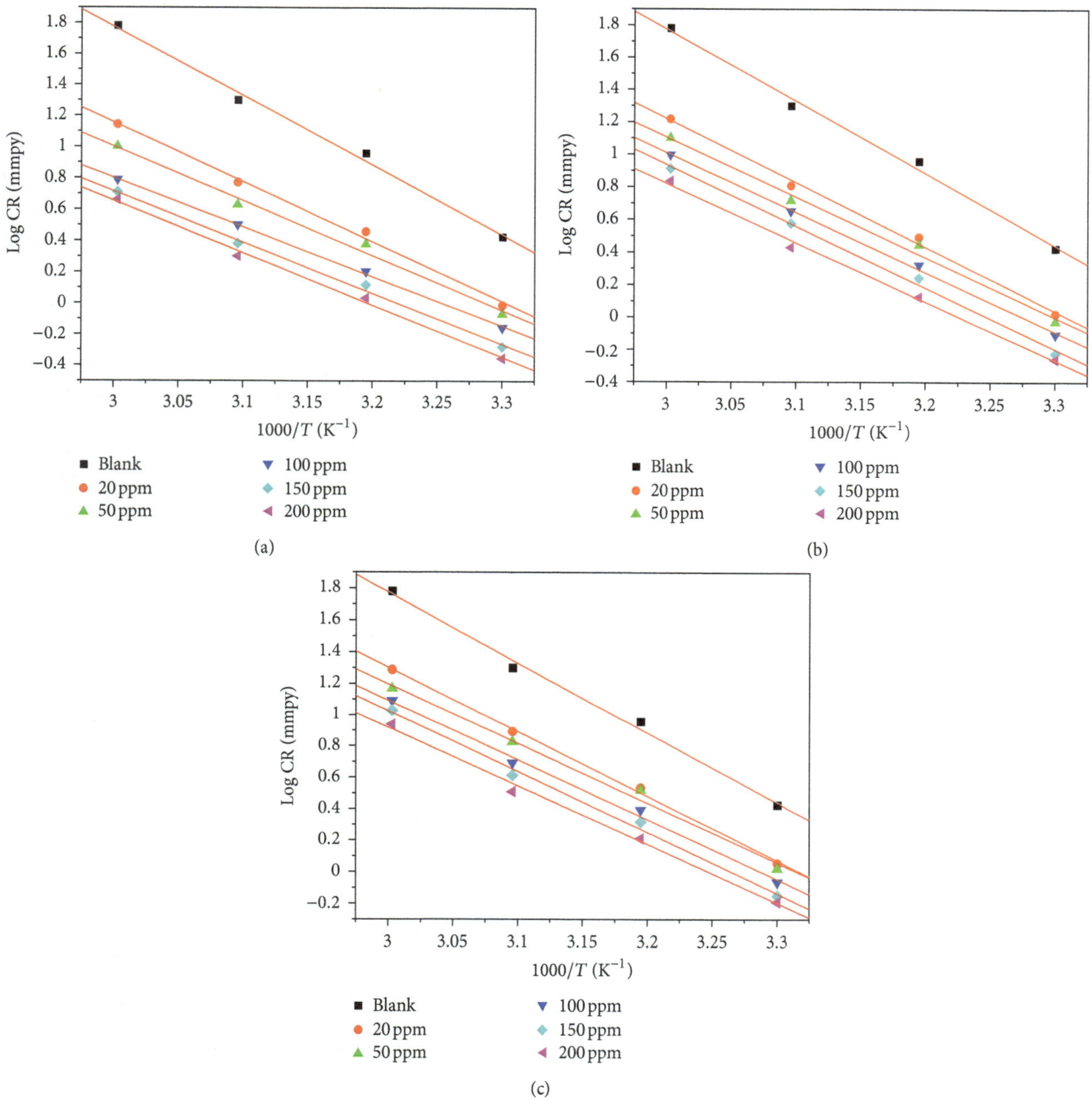

FIGURE 3: Arrhenius plots of log CR versus 1000/T for N80 steel corrosion in 15% HCl (a) AMPT, (b) APT, and (c) ACPT.

of thiadiazole derivatives suggests anodic control over the reaction.

3.3. Electrochemical Impedance Spectroscopy (EIS). Nyquist plots of N80 steel in 15% HCl solution in the absence and the presence of different concentrations of AMPT, APT, and ACPT at 303 K are shown in Figures 7(a), 7(b), and 7(c). It follows from Figures 7(a), 7(b), and 7(c) that a high frequency (HF) depressed charge-transfer semicircle was observed. The experimental data obtained from these plots are fitted by the equivalent electrical circuit shown in Figure 8. Such an equivalent circuit was also discussed by several researchers, who obtained similar depressed semicircles with single time constant due to surface heterogeneity of corrosion product covering the surface at random sites [31–33]. The charge-transfer resistance (R_{ct}) values were calculated from the difference in impedance at low and high frequencies. The values of electrochemical double-layer capacitance (C_{dl}) were calculated at the frequency f_{max}, at which the imaginary component of the impedance is maximal ($-Z_i$) by the following equation:

$$C_{dl} = \frac{1}{2\pi f_{max} R_{ct}}. \tag{7}$$

The constant phase element of double-layer (CPE) is mathematically expressed as

$$Z_{CPE} = Y_0^{-1} (i)^{-n}, \tag{8}$$

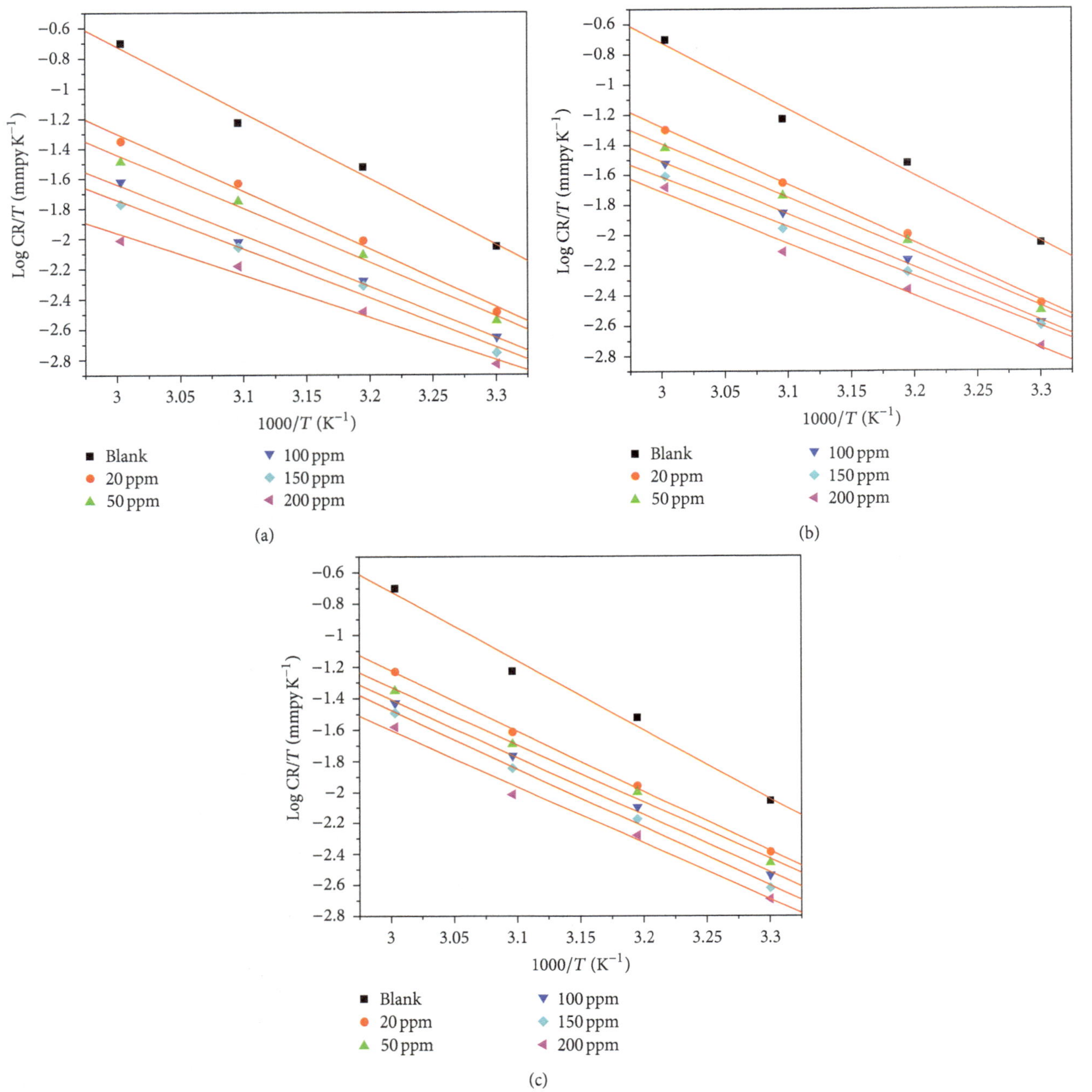

FIGURE 4: Transition state plot of log CR/T versus 1000/T for N80 steel in 15% HCl at different concentrations of (a) AMPT, (b) APT, and (c) ACPT.

where Y_0 is a proportionality factor and "n" has the meaning of phase shift. The value of "n" represents the deviation from the ideal behavior and it lies between 0 and 1.

The charge transfer resistance (R_{ct}), double-layer capacitance (C_{dl}) obtained from the Nyquist plots, and the calculated inhibition efficiency values (IE%) are shown in Table 7. It is clear that the value of R_{ct} increases with increasing the concentration of the inhibitor, indicating that the corrosion rate decreases in the presence of the inhibitor [34]. It is also clear that the value of C_{dl} decreases on the addition of inhibitors, indicating a decrease in the local dielectric constant and/or an increase in the thickness of the electrical double layer, suggesting that the inhibitor molecules function by the formation of the protective layer at the metal surface [35]. So, the changes in R_{ct} and C_{dl} values were caused by the steady replacement of the water molecules by the adsorption of inhibitor on mild steel surface, reducing the extent of dissolution [36]. The maximum inhibition efficiency was obtained for AMPT at 200 ppm concentration. Both electrochemical methods (polarization measurement and EIS study) offered nearly the same trend.

3.4. Adsorption Isotherm. Basic information on the interaction between the organic inhibitors and the N80 steel

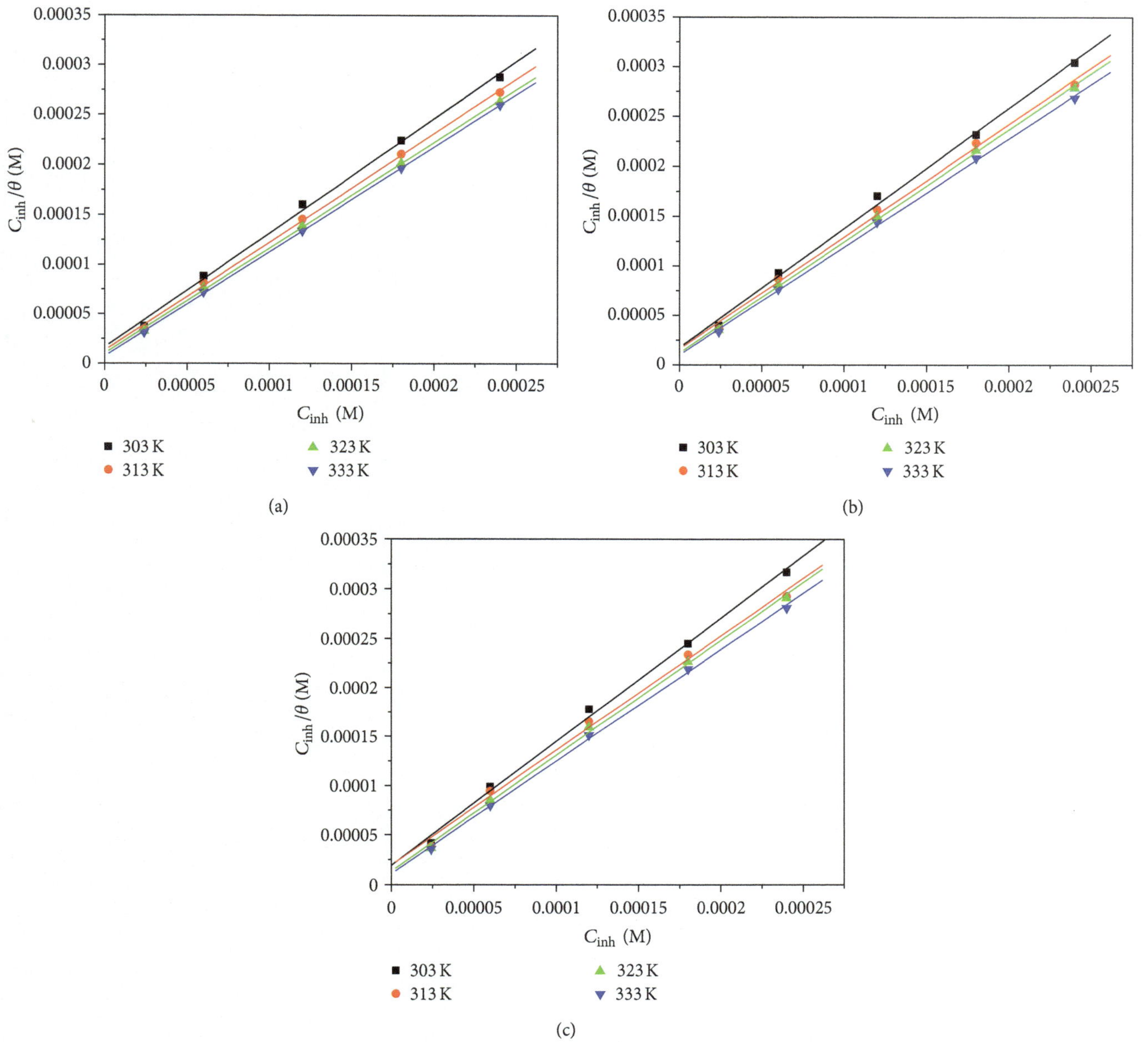

FIGURE 5: Langmuir plots of (C_{inh}/θ) versus C_{inh} for (a) AMPT, (b) APT, and (c) ACPT.

surface is obtained from various isotherms. Adsorption of these organic inhibitors has displaced the water molecule from the metal surface. The most commonly used adsorption isotherms are Langmuir, Temkin, and Frumkin. By far, the best fit was obtained with the Langmuir adsorption isotherm. The degree of surface coverage (θ) for different concentrations of inhibitor in 15% hydrochloric acid has been evaluated by the weight loss value. According to the Langmuir adsorption isotherm, the surface coverage (θ) is related to the inhibition concentration by the following equation:

$$\frac{C_{inh}}{\theta} = \frac{1}{K_{ads}} + C_{inh}, \tag{9}$$

where C_{inh} is the inhibitor concentration and K_{ads} is the equilibrium constant for adsorption-desorption process.

A plot of (C_{inh}/θ) versus C_{inh} (Figure 5) gives a straight line with an average correlation coefficient (R^2), ($R^2 = 0.998$, 0.997, and 0.999) for AMPT, APT, and ACPT, respectively, at 303 K which suggests that the adsorption obeys Langmuir adsorption isotherm. From the intercept, K_{ads} is calculated, and by using the value of K_{ads} the value of ΔG^0_{ads} was calculate by the following equation:

$$\Delta G^0_{ads} = -RT \ln(55.5 K_{ads}), \tag{10}$$

where R is the gas constant and T is the absolute temperature (K). The value of 55.5 is the molar concentration of water in solution in mol L^{-1}. In the present study a large value of K_{ads} was found for all the three studied inhibitors indicating the strong adsorption of inhibitor molecules at the surface of N80 steel. The values of ΔG^0_{ads} were −37.5, −35.2, and

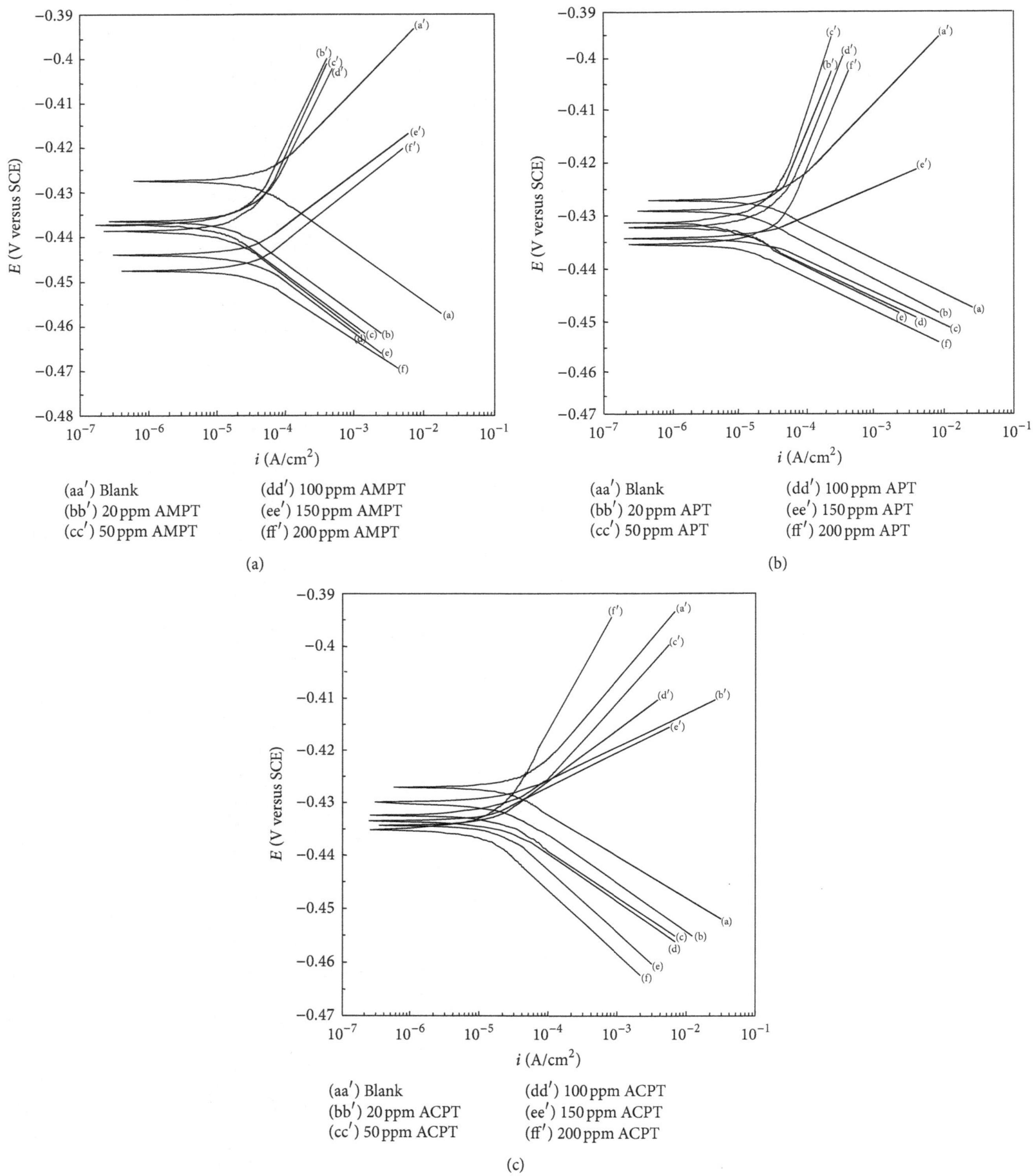

FIGURE 6: Potentiodynamic polarization curves for N80 steel in 15% HCl in the presence and the absence of inhibitors at 303 K. (a) AMPT, (b) APT, and (c) ACPT.

−32.1 kJ/mol for AMPT, APT, and ACPT, respectively. The negative values of ΔG^0_{ads} indicate a spontaneous adsorption process and stability of the adsorbed film of (AMPT, APT, and ACPT) on N80 surface [37]. It is generally accepted that for the values of ΔG^0_{ads} up to −20 kJ/mol, the types of adsorption were regarded as physisorption; the inhibition acts due to the electrostatic interactions between the charged molecules and the charged metallic surfaces, while the values around −40 kJ/mol or smaller were seen as chemisorption, which is due to the charge sharing or a transfer from the

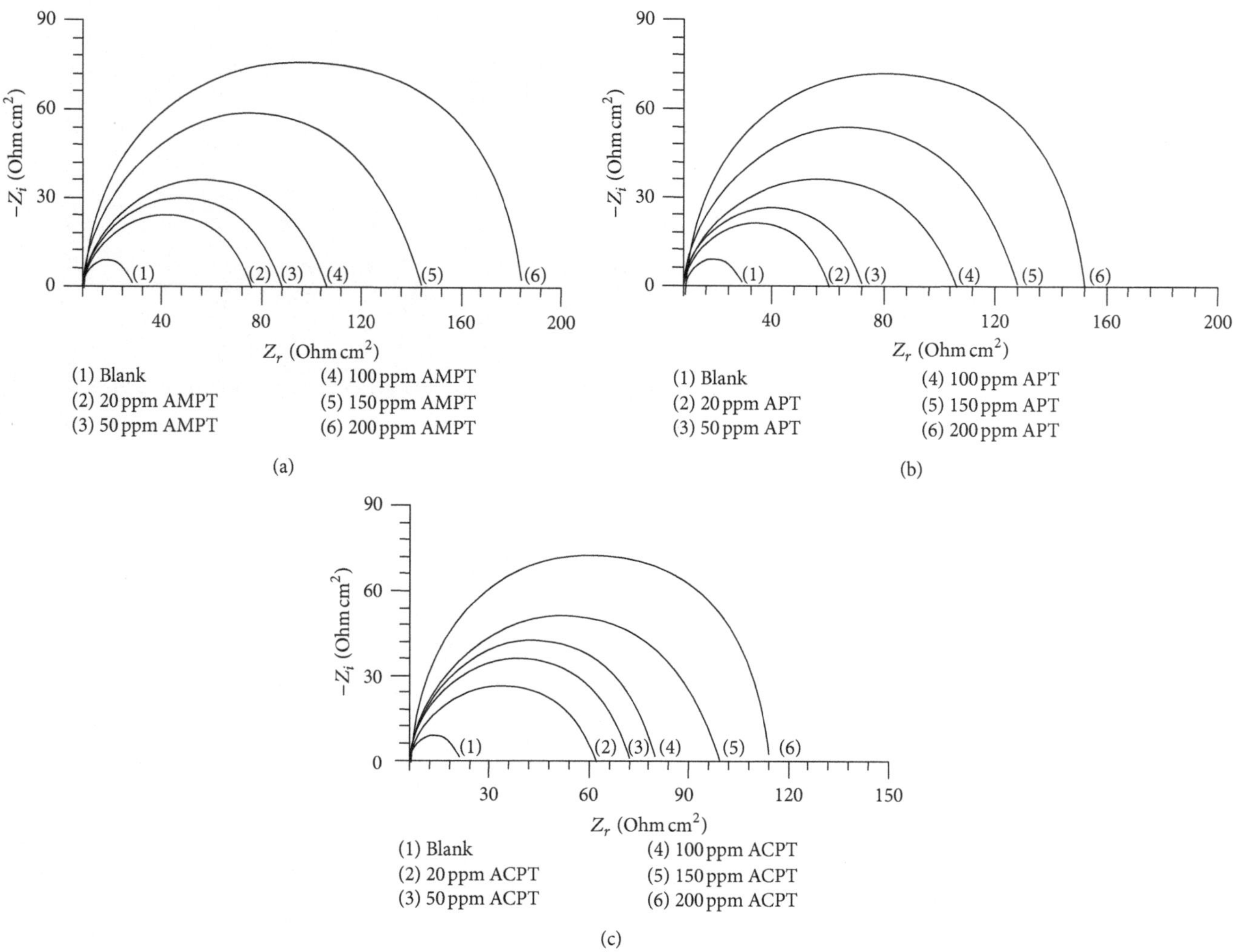

FIGURE 7: Nyquist plot for N80 steel in 15% HCl acid containing various concentrations of (a) AMPT, (b) APT, and (c) ACPT. (1) 0 ppm (2) 50 ppm (3) 100 ppm (4) 150 ppm, and (5) 200 ppm at 303 K.

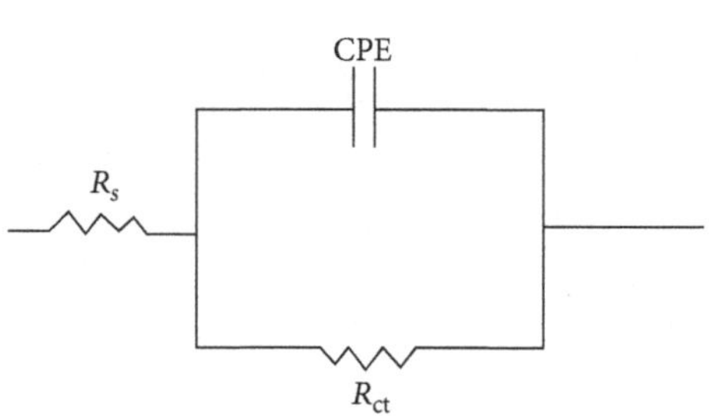

FIGURE 8: The electrochemical equivalent circuit used to fit the impedance measurements.

inhibitor's molecules to the metal surface to form a covalent bond [38, 39]. The values of ΔG^0_{ads} in our measurements range from −30 to −40 kJ·mol^{-1}; it is suggested that the adsorption of these molecules involves chemisorption.

The inhibition efficiency afforded by all the three inhibitors may be attributed to the presence of electron rich (N, S) atoms and delocalized π electrons. The possible coordinating centres are unshared electron pair of nitrogen of $-NH_2$ group, nitrogen of thiadiazole ring, C=N group, and π-electrons of phenyl and thiadiazole rings. The participation of phenyl ring during the adsorption of the inhibitors may be shown by higher % IE value in case of methoxy substituents having +I effect and lower values for chlorosubstituent with −I effect.

3.5. Scanning Electron Microscopy (SEM). The surface morphology of mild steel specimens exposed to 15% HCl solutions in the absence and presence of 200 ppm of AMPT are shown in Figures 9(a), 9(b), and 9(c). Figure 9(a) shows SEM image of polished bare N80 steel surface. Figure 9(b) reveals that the surface is severely corroded, and there is formation of different forms of corrosion products (iron oxides) on the surface in the absence of the inhibitor, and the entire surface is covered by a scale-like black corrosion product. The morphological feature of the inhibited surface with the addition of 200 ppm of AMPT is shown in Figure 9(c). On comparing these micrographs, it appears, that in the presence of inhibitors, the surface of the test material has improved remarkably with respect to its smoothness. The smoothening

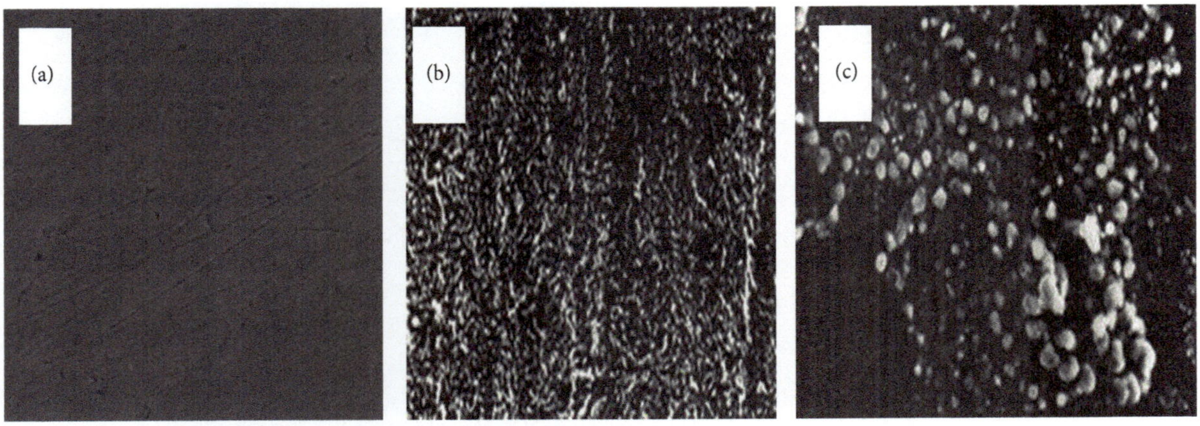

FIGURE 9: SEM image of N80 steel in 4 N HCl solution after 6 h immersion at 303 K (a) before immersion (polished) and (b) after immersion without inhibitor and (c) with inhibitor (AMPT).

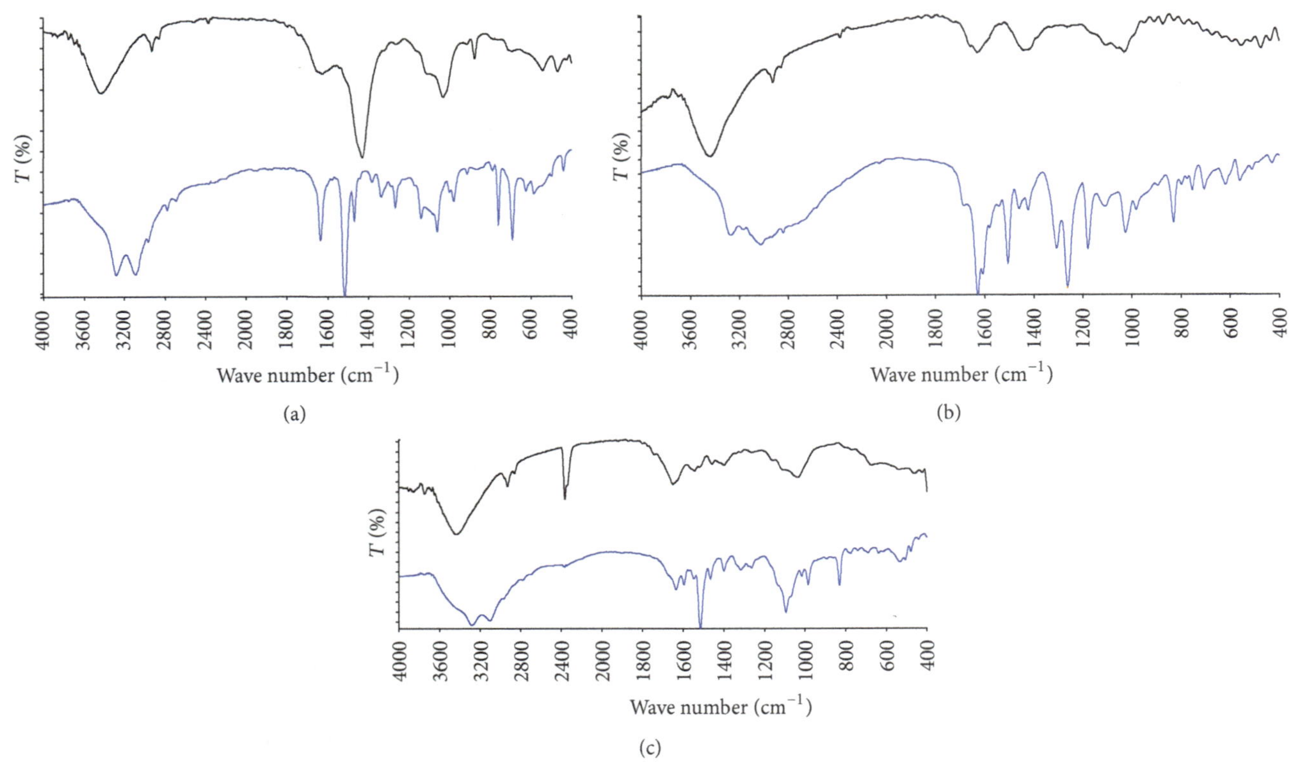

FIGURE 10: FTIR spectrum of the inhibitor (below), FT-IR spectrum of inhibitor adsorbed on metal surface (above). (a) AMPT, (b) APT, and (c) ACPT.

of the surface would have been caused by the adsorption of inhibitor molecules on it. But the closer look at such sites reveals that the inhomogeneities are due to the structural defects of the metal substrate and that these sites are also covered by the inhibitor film. This observation also accounts for the high inhibition efficiency values obtained during the weight loss studies of the inhibitor system.

3.6. FTIR Spectroscopic Analysis of Corrosion Product. The FTIR spectra of pure inhibitors and the surface films formed on N80 steel coupons immersed in 15% HCl containing optimum concentration of inhibitors (AMPT, APT, and ACPT) for 6 h are presented in Figures 10(a), 10(b), and 10(c). The FTIR spectra of pure AMPT show band at 3290 cm^{-1}, 3135 cm^{-1}, APT at 3355, 3160 cm^{-1} and ACPT at 3280, 3120 cm^{-1}, which indicate the asymmetrical and symmetrical N–H stretching vibrations for the amino group. Bands at 3280 cm^{-1}, 3125 cm^{-1} for AMPT, 3345, 3148 cm^{-1} for APT, and 3272, 3113 cm^{-1} for ACPT, respectively indicate the asymmetrical and symmetrical N–H stretching vibration for amino group. Strong peaks observed at 1640, 1630, and 1635 cm^{-1} are attributed to the N–H in plane bending

TABLE 7: Electrochemical impedance parameters and percentage inhibition efficiency for N80 steel in 15% HCl in the absence and the presence of the inhibitors at different concentrations at 303 K.

Inhibitor	Concentration (ppm)	R_{ct} (Ω cm^2)	C_{dl} (μF cm^2)	% IE
Blank	—	25	252	
AMPT	20	74	158	66.32
	50	85	122	70.58
	100	110	87	77.25
	150	142	73	82.37
	200	178	67	85.97
APT	20	66	82	62.42
	50	75	66	67.71
	100	102	53	75.47
	150	126	32	80.13
	200	150	26	83.31
ACPT	20	63	71	60.52
	50	74	58	66.12
	100	80	41	68.58
	150	107	24	76.53
	200	134	20	81.03

vibrations of the amino group in pure AMPT, APT, and ACPT, respectively. Strong peaks observed at 1626, 1622, and 1620 cm^{-1} are attributed to the N–H in plane bending vibrations of the amino group in AMPT, APT, and ACPT, respectively. Strong bands at 1530, 1520, and 1525 cm^{-1} are attributed to the stretching vibration of the C=N group in the ring in pure AMPT, APT, and ACPT, respectively. Strong bands at 1523, 1512, and 1514 cm^{-1} are attributed to the stretching vibration of the C=N group in the ring in surface product of AMPT, APT, ACPT respectively. Peaks observed at 695, 690, and 680 cm^{-1} are attributed to the ring C–S vibration for pur AMPT, APT, and ACPT, respectively. Peaks observed at 690, 684, and 670 cm^{-1} are attributed to the ring C–S vibration for surface product of AMPT, APT, ACPT respectively. The presence of all the functional groups present in pure AMPT, APT, and ACPT in the spectrum of the surface product with a negative shift indicates the adsorption of inhibitors AMPT, APT, and ACPT at the surface of N80 steel.

4. Conclusions

(1) The synthesized thiadiazoles show good inhibition efficiencies for the corrosion of N80 steel in 15 wt% HCl solutions and the inhibition efficiency increases with increasing the concentration of these inhibitors and with the temperature. The increasing order of inhibiting performance of the inhibitors is in the order of: AMPT > APT > ACPT.

(2) Thiadiazoles inhibitors were found to be mixed-type inhibitors from Tafel polarization measurements. While one equivalent structure model was selected to fit the experimental data of the impedance diagram (EIS). And the results are good in agreement with weight loss measurements.

(3) It is suggested from the results obtained from Langmuir adsorption isotherm that the mechanism of corrosion inhibition is occurring mainly through adsorption process.

(4) Surface morphological studies such as FTIR and SEM analysis showed that a film of inhibitor is formed on the electrode surface.

Acknowledgment

One of the authors, Sumit Kumar, gratefully acknowledges Council for Scientific and Industrial Research (CSIR), New Delhi India, for research fellowship.

References

[1] I. Ahamad and M. A. Quraishi, "Bis (benzimidazol-2-yl) disulphide: an efficient water soluble inhibitor for corrosion of mild steel in acid media," *Corrosion Science*, vol. 51, no. 9, pp. 2006–2013, 2009.

[2] Q. B. Zhang and Y. X. Hua, "Corrosion inhibition of mild steel by alkylimidazolium ionic liquids in hydrochloric acid," *Electrochimica Acta*, vol. 54, no. 6, pp. 1881–1887, 2009.

[3] W. Li, Q. He, C. Pei, and B. Hou, "Experimental and theoretical investigation of the adsorption behaviour of new triazole derivatives as inhibitors for mild steel corrosion in acid media," *Electrochimica Acta*, vol. 52, no. 22, pp. 6386–6394, 2007.

[4] R. Solmaz, G. Kardas, B. Yazıcı, and M. Erbil, "Inhibition effect of rhodanine for corrosion of mild steel in hydrochloric acid solution," *Protection of Metals*, vol. 41, no. 6, pp. 581–585, 2005.

[5] G. Kardas, "The inhibition effect of 2-thiobarbituric acid on the corrosion performance of mild steel in HCl solution," *Fiziko-Khimicheskaya Mekhanika Materialov*, vol. 41, no. 3, pp. 337–343, 2005.

[6] M. Yadav, D. Behera, and U. Sharma, "Development of corrosion inhibitors used in acidization of petroleum oil well," *Der Chemica Sinica*, vol. 3, pp. 262–268, 2012.

[7] M. Yadav, D. Behera, and U. Sharma, "Nontoxic corrosion inhibitors for N80 steel in hydrochloric acid," *Arabian Journal of Chemistry*, 2012.

[8] M. A. Amin, S. S. Abd El-Rehim, E. E. F. El-Sherbini, and R. S. Bayoumi, "The inhibition of low carbon steel corrosion in hydrochloric acid solutions by succinic acid. Part I. Weight loss, polarization, EIS, PZC, EDX and SEM studies," *Electrochimica Acta*, vol. 52, no. 11, pp. 3588–3600, 2007.

[9] S. T. Selvi, V. Raman, and N. Rajendran, "Corrosion inhibition of mild steel by benzotriazole derivatives in acidic medium," *Journal of Applied Electrochemistry*, vol. 33, no. 12, pp. 1175–1182, 2003.

[10] F. Bentiss, M. Lebrini, H. Vezin, and M. Lagrenée, "Experimental and theoretical study of 3-pyridyl-substituted 1,2,4-thiadiazole and 1,3,4-thiadiazole as corrosion inhibitors of mild steel in acidic media," *Materials Chemistry and Physics*, vol. 87, no. 1, pp. 18–23, 2004.

[11] S. K. Shukla and E. E. Ebenso, "Effect of condensation product of thiosemicarbazide and phenyl isothiocynate on corrosion of

mild steel in sulphuric acid medium," *International Journal of Electrochemcal Science*, vol. 7, pp. 12134–12145, 2012.

[12] I. A. Ammar and S. Darwish, "Effect of some ions on inhibition of the acid corrosion of fe by thiourea," *Corrosion Science*, vol. 7, no. 9, pp. 579–596, 1967.

[13] D. D. N. Singh, M. M. Singh, R. S. Chaudhary, and C. V. Agarwal, "Inhibitive effects of isatin, thiosemicarbazide and isatin-3-(3-thiosemicarbazone) on the corrosion of aluminium alloys in nitric acid," *Journal of Applied Electrochemistry*, vol. 10, no. 5, pp. 587–592, 1980.

[14] G. Trabanelli, G. Brunoro, C. Monticells, and M. Foganolo, in *Proceedings of the 9th ICMC*, Toronto, Canada, June 1984.

[15] G. D. M. Z. Zhon, R. Tong, and T. Notoxa, *Bulletin of the Electrochemical Society*, vol. 7, p. 60, 1991.

[16] R. M. Souto, V. Fox, M. M. Laz, M. Pérez, and R. S. González, "Some experiments regarding the corrosion inhibition of copper by benzotriazole and potassium ethyl xanthate," *Journal of Electroanalytical Chemistry*, vol. 411, no. 1-2, pp. 161–165, 1996.

[17] E. Otero and J. M. Bastidas, "Cleaning of two hundred year-old copper works of art using citric acid with and without benzotriazole and 2-amino-5-mercapto-1,3,4-thiadiazole," *Materials and Corrosion*, vol. 47, no. 3, pp. 133–138, 1996.

[18] S. S. Mahmoud and E. G. A. Malidy, *Egyptian Journal of Chemistry*, vol. 39, p. 365, 1996.

[19] O. O. Xometl, N. V. Likhanova, N. Nava et al., "Thiadiazoles as corrosion inhibitors for carbon steel in H_2SO_4 solutions," *International Journal of Electrochemcal Science*, vol. 8, pp. 735–752, 2013.

[20] X. J. Raj and N. Rajendran, "Corrosion inhibition effect of substituted thiadiazoles on brass," *International Journal of Electrochemical Science*, vol. 6, no. 2, pp. 348–366, 2011.

[21] A. Shafiee, E. Naimi, P. Mansobi, A. Foroumadi, and M. Shekari, "Syntheses of substituted-oxazolo-1,3,4-thiadiazoles, 1,3,4-oxadiazoles, and 1,2,4-triazoles," *Journal of Heterocyclic Chemistry*, vol. 32, no. 4, pp. 1235–1239, 1995.

[22] N. A. Negm, A. A. Hafiz, and M. Y. El Awady, "Influence of structure of the cationic polytriethanolammonium bromide derivatives. II. Corrosion inhibition," *Egyptian Journal of Chemistry*, vol. 48, no. 2, pp. 201–210, 2005.

[23] A. U. Ezeoke, O. G. Adeyemi, O. A. Akerele, and N. O. Obi-Egbedi, "Computational and experimental studies of 4-Aminoantipyrine as corrosion inhibitor for mild steel in sulphuric acid solution," *International Journal of Electrochemical Science*, vol. 7, no. 1, pp. 534–553, 2012.

[24] E. E. Ebenso and I. B. Obot, "Inhibitive properties, thermodynamic characterization and quantum chemical studies of secnidazole on mild steel corrosion in acidic medium," *International Journal of Electrochemical Science*, vol. 5, no. 12, pp. 2012–2035, 2010.

[25] K. Jüttner, "Electrochemical impedance spectroscopy (EIS) of corrosion processes on inhomogeneous surfaces," *Electrochimica Acta*, vol. 35, no. 10, pp. 1501–1508, 1990.

[26] I. Dehri and M. Özcan, "The effect of temperature on the corrosion of mild steel in acidic media in the presence of some sulphur-containing organic compounds," *Materials Chemistry and Physics*, vol. 98, no. 2-3, pp. 316–323, 2006.

[27] G. Moretti, G. Quartarone, A. Tassan, and A. Zingales, "Pitting corrosion behavior of superferritic stainless steel in waters containing chloride," *Werkstoffe und Korrosion*, vol. 44, no. 1, pp. 24–30, 1993.

[28] V. R. Saliyan and A. V. Adhikari, "Inhibition of corrosion of mild steel in acid media by N'-benzylidene-3- (quinolin-4-ylthio)propanohydrazide," *Bulletin of Materials Science*, vol. 31, no. 4, pp. 699–711, 2008.

[29] D. Jayaperumal, "Effects of alcohol-based inhibitors on corrosion of mild steel in hydrochloric acid," *Materials Chemistry and Physics*, vol. 119, no. 3, pp. 478–484, 2010.

[30] W. H. Li, Q. He, S. T. Zhang, C. L. Pei, and B. R. Hou, "Some new triazole derivatives as inhibitors for mild steel corrosion in acidic medium," *Journal of Applied Electrochemistry*, vol. 38, no. 3, pp. 289–295, 2008.

[31] M. S. Morad, "An electrochemical study on the inhibiting action of some organic phosphonium compounds on the corrosion of mild steel in aerated acid solutions," *Corrosion Science*, vol. 42, no. 8, pp. 1307–1326, 2000.

[32] G. Gunasekaran and L. R. Chauhan, "Eco friendly inhibitor for corrosion inhibition of mild steel in phosphoric acid medium," *Electrochimica Acta*, vol. 49, no. 25, pp. 4387–4395, 2004.

[33] A. Alagta, I. Felhösi, J. Telegdi, I. Bertóti, and E. Kálmán, "Effect of metal ions on corrosion inhibition of pimeloyl-1,5-di-hydroxamic acid for steel in neutral solution," *Corrosion Science*, vol. 49, no. 6, pp. 2754–2766, 2007.

[34] R. Solmaz, "Investigation of the inhibition effect of 5-((E)-4-phenylbuta-1,3-dienylideneamino)-1,3,4-thiadiazole-2-thiol Schiff base on mild steel corrosion in hydrochloric acid," *Corrosion Science*, vol. 52, no. 10, pp. 3321–3330, 2010.

[35] I. Ahamad, R. Prasad, and M. A. Quraishi, "Adsorption and inhibitive properties of some new Mannich bases of Isatin derivatives on corrosion of mild steel in acidic media," *Corrosion Science*, vol. 52, no. 4, pp. 1472–1481, 2010.

[36] B. Trachli, M. Keddam, H. Takenouti, and A. Srhiri, "Protective effect of electropolymerized 3-amino 1,2,4-triazole towards corrosion of copper in 0.5 M NaCl," *Corrosion Science*, vol. 44, no. 5, pp. 997–1008, 2002.

[37] J. D. Talati and D. K. Gandhi, "N-heterocyclic compounds as corrosion inhibitors for aluminium-copper alloy in hydrochloric acid," *Corrosion Science*, vol. 23, no. 12, pp. 1315–1332, 1983.

[38] Z. Szklarska-Smialowska and J. Mankowski, "Crevice corrosion of stainless steels in sodium chloride solution," *Corrosion Science*, vol. 18, no. 11, pp. 953–960, 1978.

[39] A. Yurt, S. Ulutas, and H. Dal, "Electrochemical and theoretical investigation on the corrosion of aluminium in acidic solution containing some Schiff bases," *Applied Surface Science*, vol. 253, no. 2, pp. 919–925, 2006.

Effect of Mechanical Milling and Cold Pressing on Co Powder

A. S. Bolokang,[1] M. J. Phasha,[2] D. E. Motaung,[3] and S. Bhero[1]

[1] *Department of Engineering Metallurgy, University of Johannesburg, P.O. Box 17011, Doornfontein 2028, South Africa*
[2] *Council for Scientific and Industrial Research (CSIR), Materials Science and Manufacturing, Meiring Naude Road, Brummeria, P.O. Box 395, Pretoria 0001, South Africa*
[3] *DST/CSIR Nanotechnology Innovation Centre, National Centre for Nano-Structured Materials, Council for Scientific Industrial Research, P.O. Box 395, Pretoria 0001, South Africa*

Correspondence should be addressed to A. S. Bolokang, amogelangblkng6@gmail.com

Academic Editor: Qian Zhao

Cold pressing (CP) of the amorphous-like Co powder suppressed most of the XRD peaks, in particular the peak along (100) plane. The DSC curve of unmilled CP Co powder has shown a distinct sharp exothermic peak at 615°C. Upon annealing at 700°C, only the FCC phase with lattice parameter of 3.51 Å was detected by XRD. Our results implied that the exotherm at 615°C corresponds to compaction-pressure-assisted HCP to FCC first-order phase transition. The XRD analysis of 30 h milled powder revealed for the first time the FCC phase with a = 3.80 Å. However, due to presence of (100) and (210) peaks, this phase is thought to be FCT with lattice parameters $a = b$ = 3.80 and c = 3.07 Å. Consequently, the high-energy milling carried out in the current work induced for the first time HCP to FCT transition in Co. Upon CP of milled powder, the lattice parameter a shrunk from 3.80 to 3.75 Å. However, during annealing of the CP milled Co powder at 750°C, the FCT to FCC transition occurred, yielding the FCC phase with a = 3.51 Å.

1. Introduction

Cobalt (Co) is a transition metal used in electronics, magnetic recording [1], and hard materials [2–4]. Through thermal [5] and mechanical treatment [6–9], Co metal undergoes the allotropic HCP to FCC phase transformation. The current literature shows that metastable FCC phase is induced by ball milling (BM) [6], cold pressing [10], and thermal treatments [11]. It has been shown in the previous investigations that Co milled with W, V, and C powders forms the complex FCC Co-rich carbide with large lattice parameters [12]. Apart from the FCC phase, Co-based alloys undergo amorphization via electrodeposition [13, 14] and mechanical alloying (MA) [15–17] and during irradiation processes [18, 19]. It is worth noting that in the case of milling, the XRD peaks of nonmetallic elements such as Si in a mixture with Co vanish rapidly when subjected to BM and ultimately lead to amorphization [17]. Similar behaviour occurs when a mixture of Co and metallic elements is subjected to milling [15, 16, 20]. Amorphous metallic alloys have a promising combination of high magnetic, high strength, and high thermal properties [13]. Amorphization in nonmetallic elements, such as gallium [19], silicon [21], germanium [22], and selenium [23], is easily achievable. However, some studies reported amorphous phases in metallic titanium [24], nickel [25], and cobalt [26, 27] produced via nonequilibrium processing. Among the nonequilibrium techniques, mechanical milling (MM) is a well-known process for producing a wide range of novel materials with unique properties. Of particular interest, MM is capable of forming amorphous alloys [28–31] and composites [32] as well as inducing amorphization in pure elements [23]. On contrary, some amorphous materials were found to undergo the "reverse" process under MM, namely, milling-induced crystallization (MIC) [33–37]. Despite numerous investigations of the phenomenon and mechanisms of MIC, its origin is still surrounded by controversy. The object of the current study to investigate the effects of MM and cold pressing (CP) on the Co powder, and how both processes affect the thermal behaviour.

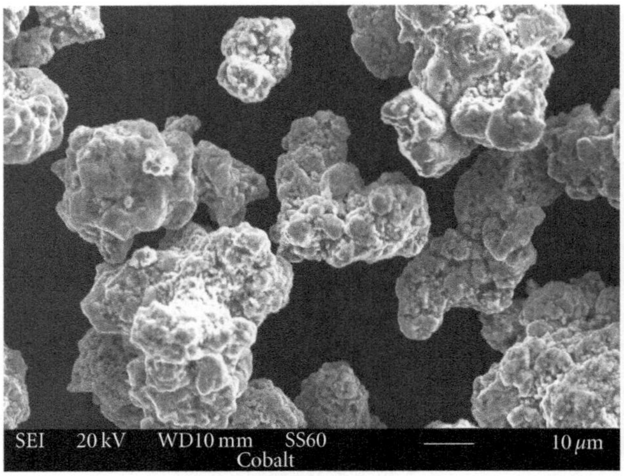

(a)

(b)

FIGURE 1: SEM micrographs of (a) unmilled (b) 30 h milled cobalt powders.

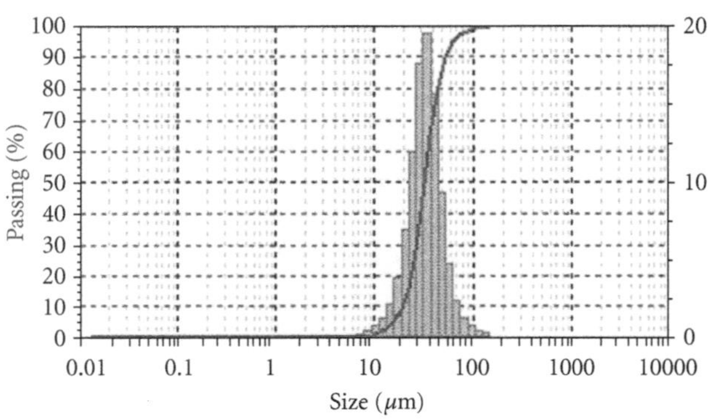

(a)

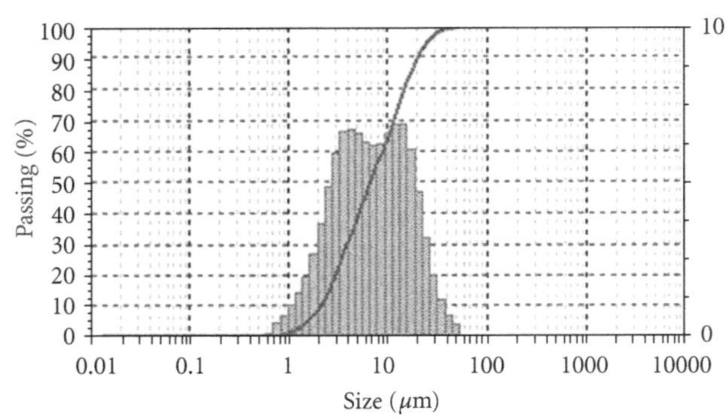

(b)

FIGURE 2: Particle size distribution in (a) unmilled and (b) 30 h milled Co powders.

2. Experimental Work

In this work, Co powder of 99.8% purity was used. Pure Co powder charge was milled under argon atmosphere at milling speed of 770 rpm for 30 hours (h). Milling was performed in the stainless steel milling medium consisting of 5 mm diameter balls and vial at ball-to-powder ratio of 20 : 1. Milling vial was equipped with cooling system to avoid heating during milling. A small powder sample was used for crystal structure analysis and morphology. The powder morphology was evaluated by LEO 1525 field-emission scanning electron microscope (FE-SEM coupled with a Robinson Backscatter Electron Detector (RBSD) and an Oxford Link Pentafet energy dispersive X-ray spectroscopy (EDX) detector. Phase evolution was traced with a Phillips PW 1830 X-ray diffraction (XRD) machine fitted with Cu Kα radiation, and 0.02 step size scanned from 20° to 90° (2θ). The Microtrac Bluewave particle analyzer was employed to determine the particle size of unmilled and milled Co powders. The thermal analyses of the samples were carried out using a differential scanning calorimetry (DSC) operating at 20°C/min heating rate under argon atmosphere. Annealing was conducted in a tube furnace at heating rate of 20°C/min under argon gas flowing at 20 ml/L standard rate.

3. Results and Discussions

3.1. Powder Characterization. Figure 1(a) shows the SEM micrographs indicating irregular lumps of unmilled Co powder particles. These particles appear to be the agglomerates formed from fine particles. Milling for 30 h fractured these powder lumps into ultrafine particles as displayed in Figure 1(b). In order to validate the particle size reduction, the particle size analyses are presented in Figures 2(a) and 2(b) for unmilled and milled powders, respectively. The unmilled powder is comprised of particles with diameters 19, 32 and 52 μm for cumulative D_{10}, D_{50}, and D_{90}, while particles of diameters 2, 7, and 19 μm constitute the milled powder. Though not presented here, the EDX analysis did not detect any sign of contamination by any of the interstitial elements.

The XRD patterns of the unmilled and cold-pressed Co powders are shown in Figures 3(a) and 3(b), respectively.

TABLE 1: XRD data of unmilled free, unmilled cold pressed (CP), 30 h milled-free, 30 h milled CP and 30 h milled CP 400°C-annealed Co powder.

Material condition	Space group and number	Phases	Lattice parameter (Å)	
			a	*c*
Unmilled powder	P63/mmc no. 194	HCP	2.51	4.07
Unmilled CP	P63/mmc no. 194	HCP	2.50	4.06
Milled powder	Fm-3m no. 225	FCC	3.80	—
Milled CP	Fm-3m no. 225	FCC	3.75	—
Milled CP annealed	Fm-3m no. 225	FCC	3.80	—

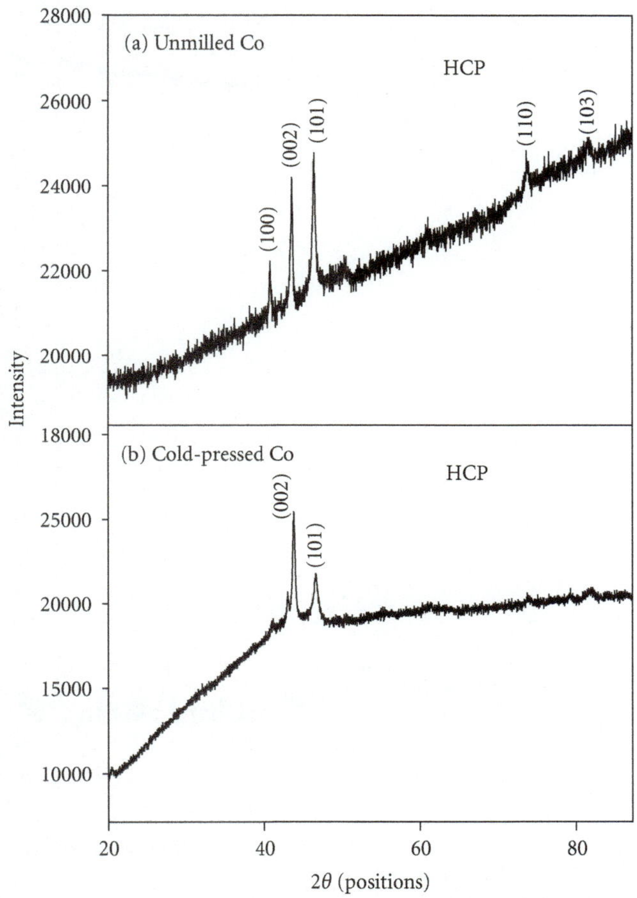

FIGURE 3: XRD patterns of (a) unmilled and (b) cold-pressed Co powders.

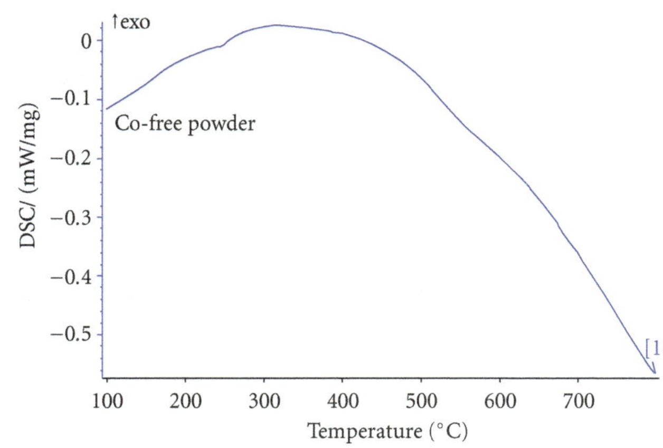

FIGURE 4: DSC curve of unmilled free Co powder.

Three intense HCP peaks belonging to (100), (002), and (101) planes as well as the two weak peaks along (110) and (103) planes are detected by XRD. The weak XRD peaks are an indication of the nanocrystalline-amorphous nature of the starting Co powder used in the current work. The lattice parameters of HCP unmilled Co powder are listed in Table 1.

As shown in Figure 3(b), the subsequent cold pressing (CP) of the amorphous-like Co powder suppressed most of the XRD peaks observed in Figure 3(a). In addition, the XRD pattern changed from a linear type with approximately constant positive slope to nearly bending behaviour, while the XRD peak along (100) plane almost disappears. This significant decrease in peak intensity along (100), (110), and (103) planes could be indicative of lattice planes that are more affected by cold pressing due to compressive shear stress. Consequently, the *a* and *c* lattice parameters of HCP Co were slightly compressed to 2.50 and 4.06 Å, respectively. Since the axial *c*/*a* ratio decreased from 1.6258 to 1.6222, it follows that the HCP crystal lattice was compressed more along the *c*-axis via prismatic slip as evidenced by suppressed reflection along the (10-10) plane.

The DSC curve of unmilled free Co powder is presented in Figure 4, showing a very broad exothermic peak above 250°C possibly due to structural relaxation of strain [38] introduced by rapid cooling during atomization process. Unless this strain is released, it will be difficult to observe the HCP to FCC allotropic transition occurring above 420°C. From the DSC curve of unmilled CP Co powder shown in Figure 5, a distinct sharp exothermic peak is observed at 615°C. This figure indicates that the effect of cold pressing is evident upon thermal treatment. The exotherm corresponds to a crystallization process or alternatively because of an annihilation process of pores inside the compacted sample [38–40]. However, it is also possible that the exotherm is an indication of pressure-assisted HCP to FCC martensitic transformation upon heating as opposed to usual thermal-induced transition [41]. In order to clarify this notion, the unmilled powder compact was annealed at 700°C for an hour. As shown in Figure 6, the XRD pattern of the annealed sample has revealed only the FCC phase with lattice parameter of 3.51 Å. This result is attributed to combination

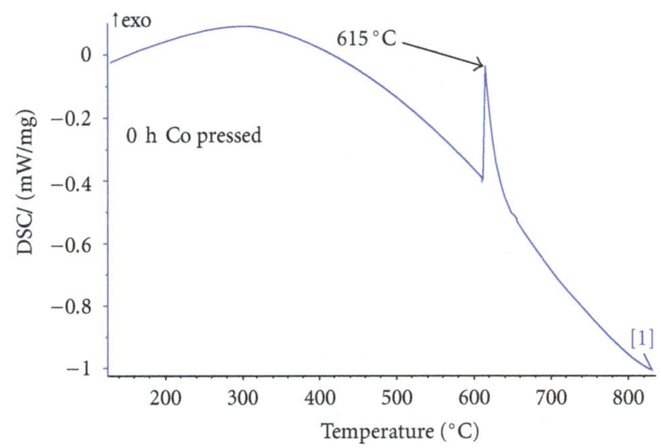

FIGURE 5: DSC curve of unmilled and cold pressed Co powder.

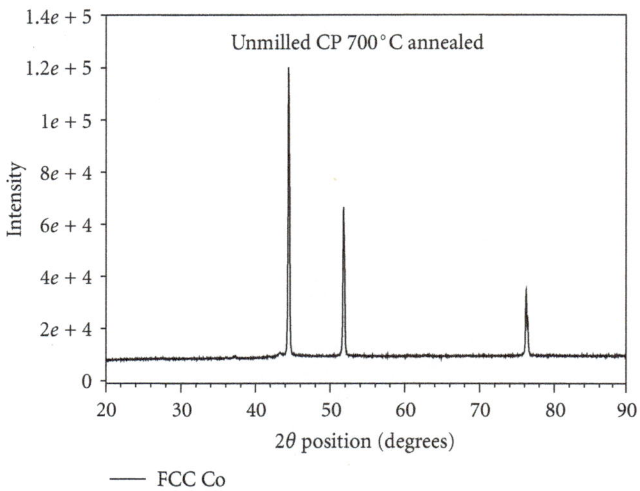

FIGURE 6: The XRD pattern of the CP sample annealed at 700°C.

of pressure during CP, the strain in the amorphous-type starting powder, and the heat upon annealing. The resulting FCC phase reveals the exotherm at 615°C as the typical signature for the pressure-assisted first-order phase transition.

Upon extensive 30 h milling, a FCC phase is obtained as shown by the XRD pattern with short and broader peaks in Figure 7(a) due to grain refinement. Although allotropic HCP to FCC phase transformation via MM has been reported before [6–9], the current study finds different results. In addition to the (100) and (210) peaks, the lattice parameter obtained in the current work is 3.80 Å, larger than the previously reported value of 3.54 Å [6]. The larger lattice parameter might imply that the HCP phase actually transformed to FCT (face-centered tetragonal) with lattice parameters $a = b = 3.80$, $c = 3.07$ Å instead of expanded FCC. This possibility would then account for the appearance of (100) and (210) peaks. Using the Scherrer equation, the estimated average crystallite size of the powder was reduced from about 84 nm (before MM) to 45 nm (after MM). Therefore, it should be noted that in current work, the agglomerates of fine-grained powder were used under high

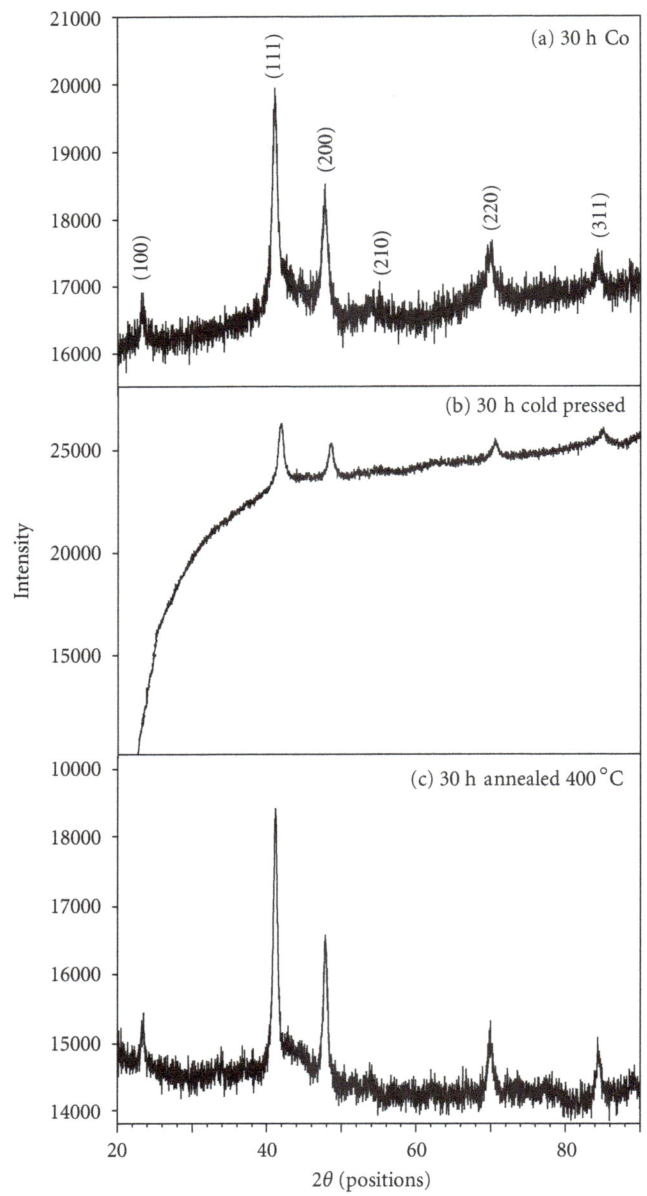

FIGURE 7: XRD patterns of (a) 30-h milled, (b) cold pressed and (c) 400°C annealed Co powders.

milling speed while in previous studies the starting powder was micron sized.

When the milled powder was subjected to cold pressing, the XRD peaks became much shorter and the (100) and (210) peaks disappeared, as shown in Figure 7(b). Similar to what occurred in unmilled HCP Co, the XRD peak of (100) plane in FCC disappeared due to cold pressing. At the same time, the compression of the lattice parameter from 3.80 to 3.75 Å took place upon cold pressing. In order to investigate the thermal behaviour, a DSC analysis was performed on the pressed 30 h milled sample.

The DSC curve in Figure 8 shows endothermic peak around 297°C upon heating the 30 h milled and pressed Co powder, followed by a small and large exothermic peaks at 426 and 707°C, respectively. It is well known that endothermic peaks show phase transformation in certain pure metals

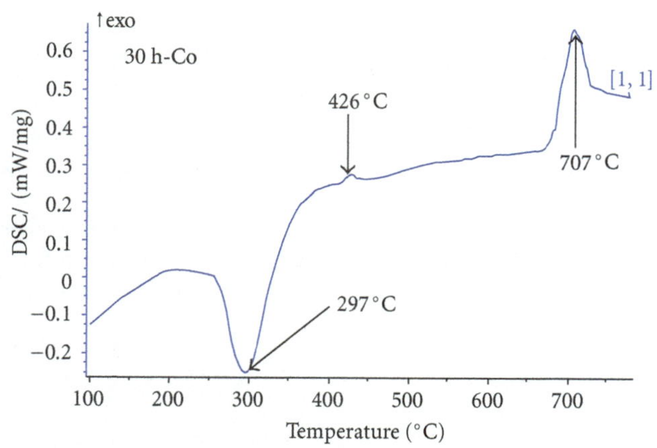

FIGURE 8: DSC curve of 30 h milled and cold pressed Co.

[42, 43] including Co [5]. However, since our powder has no HCP phase after milling, this endothermic peak at 297°C cannot be related to allotropic HCP to FCC phase transition. To investigate the cause of endothermic peak at 297°C, the pressed sample was annealed at 400°C for 1 hour. It became evident in Figure 7(c) that annealing of the cold-pressed powder changed the XRD pattern to appear similar to that of free flowing milled powder shown in Figure 7(a). It follows that the annealing reversed the cold pressing effect. The (100) plane that was suppressed by cold pressing reappeared, while the lattice parameter also reverted to 3.80 Å. Therefore, it is logical to conclude that the endothermic peak in Figure 8 is a result of washed-out exothermic effect (relaxation of CP mechanical stress), which is a characteristic of many mechanically milled powders [44–46]. On the other hand, the exothermic peaks occurring at 426 and 707°C are due to reordering of the crystallite phase during crystallization process. The difference in temperature might be due to the inhomogeneity of the milling process leading to powder with varying particle sizes as shown in Figure 2(b). However, as was shown for unmilled powder compact, it is possible that the peak at exotherm at 707°C is related to FCT to FCC transition. Consequently, the CP sample was annealed at 750°C, and the corresponding XRD pattern is similar to that in Figure 6. The results therefore validate our hypothesis for possible FCT-FCC transition in milled Co.

Co is an HCP ferromagnetic transition metal placed with FCC metals in the periodic table of elements. It has been shown that Co can either exist as HCP ($a = 2.51$; $c = 4.07$ Å) or FCC with lattice parameter 3.54 Å depending on the grain size [11, 47]. Besides rapid cooling, several authors reported the HCP to FCC phase transformation in Co by MM. Huang et al. [6] showed a close range of variation of FCC Co lattice parameters from 3.53 to 3.56 Å via powder milling. It has been reported that Co can also exist in amorphous structure [26]. The XRD pattern of the cold-pressed sample shows the suppression of the peak belonging to (100) plane, while the lattice parameters become slightly smaller. Our current results support the work of Kirin et al. [10] on effectiveness of cold pressing. They have illustrated phase transformation in Co from FCC to HCP by pressing of the powder. In addition to that, cold pressing can even induce intermetallics reaction [48]. The extensive milling carried out in the current work induced HCP to FCT transition due to mechanical deformation caused by high-speed milling. Several studies illustrated that MM induces crystallization on amorphous materials [49–51]. For example, Bednarčík et al. [50] have milled amorphous CoFeSiB for 800 h to cause crystallization of a mixture of nanocrystalline FCC Co and amorphous phase. An endothermic peak on the DSC curve in Figure 8 corresponds with the volume change of 52.70 mm^3 calculated from the FCC structure after cold pressing from the 55.01 mm^3 obtained after annealing. In our recent work, similar type of endothermic peak accompanied by volume expansion at the Curie temperature of Ni was reported [52]. These types of endothermic peaks that occur in certain pure metals such Ti (HCP to BCC) are assumed to be the resultant of external volume effect, and transformation takes place as a reconstruction of the crystal lattice by both shearing and diffusion process [53].

4. Conclusions

The large lumps of particles in the starting Co powder were fractured into ultrafine particles after 30 h MM, as evidenced by SEM micrographs and particle size distribution analyses. Cold pressing (CP) of the amorphous-like Co powder suppressed most of the XRD peaks, in particular the peak along (100) plane. Since the axial c/a ratio decreased from 1.6258 to 1.6222 due to CP, it follows that the HCP crystal lattice was compressed more along the c-axis via prismatic slip as evidenced by suppressed reflection along the (10-10) plane. The DSC curve of unmilled CP Co powder has shown a distinct sharp exothermic peak at 615°C. Upon annealing at 700°C, only the FCC phase with lattice parameter of 3.51 Å was detected by XRD. Our results implied that the exotherm at 615°C corresponds to compaction pressure-assisted HCP to FCC first-order phase transition. The XRD analysis of 30 h milled powder revealed for the first time the FCC phase with $a = 3.80$ Å. However, due to presence of (100) and (210) peaks, this phase is thought to be FCT with lattice parameters $a = b = 3.80$ and $c = 3.07$ Å. Consequently, the high-energy milling carried out in the current work induced for the first time HCP to FCT transition in Co. Similar to unmilled HCP Co, the XRD peak belonging to the (100) plane of FCT phase disappeared after CP. At the same time, the lattice parameter a shrunk from 3.80 to 3.75 Å due to CP. However, during annealing of the CP milled Co powder at 750°C, the FCT to FCC transition occurred. The lattice parameter of the FCC phase is equal to 3.51 Å.

Acknowledgments

This work was supported by the Department of Science and Technology (DST) of South Africa and the Council for Scientific and Industrial Research (CSIR), Metals and Metals Processing (MMP) Division of the Materials Science and Manufacturing (MSM) Unit.

References

[1] M. P. Pileni, "Mesoscopic domains of cobalt nanocrystals," *Pure and Applied Chemistry*, vol. 74, no. 9, pp. 1707–1718, 2002.

[2] Z. Fang, P. Maheshwari, X. Wang, H. Y. Sohn, A. Griffo, and R. Riley, "An experimental study of the sintering of nanocrystalline WC-Co powders," *International Journal of Refractory Metals and Hard Materials*, vol. 23, no. 4–6, pp. 249–257, 2005.

[3] X. Q. Ou, M. Song, T. T. Shen, D. H. Xiao, and Y. H. He, "Fabrication and mechanical properties of ultrafine grained WC-10Co-0.45Cr3C2-0.25VC alloys," *International Journal of Refractory Metals and Hard Materials*, vol. 29, no. 2, pp. 260–267, 2011.

[4] M. H. Enayati, G. R. Aryanpour, and A. Ebnonnasir, "Production of nanostructured WC-Co powder by ball milling," *International Journal of Refractory Metals and Hard Materials*, vol. 27, no. 1, pp. 159–163, 2009.

[5] H. Matsumoto, "Variation in transformation hysteresis in pure cobalt with transformation cycles," *Journal of Alloys and Compounds*, vol. 223, no. 1, pp. L1–L3, 1995.

[6] J. Y. Huang, Y. K. Wu, H. Q. Ye, and K. Lu, "Allotropic transformation of cobalt induced by ball milling," *Nanostructured Materials*, vol. 6, no. 5–8, pp. 723–726, 1995.

[7] J. Sort, J. Nogués, S. Suriñach, J. S. Muñoz, and M. D. Baró, "Microstructural aspects of the hcp-fcc allotropic phase transformation induced in cobalt by ball milling," *Philosophical Magazine*, vol. 83, no. 4, pp. 439–455, 2003.

[8] J. Sort, J. Nogués, S. Surinach, J. S. Muñoz, and M. D. Baró, "Correlation between stacking fault formation, allotropic phase transformations and magnetic properties of ball-milled cobalt," *Materials Science and Engineering A*, vol. 375–377, no. 1-2, pp. 869–873, 2004.

[9] F. Delogu, "Kinetics of allotropic phase transformation in cobalt powders undergoing mechanical processing," *Scripta Materialia*, vol. 58, no. 2, pp. 126–129, 2008.

[10] A. Kirin, A. Bonefačić, and D. Duževič, "Phase transformation in pressed cobalt powder," *Journal of Physics F*, vol. 14, no. 11, pp. 2781–2786, 1984.

[11] E. A. Owen and D. Madoc Jones, "Effect of grain size on the crystal structure of cobalt," *Proceedings of the Physical Society B*, vol. 67, no. 6, article 302, pp. 456–466, 1954.

[12] A. S. Bolokang, M. J. Phasha, C. Oliphant, and D. Motaung, "XRD analysis and microstructure of milled and sintered V, W, C, and Co powders," *International Journal of Refractory Metals and Hard Materials*, vol. 29, no. 1, pp. 108–111, 2011.

[13] S. S. Grabchikov and A. M. Yaskovich, "Effect of the structure of amorphous electrodeposited Ni-W and Ni-Co-W alloys on their crystallization," *Russian Metallurgy*, vol. 2006, no. 1, pp. 56–60, 2006.

[14] M. A. Sheikholeslam, M. H. Enayati, and K. Raeissi, "Characterization of nanocrystalline and amorphous cobalt-phosphorous electrodeposits," *Materials Letters*, vol. 62, no. 21-22, pp. 3629–3631, 2008.

[15] A. M. Bolarín-Miró, F. S.-D. Jesús, G. Torres-Villaseñor, C. A. Cortés-Escobedo, J. A. Betancourt-Cantera, and J. I. Betancourt-Reyes, "Amorphization of Co-base alloy by mechanical alloying," *Journal of Non-Crystalline Solids*, vol. 357, no. 7, pp. 1705–1709, 2011.

[16] S. Louidi, F. Z. Bentayeb, W. Tebib, J. J. Suñol, A. M. Mercier, and J. M. Grenèche, "Amorphisation of Cr-10Co mixture by mechanical alloying," *Journal of Non-Crystalline Solids*, vol. 356, no. 20–22, pp. 1052–1056, 2010.

[17] G. F. Zhou and H. Bakker, "Atomically disordered nanocrystalline Co_2Si by high-energy ball milling," *Journal of Physics: Condensed Matter*, vol. 6, no. 22, article 004, pp. 4043–4052, 1994.

[18] Y. Limoge and A. Barbu, "Amorphization mechanism in metallic crystalline solids under irradiation," *Physical Review B*, vol. 30, no. 4, pp. 2212–2215, 1984.

[19] M. Holz, P. Ziemann, and W. Buckel, "Direct evidence for amorphization of pure gallium by low-temperature ion irradiation," *Physical Review Letters*, vol. 51, no. 17, pp. 1584–1587, 1983.

[20] M. A. Xueming and J. I. Gang, "Nanostructured WC-Co alloy prepared by mechanical alloying," *Journal of Alloys and Compounds*, vol. 245, no. 1-2, pp. L30–L32, 1996.

[21] S. Takeda and J. Yamasaki, "Amorphization in silicon by electron irradiation," *Physical Review Letters*, vol. 83, no. 2, pp. 320–323, 1999.

[22] G. Patriarche, E. Le Bourhis, M. M. O. Khayyat, and M. M. Chaudhri, "Indentation-induced crystallization and phase transformation of amorphous germanium," *Journal of Applied Physics*, vol. 96, no. 3, pp. 1464–1468, 2004.

[23] G. J. Fan, F. Q. Guo, Z. Q. Hu, M. X. Quan, and K. Lu, "Amorphization of selenium induced by high-energy ball milling," *Physical Review B*, vol. 55, no. 17, pp. 11010–11013, 1997.

[24] Y. Wang, Y. Z. Fang, T. Kikegawa et al., "Amorphouslike diffraction pattern in solid metallic titanium," *Physical Review Letters*, vol. 95, no. 15, Article ID 155501, 2005.

[25] Y. Koltypin, G. Katabi, X. Cao, R. Prozorov, and A. Gedanken, "Sonochemical preparation of amorphous nickel," *Journal of Non-Crystalline Solids*, vol. 201, no. 1-2, pp. 159–162, 1996.

[26] C. Xie, J. Hu, R. Wu, and H. Xia, "Structure transition comparison between the amorphous nanosize particles and coarse-grained polycrystalline of cobalt," *Nanostructured Materials*, vol. 11, no. 8, pp. 1061–1066, 1999.

[27] Y. Song, L. L. Henrys, and W. Yang, "Stable amorphous cobalt nanoparticles formed by an in situ rapidly cooling microfluidic process," *Langmuir*, vol. 25, no. 17, pp. 10209–10217, 2009.

[28] C. C. Koch, O. B. Cavin, C. G. McKamey, and J. O. Scarbrough, "Preparation of "amorphous" $Ni_{60}Nb_{40}$ by mechanical alloying," *Applied Physics Letters*, vol. 43, no. 11, pp. 1017–1019, 1983.

[29] L. Schultz, "Formation of amorphous metals by mechanical alloying," *Materials Science and Engineering*, vol. 97, pp. 15–23, 1988.

[30] J. Eckert, L. Schultz, and K. Urban, "Comparison of solid-state amorphization by mechanical alloying and interdiffusion in NiZr," *Materials Science and Engineering A*, vol. 134, pp. 1389–1393, 1991.

[31] I. Börner and J. Eckert, "Phase formation and properties of mechanically alloyed amorphous $Al_{85}Y_8Ni_5Co_2$," *Scripta Materialia*, vol. 45, no. 2, pp. 237–244, 2001.

[32] J. T. Hsieh, C. K. Lin, J. S. Chen, R. R. Jeng, Y. L. Lin, and P. Y. Lee, "Preparation and thermal stability of mechanically alloyed amorphous Ni-Zr-Ti-Si composites," *Materials Science and Engineering A*, vol. 375–377, no. 1-2, pp. 820–824, 2004.

[33] C. Bansal, B. Fultz, and W. L. Johnson, "Crystallization of FeBSi metallic glass during ball milling," *Nanostructured Materials*, vol. 4, no. 8, pp. 919–925, 1994.

[34] M. L. Trudeau, "Deformation induced crystallization due to instability in amorphous FeZr alloys," *Applied Physics Letters*, vol. 64, no. 26, pp. 3661–3663, 1994.

[35] B. Huang, R. J. Perez, P. J. Crawford, A. A. Sharif, S. R. Nutt, and E. J. Lavernia, "Mechanically induced crystallization of metglas $Fe_{78}B_{13}Si_9$ during cryogenic high energy ball milling," *Nanostructured Materials*, vol. 5, no. 5, pp. 545–553, 1995.
[36] J. Xu and M. Atzmon, "Temperature dependence of deformation-assisted crystallization in amorphous $Fe_{78}B_{13}Si_9$," *Applied Physics Letters*, vol. 73, no. 13, pp. 1805–1807, 1998.
[37] Y. S. Kwon, J. S. Kim, I. V. Povstugar, E. P. Yelsukov, and P. P. Choi, "Role of local heating in crystallization of amorphous alloys under ball milling: an experiment on $Fe_{90}Zr_{10}$," *Physical Review B*, vol. 75, no. 14, Article ID 144112, 2007.
[38] J. J. Suñol, A. González, and J. Saurina, "Thermal analysis of two Fe-X-B (X=Nb, ZrNi) alloys prepared by mechanical alloying," *Journal of Thermal Analysis and Calorimetry*, vol. 72, no. 1, pp. 329–335, 2003.
[39] X. J. Gu, F. Ye, F. Zhou, and K. Lu, "Pressure effect on crystallization of mechanically alloyed amorphous $Al_{85}Fe_{15}$ alloy," *Materials Science and Engineering A*, vol. 278, no. 1-2, pp. 61–65, 2000.
[40] F. Ye and K. Lu, "Pressure effect on crystallization kinetics of an Al-La-Ni amorphous alloy," *Acta Materialia*, vol. 47, no. 8, pp. 2449–2454, 1999.
[41] C. Mangler, C. Gammer, H. P. Karnthaler, and C. Rentenberger, "Structural modifications during heating of bulk nanocrystalline FeAl produced by high-pressure torsion," *Acta Materialia*, vol. 58, no. 17, pp. 5631–5638, 2010.
[42] C. Papandrea and L. Battezzati, "A study of the $\alpha \leftrightarrow \gamma$ transformation in pure iron: rate variations revealed by means of thermal analysis," *Philosophical Magazine*, vol. 87, no. 10, pp. 1601–1618, 2007.
[43] A. K. Rai, S. Raju, B. Jeyaganesh, E. Mohandas, R. Sudha, and V. Ganesan, "Effect of heating and cooling rate on the kinetics of allotropic phase changes in uranium: a differential scanning calorimetry study," *Journal of Nuclear Materials*, vol. 383, no. 3, pp. 215–225, 2009.
[44] J. J. Sunol, A. González, T. Pradell, P. Bruna, M. T. Clavaguera-Mora, and N. Clavaguera, "Thermal and structural changes induced by mechanical alloying in melt-spun Fe-Ni based amorphous alloys," *Materials Science and Engineering A*, vol. 375–377, no. 1-2, pp. 881–887, 2004.
[45] M. T. Clavaguera-Mora, J. J. Suñol, and N. Clavaguera, "Relaxation Kinetics of Mechanically Alloyed Powders. Fe-Ni-Si-P: A Case Study," *Journal of Metastable and Nanocrystalline Materials*, vol. 10, pp. 459–466, 2001.
[46] V. V. Boldyrev and K. Tkáčová, "Mechanochemistry of solids: past, present, and prospects," *Journal of Materials Synthesis and Processing*, vol. 8, no. 3-4, pp. 121–132, 2000.
[47] S. Kajiwara, S. Ohno, K. Honma, and M. Uda, "New crystal structure of pure cobalt formed in ultrafine particles," *Philosophical Magazine Letters*, vol. 55, no. 5, pp. 215–219, 1987.
[48] S. D. De la Torre, K. N. Ishihara, and P. H. Shingu, "Synthesis of SnTe by repeated cold-pressing," *Materials Science and Engineering A*, vol. 266, no. 1-2, pp. 37–43, 1999.
[49] Y. Birol, "Crystallization of a $Fe_{36}Ni_{36}B_{28}$ metallic glass during ball-milling," *Scripta Materialia*, vol. 34, no. 7, pp. 1081–1085, 1996.
[50] J. Bednarčík, J. Kováč, P. Kollár et al., "Crystallization of CoFeSiB metallic glass induced by long-time ball milling," *Journal of Non-Crystalline Solids*, vol. 337, no. 1, pp. 42–47, 2004.
[51] M. Pękała, M. Jachimowicz, V. I. Fadeeva, and H. Matyja, "Phase transformations in Co-B-Si alloys induced by high-energy ball milling," *Journal of Non-Crystalline Solids*, vol. 287, no. 1–3, pp. 360–365, 2001.
[52] A. S. Bolokang and M. J. Phasha, "Thermal analysis on the curie temperature of nanocrystalline Ni produced by ball milling," *Advanced Powder Technology*, vol. 22, no. 4, pp. 518–521, 2011.
[53] W. Szkliniarz and G. Smolka, "Analysis of volume effects of phase transformation in titanium alloys," *Journal of Materials Processing Technology*, vol. 53, no. 1-2, pp. 413–422, 1995.

Mutual Influence of Special Components in Baotou Steel Blast Furnace Slag on the Crystallization Behavior of Glass

Yici Wang,[1,2] Qi Jiang,[2] Guoping Luo,[1,2] Wenwu Yu,[2] and Yan Ban[2]

[1] *College of Metallurgical and Ecological Engineering, University of Science and Technology Beijing, Beijing 100083, China*
[2] *College of Material and Metallurgy, Inner Mongolia University of Science and Technology, Baotou 014010, China*

Correspondence should be addressed to Yici Wang, wangyici01060@163.com

Academic Editor: Shien-Ping Feng

In the process of glass-ceramics prepared with Baotou steel blast furnace slag, quartz sand, and other raw materials by melting method, the mutual influence of the special components such as CaF_2, RE_XO_Y, TiO_2, K_2O, and Na_2O in the blast furnace slag on the crystallization behavior of parent glass was investigated using differential thermal analysis (DTA) and X-ray diffraction (XRD). The results show that the special components in slag can reduce the crystallization temperature and promote crystallization of glass phase, which belongs to surface crystallization of glass, and they cannot play the role of the nucleating agent; the major crystal phase composed of diopside, diopside containing aluminum and anorthite, is slightly different from the expected main crystal phase of diopside. Therefore, the nucleating agents of proper species and quantity must be added into the raw materials in order to obtain glass-ceramics. The results have important theoretical guidance meaning for realizing industrial production of Baotou steel blast furnace slag glass-ceramics preparation.

1. Introduction

Glass-ceramics are uniform polycrystalline materials with glasses and crystals, which are processed and formed as glasses and then converted into a crystalline ceramic by a subsequent heat treatment [1, 2]. It is of great significance to prepare glass-ceramics from blast furnace slag for improving utilization and added value of iron smelting slag, recovering heat of slag, increasing economic benefit of enterprises, and reducing environmental pollution [3–10]. Baotou steel blast furnace slag is unique in the world because it contains some special components such as CaF_2, RE_XO_Y, TiO_2, K_2O, Na_2O, and the radioactive element thorium [11]. Among them, CaF_2 and TiO_2 can act as nucleating agent to promote crystallization of glass [12–14], and K_2O and Na_2O can decrease the melting point of glass, but there are few reports so far regarding the effect of RE_XO_Y, Nb_2O_5, and ThO_2 on the crystallization behavior of glass in the process of glass-ceramics prepared. These special components make the crystallization behavior of glass-ceramics prepared from blast furnace slag become more complex, and the related theory of glass-ceramics prepared from ordinary blast furnace slag is not completely applicable to Baotou steel blast furnace slag. Therefore, the mutual influence of the special components in Baotou steel blast furnace slag on the crystallization behavior of parent glass was investigated without any nucleating agents added in this paper. The results not only are a supplement to the theory of slag glass-ceramics preparation, but also had theoretical guidance meaning and practical reference value for realizing industrial production of Baotou steel blast furnace slag glass-ceramics preparation.

2. The Chemical Composition Design of the Parent Glass

Two problems must be taken into account in chemical composition design of the parent glass: the structural stability of the parent glass and the major crystal phases after glass crystallization. Various crystal phases can be obtained due to different silicon-oxygen ratio (Si/O) from the viewpoint of crystal chemistry, and the prepared glass-ceramics should have higher mechanical strength, good wear resistance, chemical stability, and thermal stability.

TABLE 1: Chemical composition of the parent glass (wt%).

CaO	SiO_2	Al_2O_3	MgO
26	54	10	10

TABLE 2: The chemical composition of Baotou steel blast furnace slag (wt%).

Components	Mass percentage
TFe	1.20
FeO	1.20
SiO_2	35.94
CaF_2	1.52
P	—
S	1.06
K_2O	0.53
Na_2O	1.34
CaO	33.37
MgO	10.10
Al_2O_3	12.08
TiO_2	1.38
MnO	0.55
RE_XO_Y	0.78
Cu	0.004
Zn	0.002
C	0.36
As	—
Pb	—
Bi	0.0042
Sn	—
Sb	—
Nb_2O_5	0.0092
ThO_2	0.016

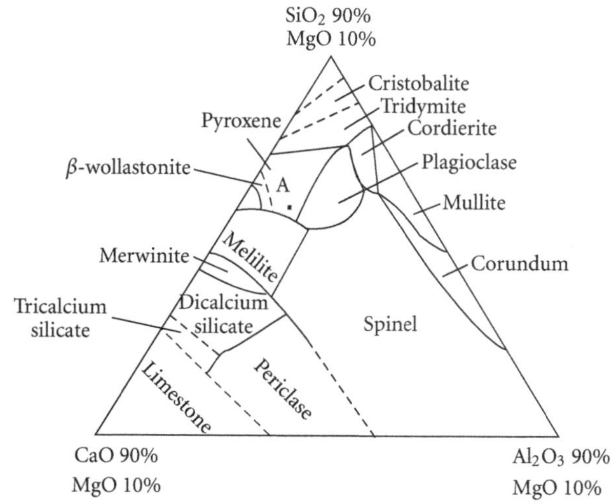

FIGURE 1: CaO-Al_2O_3-SiO_2 ternary system phase diagram containing 10% MgO.

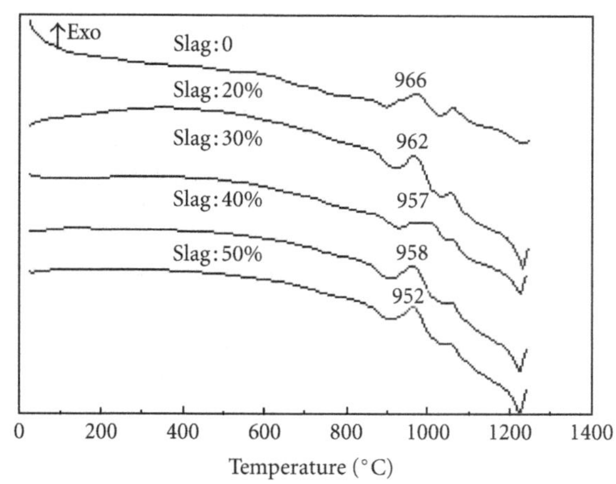

FIGURE 2: The DTA curves for each parent glass.

Consequently, according to the CaO-Al_2O_3-SiO_2 ternary system phase diagram [15] (shown in Figure 1), in which the mass percentage of MgO is 10%, the pyroxenes crystal (diopside $CaMg(SiO_3)_2$) was selected as the major crystal phase and the chemical composition of point A in Figure 1 was determined as that of the parent glass (shown in Table 1). The designed parent glass can get a lower melting temperature because point A is near the ternary eutectic point.

3. Experimental Materials, Content, and Method

3.1. Experimental Materials. Baotou steel blast furnace slag and quartz sand were used as the major raw materials, added with a small amount of pure chemical reagents CaO, MgO, and Al_2O_3, to prepare a mixture according to the chemical composition of the parent glass in Table 1. The chemical composition of Baotou steel blast furnace slag determined by means of chemical analysis is shown in Table 2, and the mass percentage of SiO_2 in the quartz sand is 97.15%.

Besides the major chemical components CaO, MgO, SiO_2, and Al_2O_3, there is a small amount of special components such as CaF_2, RE_XO_Y, K_2O, Na_2O, TiO_2, Nb_2O_5, and ThO_2 in the blast furnace slag, and the total mass percentage of these special components is about 5.58%, among which, 1.52% CaF_2, 0.78% RE_XO_Y, 0.53% K_2O, 1.34% Na_2O, 1.38% TiO_2, 0.0092% Nb_2O_5, and 0.016% ThO_2.

3.2. Experimental Content. In our experiment, the CaO-MgO-Al_2O_3-SiO_2 system glass-ceramics were prepared by melting method. And the change of mass percentage of the special components (0, 1.05%, 1.58%, 2.11%, and 2.64%) was realized by changing the proportion of the blast furnace slag in the parent glass (0, 20%, 30%, 40%, and 50%). The mutual influence of such special components as CaF_2, RE_XO_Y, K_2O, Na_2O, and TiO_2 on the crystallization behavior of the parent glass was studied under the condition of adding no nucleating agents (including nucleation and crystallization temperature of the parent glass, crystal phase

TABLE 3: Raw materials ratio of the parent glass.

Sample	Blast furnace slag (g)	Quartz sand (g)	CaO(g)	Al_2O_3 (g)	MgO (g)	B_2O_3 (g)	Special components (%)	ThO_2 (%)
1	0	55.58	26.53	10.15	10.20	2.04	0	0
2	20	47.24	19.27	7.53	7.97	2.04	1.05	0.003
3	30	43.07	15.64	6.21	6.85	2.04	1.58	0.005
4	40	38.89	12.01	4.90	5.74	2.04	2.11	0.006
5	50	34.72	8.38	3.59	4.62	2.04	2.64	0.008

composition, and microstructure). The raw materials ratio of the parent glass was shown in Table 3. In order to decrease the melting temperature of the glass, the pure chemical reagent B_2O_3 (2%) as flux was added. When the mass percentage of the blast furnace slag was less than 50% and that of ThO_2 in the parent glass was no more than 0.008% (80 ppm), that is to say, the content of the radioactive element thorium in the prepared glass-ceramics was not excessive.

3.3. Experimental Method. The particle sizes for various materials were ground to smaller than 200 mesh, weighed accurately according to the raw material ratio of the parent glass in Table 3, and mixed completely. Then, the mixture was put into a corundum crucible and melted in a furnace using Si-Mo rod as heater at 1490°C for 4 h. After that, the melt was poured into three steel molds. The moulds with the cooled melts were put into another furnace with its temperature 600°C and held for 2 h. After the annealing process, the parent glass prepared from the mixture was obtained.

The mutual influence of the special components in the blast furnace slag on the crystallization behavior of the parent glass was investigated by the method of differential thermal analysis (DTA) and X-ray diffraction (XRD).

The nucleation and crystallization temperatures of the parent glass were determined by an STA 449C comprehensive differential thermal analyzer. Because the nucleation of glass is endothermic, and its crystallization is exothermic, according to Tamman curve [16], the nucleation and crystallization of glass are both in a certain temperature range and, respectively, presented as the endothermic peak and exothermic peak on the DTA curve. Then, the parent glass was heat-treated at the nucleation and crystallization temperatures. XRD was used to identify the crystalline phases present in the glass after heat treatment, and the degree of crystallization, crystal phase composition, and their relative contents in the glass sample can be determined by observing the intensity of X-ray diffraction peaks.

4. Experimental Results and Analysis

4.1. The Impact of Special Components on the Nucleation and Crystallization Temperatures of the Parent Glass. The nucleation and crystallization temperatures of the parent glass were determined by the method of differential thermal analysis. The parent glass was ground to smaller than 200 mesh, filled into a corundum pot, then heated up to 1250°C at a rate of 10°C/min in argon. The DTA curves for each parent glass are shown in Figure 2, and the impact of special components in the blast furnace slag on the crystallization temperatures is shown in Figure 3. From Figures 2 and 3, it can be found that there are no obvious endothermic peaks corresponding to the nucleation process of glass, but there exist obvious exothermic peaks corresponding to the crystallization process of glass on the DTA curves for each sample. Due to one exothermic peak corresponding to the crystallization of a kind of mineral, two exothermic peaks on each DTA curve show that there are at least two kinds of minerals to precipitate. The area of the first peak is much larger than that of the second peak, which is caused by the precipitation of the main crystal phase, and the second peak stands for the precipitation of the minor crystalline phase. In this paper, the crystallization temperature of the main crystal phase was used as the crystallization temperature of the parent glass. When the blast furnace slag was not added into the raw materials of parent glass (the mass percentage of the special components was 0 in the parent glass), the crystallization peak temperature was 966°C; when the ratio of the blast furnace slag was increased to 50% (the mass percentage of the special components was 2.64%), it decreased to 952°C. Consequently, the crystallization peak temperature of the parent glass was gradually decreased with the increase of the blast furnace slag proportion in the parent glass, that is to say, the special components in the blast furnace could reduce the crystallization temperature of parent glass and promote the crystallization of the glass phase. The more slag addition means the more special components in the slag was added into the raw materials of

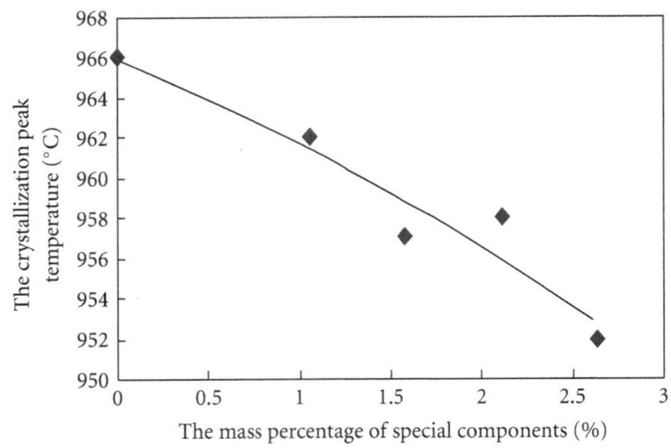

FIGURE 3: The impact of special components in the blast furnace slag on the crystallization temperatures of glass.

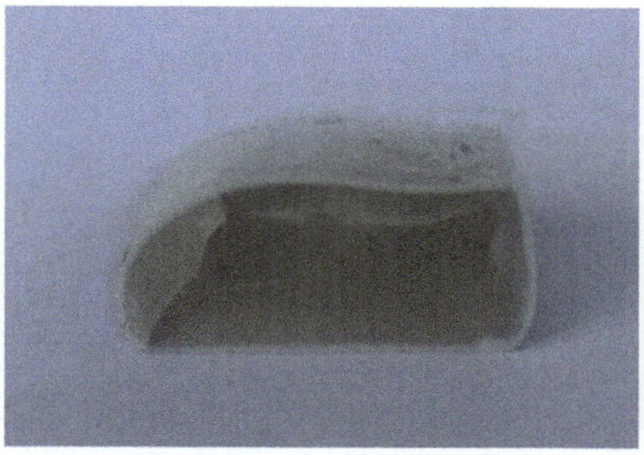

(a) Containing 20% slag (crystallization time 1 h)

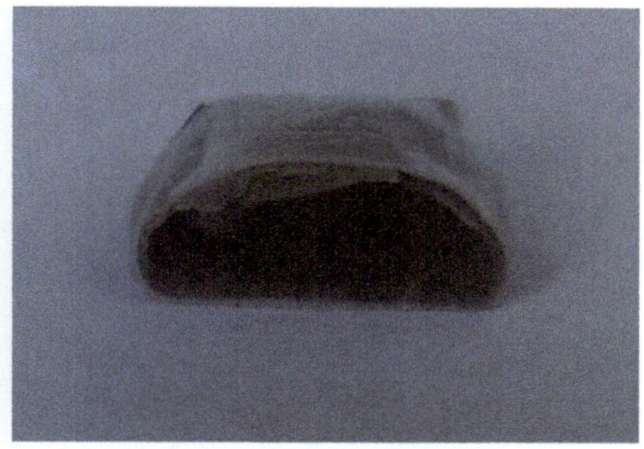

(b) Containing 40% slag (crystallization time 1 h)

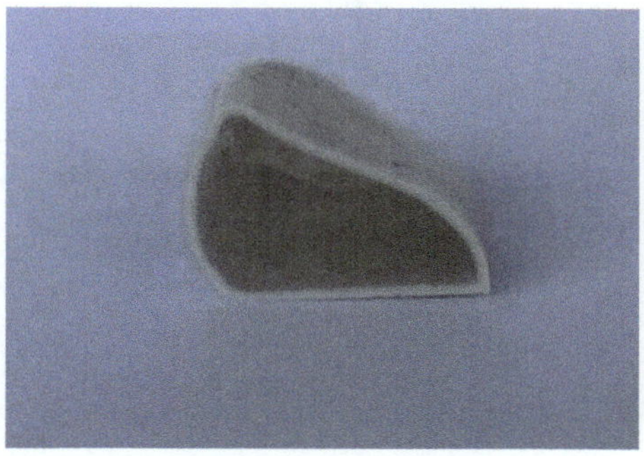

(c) Containing 20% slag (crystallization time 2 h)

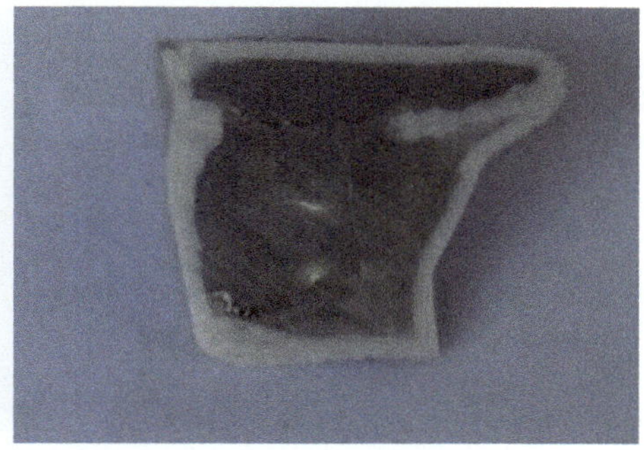

(d) containing 40% slag (crystallization time 2 h)

FIGURE 4: The crystallization state of glass samples containing 20% and 40% blast furnace slag after heat treatment.

the basic glass, in which, such special components as CaF_2 and TiO_2 are nucleating agents to promote the crystallization of glass. And the more nucleating agents can promote crystallization of glass. Therefore, it makes the crystallization temperature of glass reduce.

4.2. The Impact of Special Components on the Crystallization of the Parent Glass. According to the DTA curves of the parent glass samples, the heat treatment system of glass was selected as nucleation temperature 770°C, nucleation time 1 hour and crystallization temperature 962°C, crystallization time 1 hour. The crystallization temperature was selected as 962°C, because it was the highest for the glass samples, respectively, containing slag 20%, 30%, 40%, and 50%. Consequently, it ensures that crystals can precipitate from all the glass samples at this temperature. The glass samples containing 20% and 40% blast furnace slag after heat treatment is shown in Figures 4(a) and 4(b). From Figures 4(a) and 4(b), it can be seen that the surface of the samples is a crystal layer and all the rest part of the samples is glass when no nucleating agents were added into the raw materials of the parent glass. The surface layer has been separated from the inner glass, grinded into powders, and detected by XRD; the XRD results were shown in Figure 5, and three minerals were found. Consequently, the surface layer is crystallized. The crystallization state of glass samples containing 0%, 20%, 30%, 40%, and 50% blast furnace slag after heat treatment is similar, which belongs to the surface crystallization. In this paper, the glass samples containing 20% and 40% blast furnace slag after heat treatment were selected as the typical representative. With the increase of the blast furnace ratio, the color of the parent glass becomes darker. When the crystallization time was extended to 2 h, the other heat treatment parameters were unchanged, the crystallization state of these two glass samples is shown in Figures 4(c) and 4(d). It can be found that the crystal layer thickness of samples increases with prolonging the crystallization time and/or increasing the blast furnace slag ratio in the parent glass (that is, increasing mass percentage of the special components).

The special components in the blast furnace slag can promote the crystallization of the glass phase, which belongs to surface crystallization of glass, and they cannot play the role of nucleating agents to provide opportunities of nucleation, but the microdefects in the surface of the glass can provide heterogeneous nucleation sites and induce the crystallization during heat treatment. Consequently, the nucleating agents of proper species and quantity must be added into the raw materials in order to obtain uniform and compact grains in the glass body.

4.3. The Impact of Special Components on the Crystallization Phase after the Parent Glass Heat Treatment. The crystallization phase after the parent glass heat treatment was determined by XRD, and the results are shown in Figure 5. From Figure 5, it can be found that diopside $Ca(Mg, Al)(Si, Al)_2O_6$ and anorthite $CaAl_2Si_2O_8$ are the major minerals precipitated from the glass sample containing 20% blast furnace slag. Diopside containing aluminium $Ca(Mg_{0.5}Al_{0.5})(Al_{0.5}Si_{1.5}O_6)$ was also precipitated from the

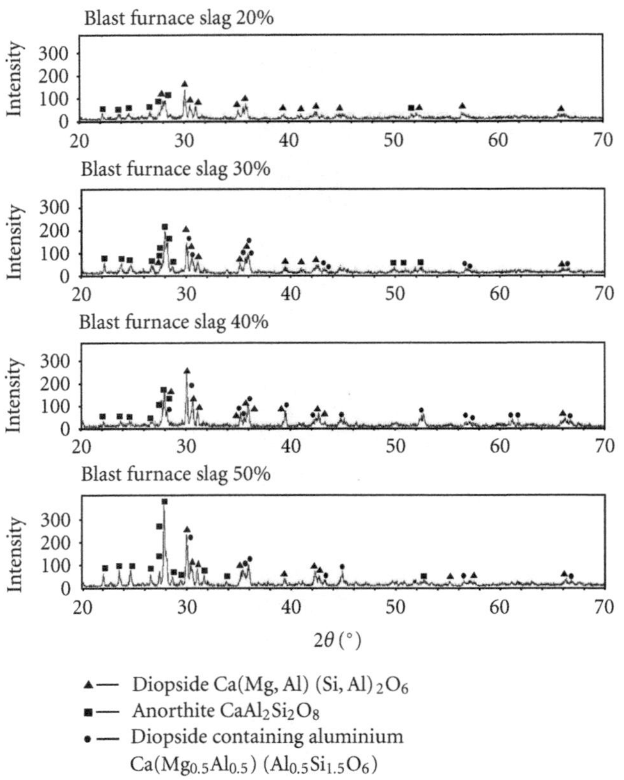

FIGURE 5: XRD patterns of the crystallization phase after the parent glass heat-treated.

glass samples containing 30%, 40%, and 50% blast furnace slag except for diopside and anorthite. And the quantity of diopside, anorthite, and diopside containing aluminum increased with the increase of the blast furnace slag content in the parent glass (i.e., the increase of special components). Therefore, the special components in the blast furnace slag can promote crystallization of glass, which is consistent with the DTA results. However, the major minerals precipitated from the parent glass were slightly different from the expectation. In order to get the expected major crystalline phase diopside, the appropriate quantity nucleating agents must be added to promote the precipitation of diopside and inhibit the precipitation of anorthite.

5. Conclusions

(1) In the process of glass-ceramics prepared with Baotou steel blast furnace slag by melting method, the special components such as CaF_2, RE_XO_Y, TiO_2, K_2O, and Na_2O in the blast furnace slag can reduce the crystallization temperature of parent glass and promote the crystallization of glass phase.

(2) The special components in the blast furnace slag can promote the crystallization of the glass phase, which belongs to surface crystallization of glass, but they cannot play the role of nucleating agents.

(3) When no nucleating agents were added, the major crystals precipitated from the parent glass are diopside, diopside containing aluminum, and anorthite, which are slightly different from the expectation.

Acknowledgments

The research work in this paper was financially supported by the Major Project of Inner Mongolia Natural Science Foundation (2011ZD06) and Research Projects of Inner Mongolia Colleges and Universities (NJZZ11141).

References

[1] H. N. Xiao, W. Q. Peng, and C. M. Deng, "Preparation of microcrystalline ceramic technology, performance, and uses," *Chinese Ceramics*, vol. 36, pp. 31–33, 2000.

[2] A. A. Francis, "Conversion of blast furnace slag into new glass-ceramic material," *Journal of the European Ceramic Society*, vol. 24, no. 9, pp. 2819–2824, 2004.

[3] Z. F. Shan, "Domestic and foreign steel slag processing technology and utilization technology development," *Industrial Safety and Dust*, pp. 27–30, 2000.

[4] M. L. Qvecoglu, B. Kuban, and H. Özer, "Characterization and crystallization kinetics of a diopside-based glass-ceramic developed from glass industry raw materials," *Journal of the European Ceramic Society*, vol. 17, no. 7, pp. 957–962, 1997.

[5] T. Kehagias, P. Komninou, P. Kavouras, K. Chrissafis, G. Nouet, and T. Karakostas, "Crystal phase separation and microstructure of a thermally treated vitrified solid waste," *Journal of the European Ceramic Society*, vol. 26, no. 7, pp. 1141–1148, 2006.

[6] Z. J. Wang, W. Ni, Y. Jia, L. P. Zhu, and X. Y. Huang, "Crystallization behavior of glass ceramics prepared from the mixture of nickel slag, blast furnace slag and quartz sand," *Journal of Non-Crystalline Solids*, vol. 356, no. 31-32, pp. 1554–1558, 2010.

[7] G. A. Khater, "The use of Saudi slag for the production of glass-ceramic materials," *Ceramics International*, vol. 28, no. 1, pp. 59–67, 2002.

[8] T. Toya, Y. Tamura, Y. Kameshima, and K. Okada, "Preparation and properties of CaO-MgO-Al_2O_3-SiO_2 glass-ceramics from kaolin clay refining waste (Kira) and dolomite," *Ceramics International*, vol. 30, no. 6, pp. 983–989, 2004.

[9] H. Liu, H. Lu, D. Chen, H. Wang, H. Xu, and R. Zhang, "Preparation and properties of glass-ceramics derived from blast-furnace slag by a ceramic-sintering process," *Ceramics International*, vol. 35, no. 8, pp. 3181–3184, 2009.

[10] S. Banijamali, H. R. Rezaei, B. Eftekhari Yekta, and V. K. Marghussian, "Sinterability, crystallization and properties of glass-ceramic tiles belonging to CaF_2-CaO-MgO-Al_2O_3-SiO_2 system," *Ceramics International*, vol. 33, no. 8, pp. 1557–1561, 2007.

[11] K. Cai, "Comprehensive utilization of Baotou iron and steel company new blast furnace slag," *Science & Technology of Baotou Steel*, vol. 34, pp. 77–80, 2008.

[12] W. Q. Peng and H. N. Xiao, "The impact of fluoride on CaO-Al_2O_3-SiO_2 system glass crystallization behavior," *Materials Development and Application*, vol. 16, pp. 16–17, 2001.

[13] B. Yu, K. Liang, A. Hu, and S. Gu, "Influence of different TiO_2 content on crystallization of CaO-MgO-P_2O_5-SiO_2 system glasses," *Materials Letters*, vol. 56, no. 4, pp. 539–542, 2002.

[14] W. Q. Peng, *CaO-MgO-Al_2O_3-SiO_2 system glass-ceramics [M.S. thesis]*, Hunan University, Hunan, China, 2000.

[15] H. N. Xiao, Y. Liu, and H. X. Shi, "Effects of content of blast slag and heat-treatment parameters on properties of slag glass-ceramics," *Science and Art of Ceramic*, pp. 42–46, 2003.

[16] E. Rowlands and P. F. James, "Analysis of steady state crystal nucleation rates in glasses," *Physics and Chemistry of Glasses*, vol. 20, pp. 2–14, 1979.

8

Unexpected Precipitation of a Re-Rich Phase in Single Crystal Superalloy CMSX-4 during Thermal Exposure

Kuiyu Cheng,[1,2] ChangYong Jo,[2] DooHyun Kim,[3] Tao Jin,[1] and Zhuangqi Hu[1]

[1] *Superalloy Division, Institute of Metal Research, Chinese Academy of Sciences, 72 Wenhua Road, Shenyang 110016, China*
[2] *High Temperature Materials Group, KIMS, 531 Changwondaero, Changwon 641-010, Republic of Korea*
[3] *Materials Test and Characterization Group, KIMS, 531 Changwondaero, Changwon 641-010, Republic of Korea*

Correspondence should be addressed to Kuiyu Cheng, kuiyucheng@hotmail.com

Academic Editor: Hao Chen

A Re-rich phase (up to 85.40% wt.Re) was occasionally found in a single crystal superalloy during thermal exposure. Interestingly, this phase is located in front of the dendrite periphery in the interdendritic region, while the Re is segregated in the dendrite core. Formation of this phase was suggested to relate to the unusual accumulation of the Re resulting from the uneven diffusion process of Re. Appearance of this phase may imply the preference of forming Re-Re atomic bond in the superalloys.

1. Introduction

The beneficial effect of Re on the high temperature properties of single-crystal nickel-based superalloys has resulted in a steady increase in the levels of Re used in these alloys [1, 2]. As a result, role and behavior of Re in superalloys have been drawing increasing attention over recent years [3–16]. Rhenium could significantly reduce the γ' coarsening rate [3] and is beneficial in producing alloys with small negative lattice misfit parameters [4]. In addition, atom probe investigations of Re-containing superalloys have reported both clustering of Re in the γ phase [5, 6] and pileup of Re at the γ/γ' interface [7, 8]. Therefore, the position and the existing way of Re plays a significant role in mechanical properties of the superalloys. However, the mechanisms for Re acting in superalloys are not well established yet.

In this study, a Re-rich phase (up to 85.40% wt.Re) was unexpectedly found in the well-known single crystal superalloy CMSX-4 during thermal exposure. Interestingly, this phase is located in front of the dendrite periphery in the interdendritic region, while the Re is segregated in the dendrite core. The possible formation mechanism of this phase was discussed as well as its significance to the superalloys in the present study. It is thought that the present study is helpful for understanding the particular role and behavior of Re in Nickel-base superalloys.

2. Experimental Procedure

The mother alloy of single crystal alloy CMSX-4 used for the present work was supplied by CMSX Company. The casting method and nominal composition of the single crystal bars have been given in previous work [9]. The single crystal bars were sliced to small pieces (12 mm × Φ14 mm) and were given a standard heat treatment (heat to 1260°C at 360°C/h, hold 0.2 h, heat to 1288°C at 30°C/h and heat to 1320°C at 2°C/h, hold 2 h + AC; 1140°C/2 h + AC; 870°C/20 h + AC). Then, the samples were thermally exposed at 950°C between 500~5000 hours. The microstructures were carefully examined by the scanning electron microscope (SEM) with an energy dispersive X-ray analysis (EDS) accessory. It should be pointed out that only the observations made on the samples which show the unexpected formation of the Re-rich phase will be presented in this paper.

The composition mapping of Re was carried out with a Shimadzu mode electron probe microanalysis (EPMA) machine to examine and analyze the Re distribution before and after standard heat treatment according to the dendrite structure.

3. Result

3.1. Appearance of the Re-Rich Phase. The only phases occurred in the CMSX-4 specimens before or after standard

Table 1: Chemical composition of the Re-rich phase in present study (wt.%).

Al	Cr	Co	Ni	W	Re
1.08	1.89	1.05	10.15	0.43	85.40

heat treatment are γ, γ', and γ/γ' eutectics. The detail of the specimen after standard heat treatment can be found in previous work [9]. Figure 1 shows the microstructure and EDS analysis of the specimen after thermal exposure of 800 hours at 950°C. As we can see, a bright phase was occasionally found in the backscattered electron image (Figure 1(a)), which indicates that this phase consists of high content of refractory alloying elements. Moreover, this phase presents faceted morphology and is distributed in front of the dendrite periphery in the interdendritic region, especially in the corners between the dendrite arms. The EDS analysis clearly proves that this phase is primarily composed of Re and a little Ni with minor of Al, Cr, Co, and W, most of which are γ forming elements (Figure 1(c)). The detail of the composition obtained by EDS is given in Table 1. It can be seen that the content of Re (up to 85.40% wt.) is much higher than those of other elements.

In the specimens thermally exposed for 1000 hours, this Re-rich phase could be hardly found at low magnification, while at high magnification pieces of this phase were detected as shown in Figure 2. The morphology of this Re-rich phase is more regular with clearly pyramid facets in this condition. Furthermore, the phase distribution and composition are consistent with those in the condition of 800 hours exposure.

It should be pointed out that, although this phase was not found in the specimens with the shorter or longer exposure times, it is hard to say whether this phase formed or not in those conditions because of the much few amount. However, it at least demonstrates that this Re-rich phase mostly formed after approximate 800 hours during thermal exposure at 950°C and gradually disappeared with increasing the thermal exposure time. In addition, since the amount of this Re-rich phase is very low in the specimen, it is difficult to find an effective way to identify this phase such as selected diffraction pattern.

3.2. Re Distribution. As this phase is strongly rich in Re, the Re distribution in the specimens before and after the standard heat treatment was analyzed by EPMA mapping analysis. The mapping analysis is conducted on a scale of one dendrite structure as shown in Figure 3. The Re is strongly segregated in the dendrite core or in the center of the secondary arm in the as-cast condition (Figure 3(a)). After the standard heat treatment, most microsegregation of the Re was homogenized with unclear dendrite structure in the EPMA maps (Figure 3(b)). Inset shows the analyzed area in the standard heat-treated specimen. Moreover, it should be emphasized that as many as EPMA maps obtained from other fields of the examined as-cast and standard heat-treated specimens are consistent with that shown in Figure 3. No particular segregation of Re was detected in the corresponding region in which the Re-rich phase precipitated during following thermal exposure.

4. Discussion

4.1. Re-Rich Phase in Superalloy. The Re-rich phase is very unusual in superalloys. A Re-rich δ-phase has been reported in as-cast and heat-treated Ni-Al-Re ternary alloys [10, 11] and later was found in variety of superalloys [12, 13]. Composition of the Re-rich phase found in this study is of some difference to theirs, with lower Re content and higher Ni. However, since Re and Ni are the main components in the Re-rich phase in this study (Table 1) and there is no other compound of Re and Ni according to the Ni-Re phase diagram [14], this Re-rich phase might be the Ni-oversaturated δ-Re phase formed by the incomplete precipitation of the Re from the matrix during the thermal exposure. This could also explain the minor contents of Al, Cr, Co, and W in this phase.

4.2. Possible Formation Mechanism of the Re-Rich Phase in CMSX-4. In this study, the alloy CMSX-4 used and other Re-containing alloys as well [13] have been melted once to make the initial ingots and again in the single crystal solidification process. Furthermore, this phase has displayed certain crystallographic morphology with matrix in the solid state as shown in Figure 2. Therefore, presence of this phase is unlikely due to insufficient time for dissolution of the initial Re additions to the alloy. Feng et al. have come up with two ways in which the Re-rich δ-phase could form in their alloys: form in the interdendritic region by peritectic reaction during the solidification or form in the dendrite core during the aging heat treatment due to the strong segregation of Re there [13]. However, neither of them is applicable in present study, since this phase was not found in the as-cast condition and the Re is segregated in the dendrite core rather than in the area where this phase formed (Figure 3). It indicates that formation of this phase in CMSX-4 alloy was controlled by a distinct mechanism with those in Feng's study. It is proposed that this Re-rich phase here is possibly induced by unusual accumulation of Re during later thermal exposure.

It is well known that alloy elements have different preference of segregation in dendrite and interdendritic area which would lead to two cross-diffusion fluxes in elevated temperature: W, Re, Cr, Mo, and Co from dendrite to the interdendritic regions and Al, Ti, Ta, and Ni in the opposite direction [15]. To simplify the diffusion process here, only the diffusion of Re is considered with two assumptions: the diffusion of Re is not affected by other alloy elements and is not permitted in the γ' phase.

Then considering the process of diffusion, the flux of matter or else can be given by the relation: Flux = (conductivity) (driving force) [17]. In our case, the conductivity is the diffusivity of Re which is determined by temperature and the volume fraction of γ phase according to the assumptions and the driving force is related to the concentration gradient of Re. On one hand, the conductivity of Re is better in the dendrite than that in the interdendritic region, since the volume fraction of the γ phase is higher in the dendrite

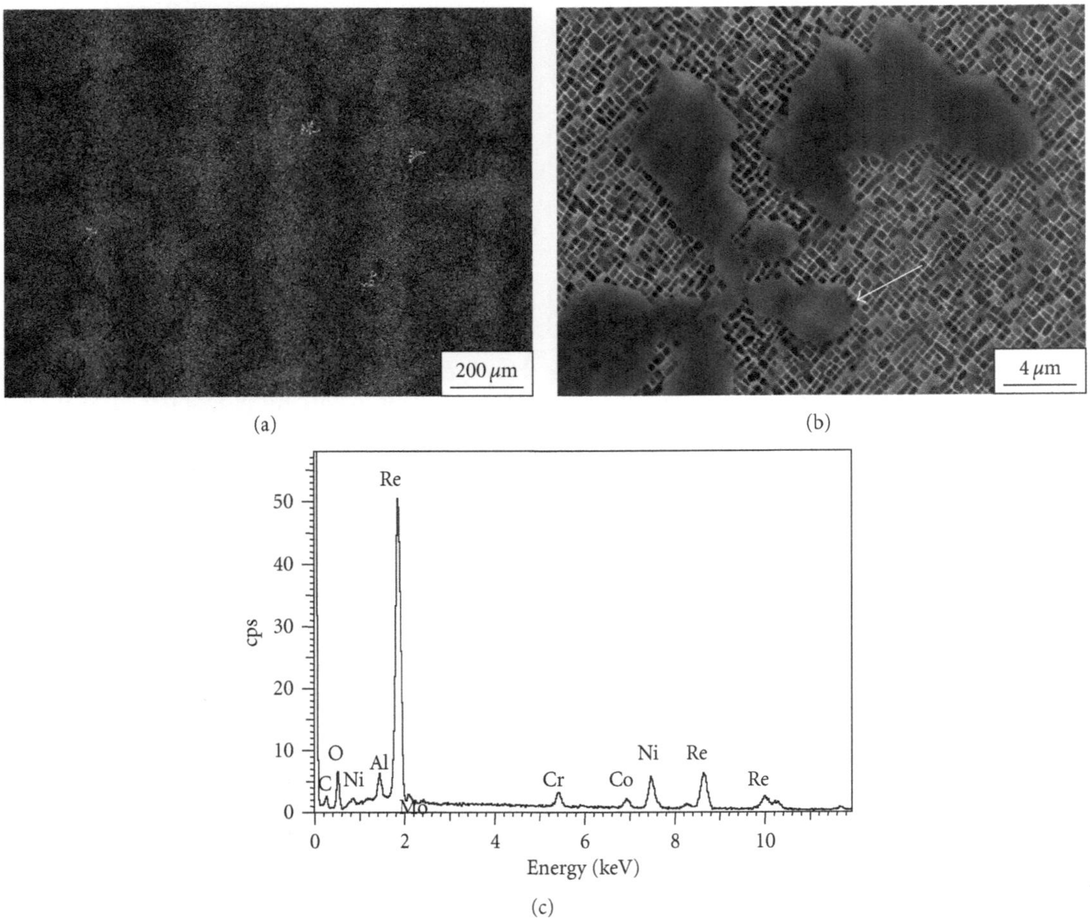

FIGURE 1: SEM micrographs and EDS analysis of the specimen after 800-hour thermal exposure: (a) BE image in low magnification, (b) SE image showing the detail of the Re-rich phase, and (c) EDS result of the faceted particle marked by the arrow in (b).

FIGURE 2: SEM micrographs of the specimen after 1000-hour thermal exposure: (a) BE image in low magnification, (b) SE image showing the detail of the Re-rich phase.

and the temperature is constant. On the other hand, it has proven in previous work [9] that the concentration along the line from dendrite core to interdendritic area of Re decreases smoothly, has a relative sharp decrease at the dendrite periphery, and then goes down smoothly again, which implies that the concentration gradient of Re is largest near the dendrite periphery between the dendrite core and the interdendritic region.

As a result, the flux exchange of Re would be relatively higher near the dendrite periphery and, at the same time, the Re might not diffuse away into the interdendritic region as fast as it diffused from the dendrite core to the dendrite

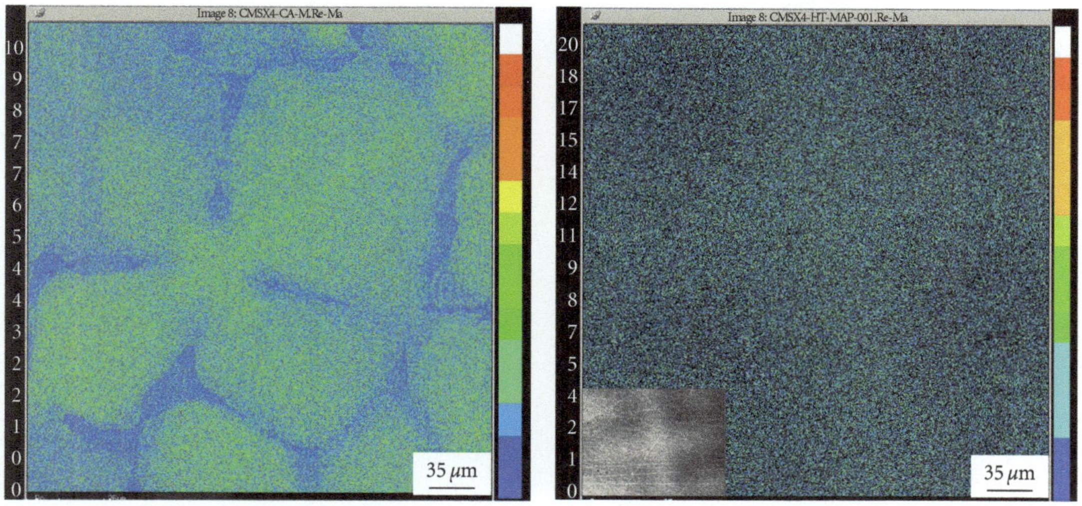

(a) (b)

FIGURE 3: EPMA mapping analysis of Re: (a) as-cast condition, (b) after standard heat treatment with the inset showing the mapping area.

periphery. This may give rise to an accumulation of Re at the dendrite periphery. Moreover, the dendrite periphery might be taken as a pseudo-boundary where a new phase may readily form. Then, there might be a possibility to form the Re-rich phase from the matrix in this area. And with increasing the thermal exposure time, this phase gradually disappeared by diffusing away of the accumulated Re.

4.3. Significance of This Re-Rich Phase. As shown preciously, the amount of this Re-rich phase is extremely low in the specimen. Therefore, it could hardly affect the mechanical properties of the superalloy (Ph.D.). Importantly, the appearance of this Re-rich phase, even a little, may to some extent illustrate the particular behavior of Re in the superalloys. It may imply the preference of forming the Re-Re bond in the superalloys rather than forming Re-X bond (X means the other alloying elements), otherwise other compounds would form but the Re-rich phase. Lately, atom probe investigations of Re-containing superalloys have reported the clustering of Re in the γ phase [5, 6], which could partly explain the obvious improvement of the Re addition to the superalloys [16]. The preference formation of the Re-Re bond may to some extent illustrate the formation of Re cluster in superalloys by means of short-range order of the Re atoms. It is tempted to say that the Rhenium atoms are prone to holding together locally rather than to being evenly distributed in the γ matrix of superalloys.

5. Conclusions

In present study, a Re-rich phase (up to 85.40% wt. Re) was unexpectedly found in the single crystal superalloy CMSX-4 after 800–1000-hour exposure at 950°C. It is surprisingly that this phase is located in front of the dendrite periphery in the interdendritic region, while the Re is segregated in the dendrite core but depleted in the interdendritic region. It was proposed that this Re-rich phase was caused by the unusual accumulation of the Re resulting from the uneven diffusion process of Re in the superalloys. Due to the scarce amount, it is incompetent to identify this Re-rich phase in this study. However, appearance of this Re-rich phase, even a little, may imply the preference of forming the Re-Re bond in the superalloys

Acknowledgment

This work was supported by the Korea Foundation for International Cooperation of Science and Technology (KICOS) through a grant provided by the Korean Ministry of Science and Technology (MOST) in K20711000001-07A0100-00110.

References

[1] T. Khan, P. Caron, and C. Duret, "The development and characterization of a high performance experimental single crystal superalloy," in *Superalloys*, M. Gell et al., Ed., pp. 145–155, The Metallurgical Society of AIME, Warrendale, PA, USA, 1984.

[2] A. D. Cetel and D. N. Duhl, "Second-generation nickel-base single crystal superalloy," in *Suoerallovs*, D. N. Dun et al., Ed., pp. 235–244, The Metallurgical Society, Warrendale, PA, USA, 1988.

[3] A. F. Giamei and D. L. Anton, "Rhenium additions to a Ni-base superalloy: effects on microstructure," *Metallurgical Transactions A*, vol. 16, no. 11, pp. 1997–2005, 1985.

[4] H. Murakami, H. Harada, and H. K. D. H. Bhadeshia, "The location of atoms in Re- and V-containing multicomponent nickel-base single-crystal superalloys," *Applied Surface Science*, vol. 76-77, pp. 177–183, 1994.

[5] D. Blavette, P. Caron, and T. Khan, "An atom probe investigation of the role of rhenium additions in improving creep resistance of Ni-base superalloys," *Scripta Metallurgica*, vol. 20, no. 10, pp. 1395–1400, 1986.

[6] N. Wanderka and U. Glatzel, "Chemical composition measurements of a nickel-base superalloy by atom probe field ion

microscopy," *Materials Science and Engineering A*, vol. 203, no. 1-2, pp. 69–74, 1995.

[7] H. Murakami, P. J. Warren, and H. Harada, "Atom-Probe Microanalysis of Some Ni-base Single Crystal Superalloys," in *Proceedings of the 3rd International Charles Parsons Turbine Conference*, p. 343, Institute of Materials, Newcastle-upon-Tyne, UK, April 1995.

[8] P. Druska, U. Steinike, and V. Sepelak, "Surface structure of mechanically activated and of mechanosynthesized zinc ferrite," *Journal of Solid State Chemistry*, vol. 146, no. 1, pp. 13–21, 1999.

[9] K. Y. Cheng, C. Y. Jo, T. Jin, and Z. Q. Hu, "Effect of stress on μ phase formation in single crystal superalloys," *Materials Science and Engineering A*, vol. 528, no. 6, pp. 2704–2710, 2011.

[10] L. A. Cornish and M. J. Witcomb, "Metallographic study of the Al-Ni-Re phase diagram," *Journal of Alloys and Compounds*, vol. 291, no. 1-2, pp. 145–166, 1999.

[11] S. Miyazaki, Y. Murata, and M. Morinaga, "Site occupancy of Re atoms in Ni3Al(γ') and γ-γ'; phase equilibria in Ni-Al-Re ternary alloys," *Tetsu-To-Hagane/Journal of the Iron and Steel Institute of Japan*, vol. 80, no. 2, pp. 161–165, 1994.

[12] Q. Feng, T. K. Nandy, and T. M. Pollock, "The Re (Ru)-rich δ-phase in Ru-containing superalloys," *Materials Science and Engineering A*, vol. 373, no. 1-2, pp. 239–249, 2004.

[13] Q. Feng, T. K. Nandy, S. Tin, and T. M. Pollock, "Solidification of high-refractory ruthenium-containing superalloys," *Acta Materialia*, vol. 51, no. 1, pp. 269–284, 2003.

[14] V. Sepelak, D. Schultze, F. Krumeich, U. Steinike, and K. D. Becker, "Mechanically induced cation redistribution in magnesium ferrite and its thermal stability," *Solid State Ionics*, vol. 141-142, pp. 677–682, 2001.

[15] B. S. Bokstein, A. I. Epishin, T. Link, V. A. Esin, A. O. Rodin, and I. L. Svetlov, "Model for the porosity growth in single-crystal nickel-base superalloys during homogenization," *Scripta Materialia*, vol. 57, no. 9, pp. 801–804, 2007.

[16] D. Blavette, P. Caron, and T. Khan, "Superalloys 1988," in *Proceedings of the 6th International Symposium on Superalloys, Seven Springs*, p. 305, TMS, Pennsylvania, Pa, USA, September 1988.

[17] R. Barrett Craig, D. Nix William, and S. Tetelman Alan, *The Principles of Engineering Materials*, Prenitice-Hall, New Jersey, NJ, USA, 1st edition, 1973.

Cathodic Cage Plasma Nitriding: An Innovative Technique

R. R. M. de Sousa,[1] F. O. de Araújo,[2] J. A. P. da Costa,[3] A. de S. Brandim,[1] R. A. de Brito,[4] and C. Alves Jr.[4]

[1] *Grupo de Pesquisa em Materiais e Metrologia, Departamento de Mecânica, Instituto Federal IF-PI, 64000040 Teresina, PI, Brazil*
[2] *Departamento de Ciências Exatas e Naturais, Universidade Federal Rural do Semi-Árido, 59625900 Mossoró, RN, Brazil*
[3] *Departamento de Física, UERN, Campus Central, 59610210 Mossoró, RN, Brazil*
[4] *LabPlasma, Departamento de Física, UFRN, Campus Universitário, 59072970 Natal, RN, Brazil*

Correspondence should be addressed to R. R. M. de Sousa, romulorms@gmail.com

Academic Editor: Thierry Grosdidier

Cylindrical samples of AISI 1020, AISI 316, and AISI 420 steels, with different heights, were simultaneously treated by a new technique of ionic nitriding, entitled cathodic cage plasma nitriding (CCPN), in order to evaluate the efficiency of this technique to produce nitrided layers with better properties compared with those obtained using conventional ionic nitriding technique. This method is able to eliminate the edge effect in the samples, promoting a better uniformity of temperature, and consequently, a smaller variation of the thickness/height relation can be obtained. The compound layers were characterized by X-ray diffraction, optical microscopy, and microhardness test profile. The results were compared with the properties of samples obtained with the conventional nitriding, for the three steel types. It was verified that samples treated by CCPN process presented, at the same temperature, a better uniformity in the thickness and absence of the edge effect.

1. Introduction

Conventional plasma nitriding has been employed a lot in the industry, due to several advantages in relation to other nitriding processes (gas nitriding and salt bath nitriding), such as higher gas economy and smaller processing time, since the diffusion velocity of nitrogen (N_2) in plasma nitriding is higher [1]. This process is used to improve superficial properties like hardness, wear, and corrosion resistances, in order to increase the useful life of nitrided components. However, it presents some inconveniences like edge effect in the sample surface, hollow cathode effect, nonuniform temperature, and arcing during the treatment, especially in pieces with complex geometry [2, 3].

When samples with different dimensions are nitrided by conventional plasma process, they present a thermal gradient between the surface and sample holder. In samples with height of 10 mm, the difference of temperature between the top surface and sample holder can reach 373 K [4].

In this work, a new plasma nitriding technique entitled *cathodic cage plasma nitriding* (CCPN) based on multiple hollow cathode effects is used and the results compared with the conventional plasma nitriding. This technique presents some advantages compared to the conventional method, including process efficiency and the production of layers with uniform thickness. The samples are totally involved by a metallic cage, where a high cathodic potential is applied, so the plasma is not formed directly on the sample surface, but in the cathodic cage. A ceramic disk (In this work, machined with aluminum oxide) is placed on the samples holder in order to electrically isolate the samples from the cathode. Besides, the temperature inside the nitriding chamber is uniform, and the thermal gradient is very small [5].

2. Experimental

Samples of AISI 1020, AISI 316, and AISI 420 steels, machined with heights of 1, 3, 5, 8, and 10 mm, respectively, were used to carry out the tests. All the specimens were previously annealed and submitted to a metallographic preparation, where they were mounted in epoxy resin, mechanically ground (using sandpapers of 320, 400, 600, and 1200) and polished (with alumina 1 and 0.3 μm). After the metallographic procedure, the samples were cleaned in an ultrasound machine with acetone, and treated by CCPN technique, in a plasma nitriding reactor. The system used in conventional nitriding is composed by a power source with maximum output voltage and DC current of 1500 V and 1.5 A, respectively. In cathodic cage technique the same configuration is used, the only modification is the use of a cathodic cage. A schematic picture of the plasma reactor is shown in Figure 1. It consists of a cylindrical chamber (diameter: 30 cm and height: 40 cm) made in stainless steel in which the negative polarized electrodes are applied and the rest of the chamber held at ground potential (anode). The cage is manufactured with 316 stainless steel (thickness: 8 mm, diameter: 112 mm, height: 25 mm and hole diameter: 8 mm). The holes have a uniform distribution, and they are positioned in a distance of 9.2 mm between the centers of adjacent holes. The lateral minimum distance between samples and the cage is 28 mm [6]. Two samples to each height (1, 3, 5, 8, and 10 mm) were placed onto the insulating aluminum oxide disk which function as the sample holder, according to Figure 1. In this configuration, the plasma is formed on the cathodic cage walls and not directly on the sample surface.

The AISI 1020 steel samples were treated for 3 h while the samples of AISI 316 and 420 steels are treated for 5 h, using pressures of 250 and 500 Pa. The samples of AISI 1020, were treated under the temperatures of 673, 773, and 823 K, the samples of AISI 316 under the temperatures of 673 K, 723, and 773 K, while for the AISI 420 steel samples 623 and 673 K. All the treatments were made with a nitriding mixture of 80% N_2 + 20% H_2, in a total flow of 20 sccm.

The phase composition and texture were analyzed by XRD, using Cu Kα radiation (wave length: 0.154 nm), in an XRD device (Shimadzu, XRD-6000). The optical microscopy (Olympus BX60M) was used to observe the morphology and thickness of the nitrided layers. Finally, the microhardness profiles were accomplished in order to analyze the thickness and uniformity of the obtained layers, beside the elimination of the edge effect. A Vickers microhardness testing device (Pantec HVS 1000) was used to measure the hardness along the surface. A load of 100 gf was used to carry out the indentations.

3. Results and Discussion

Figure 2 shows the thickness of formed layers for five samples of carbon steel AISI 1020 at the same temperature (673, 773, and 823 K) and did not present significant variation with the increase of the height for samples treated in CCPN. This fact is a consequence of the uniform temperature characteristic in this process, since in CCPN technique there are no thermal gradients between the surfaces of the sample and sample holder. The required energy to heat the samples comes from the cathodic cage, avoiding the edge effect and, consequently, promoting uniformity of the nitrided layers.

Table 1 presents the values of the thicknesses of nitrided layer for the different samples of the austenitic stainless steel AISI 316 for the two plasma nitriding techniques. A small gradient is verified in the variation of the thickness of the layers nitrided by the new technique as function of the height of the samples, especially for higher treatment temperatures (773 K), where the maximum variation observed in the thickness is of the order of 10%, unlikely for samples treated by conventional nitriding technique the nitriding layer thickness variation was of the order of 25% between two successive sample heights value. This small nitriding layer thickness variation by cathodic cage technique, rather than being a consequence of the temperature variation inside the nitriding chamber, is only due to the different distances of the samples from the cage walls. Certainly for samples of bigger height the distance from the cage lid is smaller and consequently the nitriding species reaches the sample surface with higher energy, which is actually confirmed by the larger presence of iron nitride in the XRD diffractograms.

The XRD patterns for the AISI 316 austenitic stainless steel samples are presented in Figure 3, nitrided by CCPN, showed that the peak of "S phase" (expanded austenite) is situated in the same position for all samples with different heights, presenting the same intensity. According to Sun et al. [7], when the temperature is increased, this peak is dislocated to the left, due to the interplanar spacing increase. Therefore the absence of such peak position dislocation is indicative of the small gradient of temperature for samples with different heighs, treated in CCPN technique. The same behavior is observed to γ'-Fe_4N phase.

The microhardness profiles for nitrided samples of the AISI 316 at the same temperature, but with different heights are shown in Figure 4. These values present a low variation due to the uniformity of superficial temperature, independently of the height, confirming the existence of a small thermal gradient between the top surface of the samples and the sample holder [8, 9].

Figure 5 presents a comparative of the samples of the AISI 316, AISI 1020, and AISI 420 steel which were nitrided by conventional method (a) and by the CCPN process (b), to show the visual aspect of the surface due the edge effect. Samples treated by conventional nitriding presented erosion rings, characteristics of the edge effect. In the case of cathodic cage plasma nitriding, the surface presents a uniform color in the whole surface area due to the elimination of the this effect for three types of steels used in this work.

Figure 6 presents the values of correspondent microhardness tests along the top surface for samples treated by conventional and CCPN nitriding process. Radial measurements were accomplished, following the regions 1, 2, and 3 in the sample surface for AISI 420, according to Figure 5. Samples treated by conventional process presented a variation in the values along the disk radius, and it was observed that region 2 presented a significant decreasing in these values. Samples treated by CCPN process presented uniform values along the disk radius, confirming the elimination of the edge effect. This fact testifies to the predominance of the mechanism of

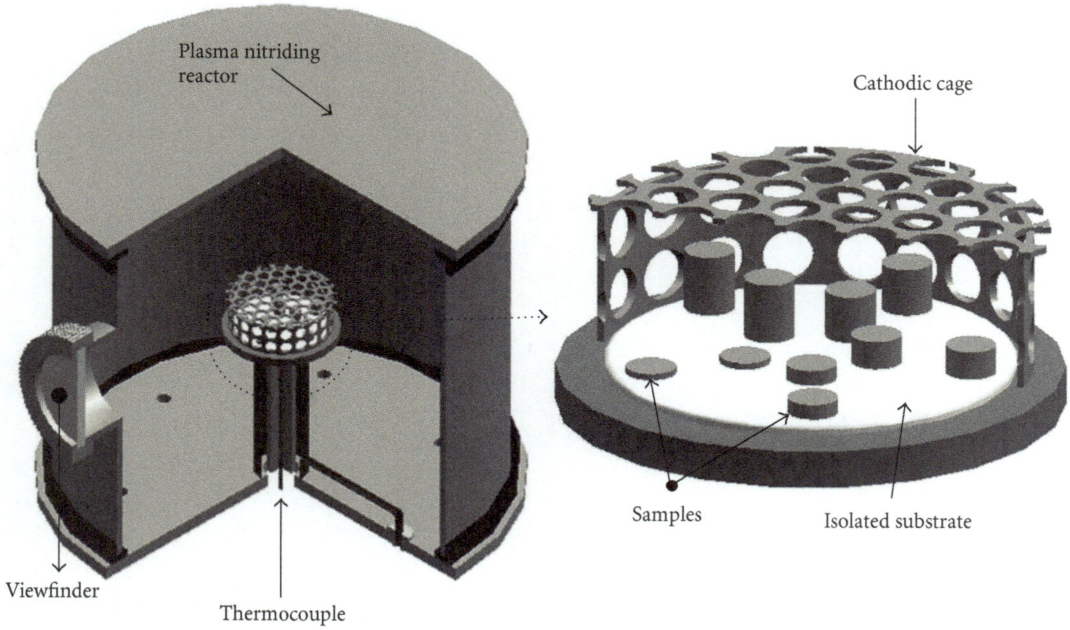

FIGURE 1: Schematic picture of the plasma reactor with the cathodic cage device, showing the disposition of samples with different heights.

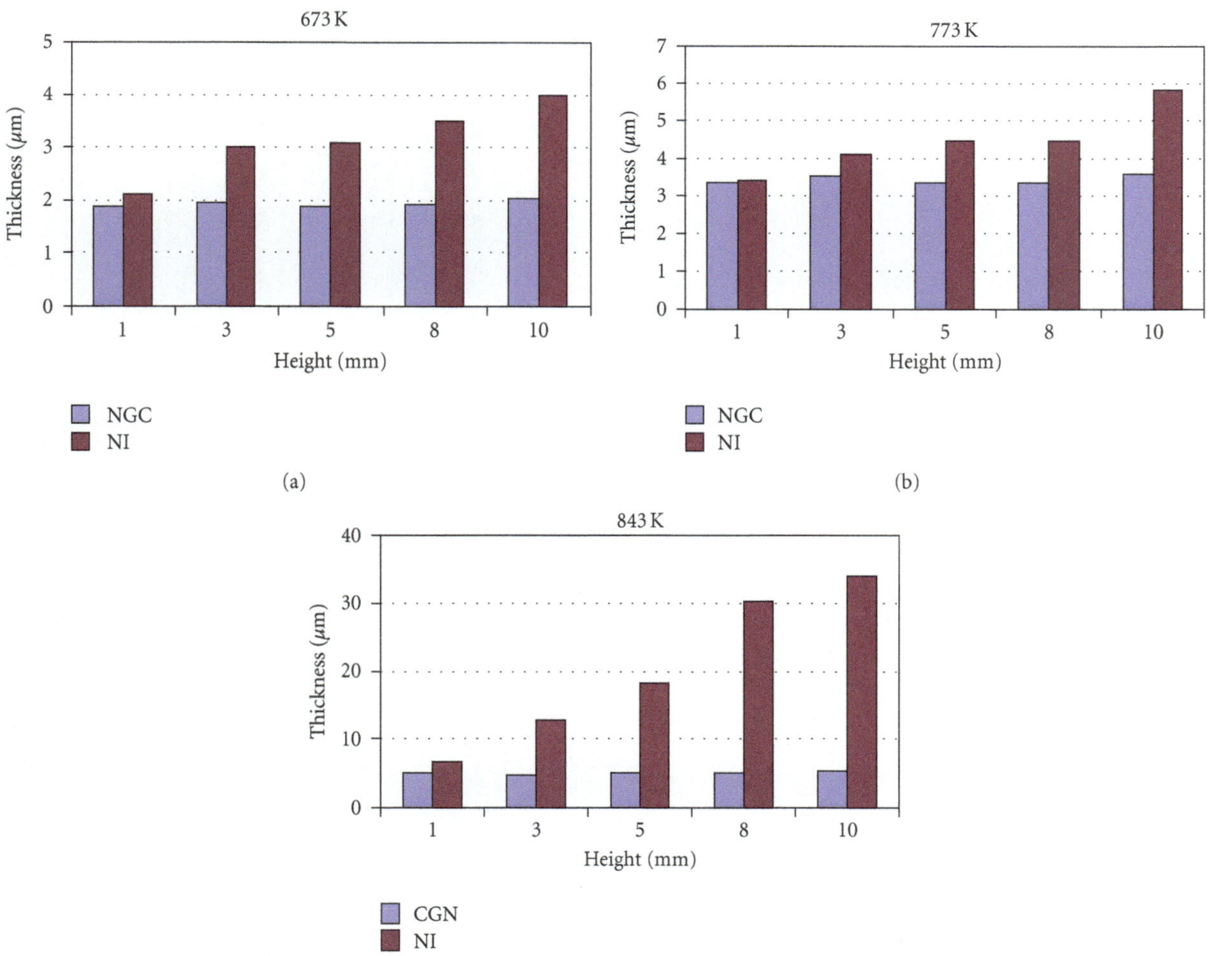

FIGURE 2: Thickness of layers to AISI 1020 steels samples with a height respectively 1, 3, 5, 8, and 10 mm nitrided by CCPN at (a) 673, (b) 773, and (c) 823 K. (3 h).

Table 1: AISI 316 thickness of the nitrided layers for samples height of 1, 3, 5 and 10 mm, at different temperatures.

Sample height (mm)		Temperature (K) 673	723	773	673	723	773
		Cathodic cage			Conventional plasma nitriding		
1	Layer thickness (μm)	7,8	22	49,1	1,3	16	24,5
3		8,5	23	50,9	2,1	17,4	28
5		10	28	52,2	3,5	23,5	36,4
10		12	32	53,4	4,4	29,5	45,8

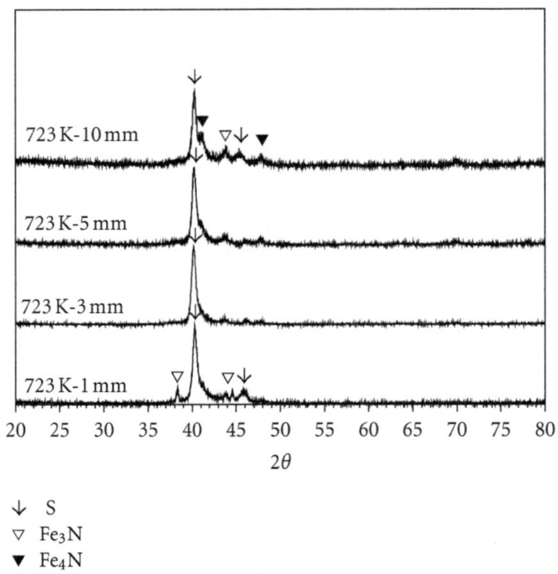

Figure 3: XRD patterns of AISI 316 steel samples with 1, 3, 5, and 10 mm and nitrided with 250 Pa, 80% of N_2, at 723 K, (5 h).

sputtering of the compounds presents in the cage, followed by deposition of such compounds on the surface of samples.

4. Conclusions

The samples made of different materials and different heights, nitrided by CCPN, present layers with considerable uniformity of thickness. This behavior is confirmed when observing the microhardness results for the surfaces of the samples. Besides, peaks of "S phase" (expanded austenite) did not present dislocation, showing a uniform temperature for austenitic stainless steel case. The visual aspect and microhardness along the top surface are uniform for the three different steel types treated, confirming the elimination of the edge effect. In this process, undesired phenomena, like arcing and hollow cathode effect, in samples containing holes and edges, will occur in the metallic cage instead of the samples surfaces.

Except for 1020 steel, the thickness of the obtained layer is higher than the observed in conventional ionic nitriding process, since, added to the deposition effect, the samples

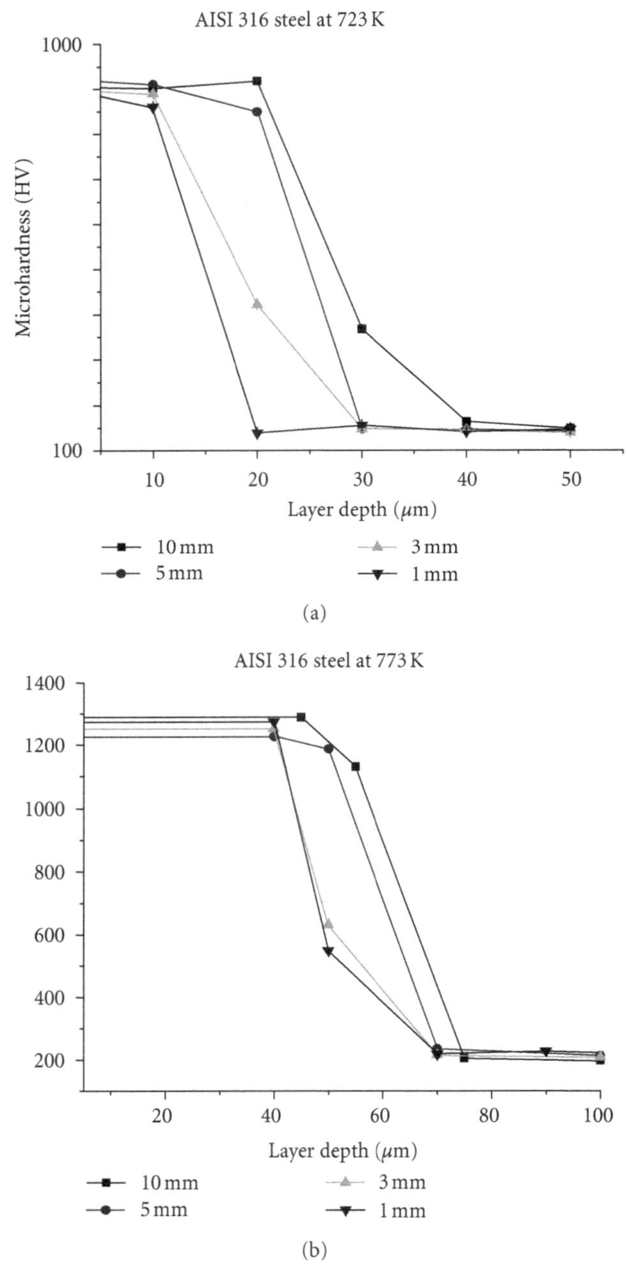

Figure 4: Profile of microhardness versus depth to the samples layer of AISI 316 stainless steel nitrided by CCN with 250 Pa, during 5 h at 723 and 773 K.

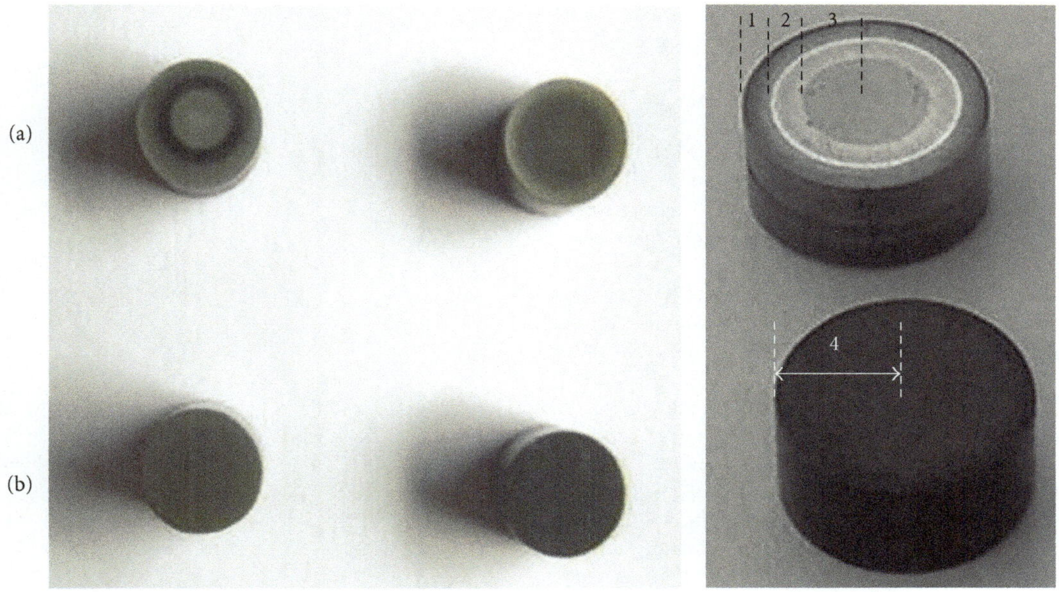

FIGURE 5: Micrographs of AISI 316, 1020, and 420 steel treated by (a) conventional nitriding and (b) cathodic cage plasma nitriding. The pressure used was 250 Pa, during 5 h (3 h for 1020), at 673 K with 80% N_2-20% H_2 mixture.

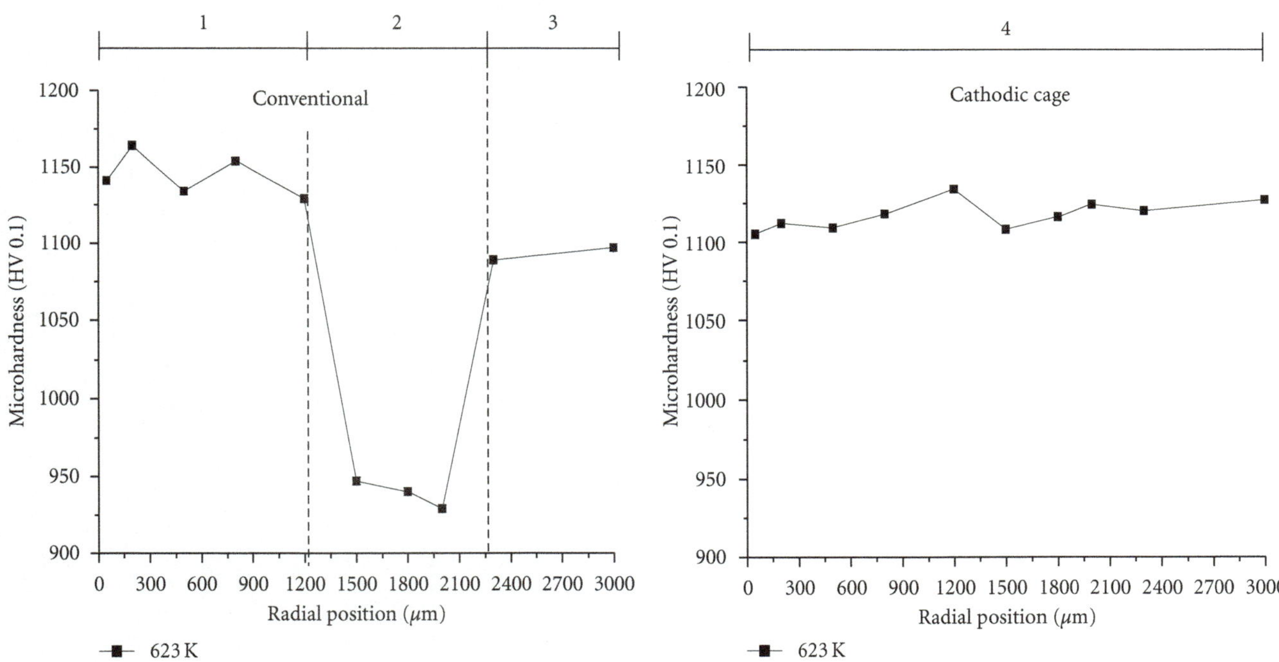

(a) (b)

FIGURE 6: Microhardness along the surface of AISI 420 stee at 673 K.

do not suffer active sputtering, which competes with the diffusion process in ionic nitriding, confirming the efficiency of the technique.

Acknowledgment

The authors are grateful to FAPEPI and CT-INFRA (MCT/CNPq) for the financial support.

References

[1] C. Alves Jr., *Plasma Nitriding: Foundations and Applications*, EDFRN, Natal, South Africa, 2001.

[2] C. X. Li, T. Bell, and H. Dong, "A study of active screen plasma nitriding," *Surface Engineering*, vol. 18, no. 3, pp. 174–181, 2002.

[3] J. Georges, "TC plasma nitriding," *Heat Treatment of Metals*, vol. 28, no. 2, pp. 33–37, 2001.

[4] R. S. Sousa, *Influence of the geometry of pieces and process parameters in layers nitrided by plasma*, Doctor thesis, UFRN, Natal, South Africa, 2005.

[5] C. Alves Jr., F. O. de Araújo, K. J. B. Ribeiro, J. A. P. da Costa, R. R. M. Sousa, and R. S. de Sousa, "Use of cathodic cage in plasma nitriding," *Surface and Coatings Technology*, vol. 201, no. 6, pp. 2450–2454, 2006.

[6] C. X. Li and T. Bell, "Corrosion properties of active screen plasma nitrided 316 austenitic stainless steel," *Corrosion Science*, vol. 46, no. 6, pp. 1527–1547, 2004.

[7] Y. Sun, X. Y. Li, and T. Bell, "X-ray diffraction characterisation of low temperature plasma nitrided austenitic stainless steels," *Journal of Materials Science*, vol. 34, no. 19, pp. 4793–4802, 1999.

[8] R. R. M. de Sousa, F. O. de Araújo, K. J. B. Ribeiro, M. W. D. Mendes, J. A. P. da Costa, and C. Alves Jr., "Cathodic cage nitriding of samples with different dimensions," *Materials Science and Engineering A*, vol. 465, no. 1-2, pp. 223–227, 2007.

[9] R. R. M. de Sousa, F. O. de Araújo, K. J. B. Ribeiro, J. A. P. da Costa, R. S. de Sousa, and C. Alves Jr., "Uniformity of temperature in cathodic cage technique in nitriding of austenitic stainless steel AISI 316," *Surface Engineering*, vol. 24, no. 4, pp. 313–318, 2008.

10

Control of Carbides and Graphite in Cast Irons Type Alloy's Microstructures for Hot Strip Mills

Sergio Villanueva Bravo,[1,2] Kaoru Yamamoto,[3] Hirofumi Miyahara,[1] and Keisaku Ogi[4]

[1] *Department of Materials Science and Engineering, Kyushu University, Fukuoka 819-0395, Japan*
[2] *Autonomous San Luis Potosi University, S. L. P., Mexico*
[3] *Kurume National College of Technology, Fukuoka 830-8555, Japan*
[4] *Oita National College of Technology, 1666 Maki, Oita 870-0152, Japan*

Correspondence should be addressed to Sergio Villanueva Bravo, svillanu@uaslp.mx

Academic Editor: Hao Chen

The carbide and graphite formation and redistribution of alloy elements during solidification were investigated on high-speed steel (HS) and Ni-hard type cast irons with Nb and V. The crystallization of hypereutectic HSS proceeds in the order of primary MC, γ + MC, $\gamma + M_6C$, $\gamma + M_7C_3$, and γ+ graphite eutectic, in hypoeutectic alloys proceeds in the order of primary γ, γ + MC, γ + graphite, $\gamma + M_6C$, and $\gamma + M_7C_3$ eutectic, and in Ni-hard proceeds in the order of primary γ, γ + MC, $\gamma + M_3C$, and γ+ graphite eutectic. The γ+ graphite eutectic solidifies with the decrease of V, Nb, and Cr and the increase of Si and C contents in residual liquid during solidification. The behavior in graphite forming tendency in the residual liquid is estimated by the parameter $\sum C_L^i m_i'$. The eutectic graphite crystallizes at the solid fraction when $\sum C_L^i m_i'$ takes a minimum value. The amount of graphite increases with the decrease in $\sum C_L^i m_i'$ of initial alloy content in both specimens. Inoculation with ferrosilicon effectively increases the graphite content in both specimens.

1. Introduction

In the roll material for hot rolling, the alloys which disperse a large amount of carbide in the matrix are widely used because they are superior in abrasion resistance. High-alloy white cast irons, in which a large amount of carbides disperse in the hardenable matrix, are widely used for abrasion resistant parts. Steel strip mills are also one of their important application fields, though the durability of high-alloy cast iron rolls is superior to conventional low-alloy ones. The scoring could sometimes shorten the life service and impair the surface quality of products [1, 2]. It is expected that the dispersed graphite flakes eliminate the scoring and sticking that could appear in alloy white cast iron. Therefore, the control of the amount and distribution of carbides and graphite is essential to get a high-quality cast iron roll. It is well known that the addition of Nb and V to white cast iron promotes the formation of MC type carbide [3]. The dispersion of MC carbide in the matrix would raise the wear resistance performance on the cast iron alloy where it is added. However, as they are the stronger carbide formers, the effects of Nb and V on the microstructure and the graphite formation must be investigated. In this study, carbide and graphite formation and redistribution of alloy elements during solidification were investigated on high-speed steel alloys (HS) and Ni-hard type cast irons.

2. Experimental Procedures

The chemical compositions of specimens tested are shown in Table 1. For HS type cast iron, different amounts of niobium (Nb), vanadium (V), tungsten (W), and cobalt (Co) were added, and V is used to disperse the larger amount of MC carbide in matrix. However, as γ + MC eutectic line for Fe-Cr-C-Nb system is located at lower concentration level of MC former than Fe-Cr-C-V system [3], the same amount of MC former addition results in a rise to more eutectic MC.

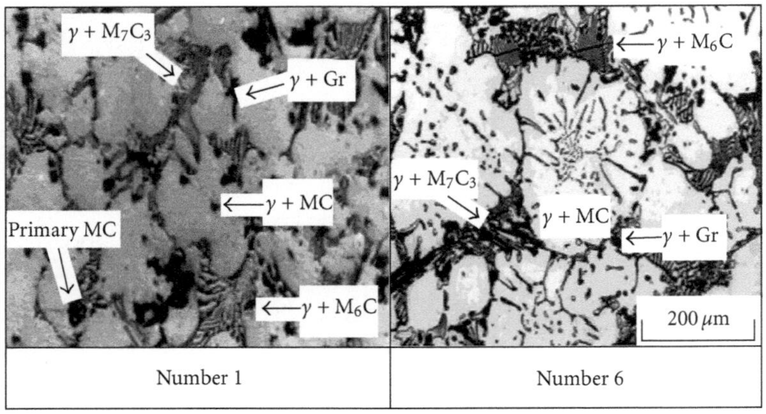

FIGURE 1: Microstructures of HS type specimens.

TABLE 1: Chemical composition for tested samples.

		Chemical composition (mass %)									Carbide
		C	Cr	V	Nb	Ni	Si	Mo	W	Co	
High-speed steel type (HS)	1	2.64	4.58	—	2.94	7.01	3.27	2.21	—	—	$MC + M_6C + M_7C_3$
	2	2.66	0.99	2.30	2.05	4.98	3.46	—	—	—	MC
	3	2.53	2.88	1.82	1.92	6.06	3.01	—	—	—	$MC + M_7C_3$
	4	2.63	2.97	2.08	1.86	5.92	3.35	5.05	—	—	$MC + M_6C + M_7C_3$
	5	2.73	3.02	3.22	0.97	6.09	3.25	5.15	4.93	—	$MC + M_6C + M_7C_3$
	6	2.79	3.05	2.95	0.99	6.10	3.23	5.16	—	5.11	$MC + M_6C + M_7C_3$
	7	2.81	2.97	2.89	0.94	6.45	3.45	5.14	—	—	$MC + M_6C + M_7C_3$
	8	2.74	3.09	2.10	0.87	5.99	4.46	5.09	—	—	$MC + M_6C + M_7C_3$
Ni-hard type	1	3.37	1.87	—	—	4.26	0.90	0.55	—	—	M_3C
	2	3.25	1.85	0.04	—	4.20	0.94	0.49	—	—	$MC + M_3C$
	3	3.16	1.81	1.86	0.53	4.24	0.93	0.46	—	—	$MC + M_3C$
	4	3.20	1.67	1.93	0.75	4.16	1.19	0.51	—	—	$MC + M_3C$
	5	3.25	1.59	0.96	0.90	4.12	1.24	0.52	—	—	$MC + M_3C$
	6	3.31	1.04	0.76	0.74	4.08	0.84	0.50	—	—	$MC + M_3C$
	7	3.35	1.24	1.94	0.35	4.32	0.88	0.54	—	—	$MC + M_3C$
	8	3.30	1.20	1.94	0.62	4.40	0.88	0.54	—	—	$MC + M_3C$
	9	3.33	0.71	0.97	1.80	4.41	0.90	0.53	—	—	$MC + M_3C$
	10	3.35	0.96	0.97	1.82	4.40	0.96	0.54	—	—	$MC + M_3C$

Furthermore, almost all Nb crystallizes in MC carbide; therefore, the effect of Nb addition on graphitization is very small. On the other hand, it is commonly recognized that NbC-γ interface sometimes shows lower bonding characteristics. Therefore, V and Nb contents in samples were controlled.

Ni-hard type cast iron series were based on alloy no. 1, and variable amounts of vanadium (V) and niobium (Nb) which are MC carbide formers were changed systematically. The amounts of Cr and Si were also controlled in some specimens. Thermal analysis was carried out for each specimen, and every specimen was melted over 1773 K in a siliconit furnace under argon atmosphere and then cooled at 10 K/min until 1173 K and quenched in oil. Distribution of carbides and graphite was analyzed in relation with the solidification sequence of the alloy. Moreover, EPMA analysis was carried out for the specimen quenched during solidification, and the relation between the behavior of alloy elements during solidification and crystallization of graphite was investigated.

Inoculation tests were also carried out for High-speed steel type cast iron and Ni-hard type cast iron. The specimens were remelted in a carbon resistance furnace, inoculated with Fe-75% Si, and then poured into the sand mold preheated at 1173 K. The amount of graphite in all specimens was examined metallographically.

3. Experimental Results and Discussions

3.1. The Graphite Formation in HS Type Cast Iron. The typical microstructures are shown in Figure 1. The specimens nos. 1 to 4 are hypereutectic, and specimens nos. 5 to 8

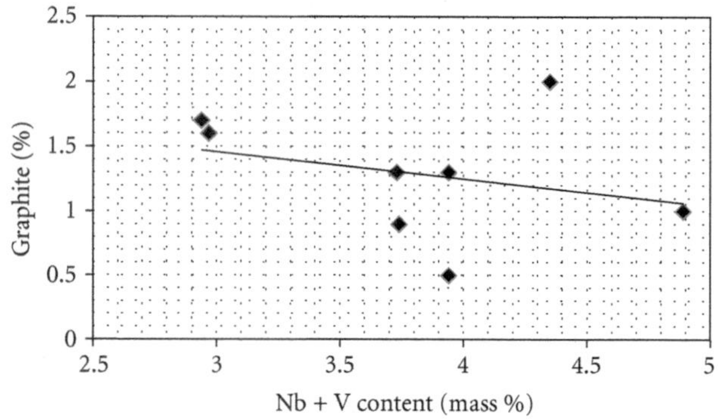

FIGURE 2: Amount of graphite-tested specimens.

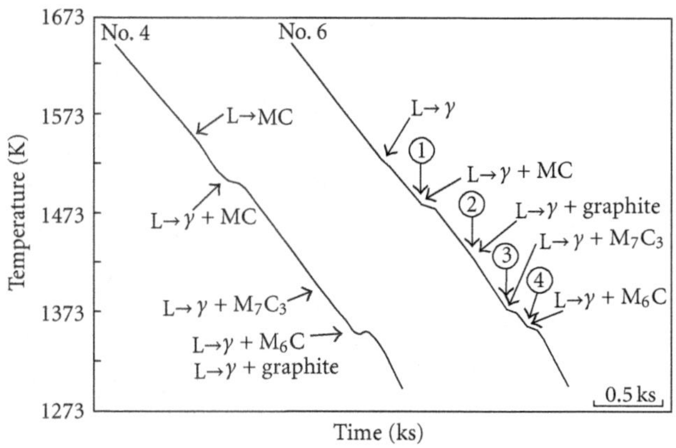

FIGURE 3: Thermal analysis curves of HSS samples.

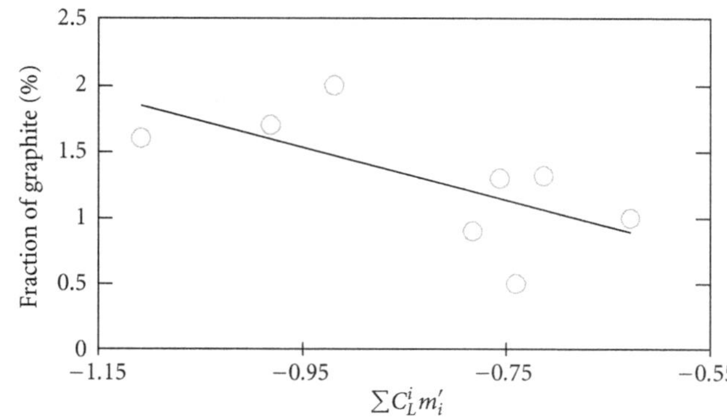

FIGURE 4: Relation between graphite fractions and $\Sigma C_L^i m_i'$.

are hypoeutectic alloys. In hypereutectic alloys, primary and eutectic MCs crystallize, and graphite flakes and particles distribute in the boundary of γ + MC eutectic cells. Meanwhile in hypoeutectic alloy, the graphite also distributes in cellar boundary. MC, M_6C, and M_7C_3 carbides crystallize, and the total amount of carbide changes from 13 to 24% depending on the chemical composition.

The amount of graphite in all specimens measured is from 0.5 to 2% showing a tendency to decrease as the amount of Nb and V increases as is shown in Figure 2, and the results are scattered due to the change on the amounts of Ni, Cr, and Mo, and the addition of W and Co in some specimens. The numbers of graphite flakes counted are from 50 to 400/mm^2.

Thermal analysis curves of hyper- (no. 4) and hypoeutectic (No. 6) alloys are shown in Figure 3. For hypereutectic alloy, the crystallization proceeds in the order of primary MC, γ + MC, γ + M_6C, γ + M_7C_3, and γ + graphite eutectic. Eutectic graphite crystallizes at the final stage of solidification because of the decreasing of carbide formers and the increasing of Ni and Si contents in residual liquid just like the case of low-Cr and high-Cr cast iron [4]. On the other hand, for hypoeutectic alloys, the solidification proceeds in the order of primary γ, γ + MC, γ + graphite, γ + M_6C, and γ + M_7C_3 eutectic. Graphite crystallizes after the γ + MC eutectic reaction.

The influence of each element on graphite formation is commonly evaluated based on the solubility of C in molten iron [5]. Therefore, the change in graphite forming tendency of residual liquid is estimated by the parameter $\Sigma C_L^i m_i'$ [5] that is shown in (1). The elements that decrease the solubility of C promote the graphitization, while the elements which decrease the solubility of C prevents the graphitization promoting the formation of carbide compounds

$$\begin{aligned}\Sigma C_L^i m_i' = {} & 0.07[\mathrm{Cr}\%] + 0.14[\mathrm{V}\%] + 0.07[\mathrm{Nb}\%] \\ & - 0.06[\mathrm{Ni}\%] - 0.31[\mathrm{Si}\%] + 0.02[\mathrm{Mo}\%] \\ & - 0.01[\mathrm{W}\%] - 0.03[\mathrm{Co}\%],\end{aligned} \quad (1)$$

where C_L^i is chemical composition of each element, m_i' is the parameter showing the influence of each element on the solubility limit of C to molten iron.

The higher values of $|m'|$ of Si and V indicate that Si promotes graphite formation and V interferes with the crystallization of graphite. The relation between the amount of graphite and the $\Sigma C_L^i m_i'$ value estimated with the initial composition is shown in Figure 4. The amount of graphite becomes larger by decreasing $\Sigma C_L^i m_i'$; thus, the amount of graphite can be predicted from the $\Sigma C_L^i m_i'$ value calculated using initial alloy contents.

3.2. The Graphite Formation in Ni-Hard Type Cast Iron. Conventional Ni-hard specimen consists of austenite (γ), eutectic M_3C, and graphite. On the other hand, the specimens with Nb and V consist of austenite (γ), eutectic MC, eutectic M_3C, and graphite. Figure 5 shows the typical microstructure and distribution of graphite of specimens solidified at 10 K/min.

Fine graphite particles crystallize in the specimen with Nb and V content. Primary MC crystallizes in the specimen contained over 0.5% Nb (alloy no. 9). As shown in Figure 6, the amount of graphite is 2.2% in the specimen without Nb and V (base alloy), when the Nb and V contents increase the graphite content on the specimen decreases. Although the data scatter slightly, due to different additions of Si and Cr a decreasing trend is observed. The cooling curves for the specimens no. 1 and no. 7. are shown in Figure 7.

The specimen no. 1 crystallizes in the order of primary austenite (γ), γ + M_3C, and γ + graphite, while in the specimen no. 7 γ + MC, eutectic crystallizes after primary γ.

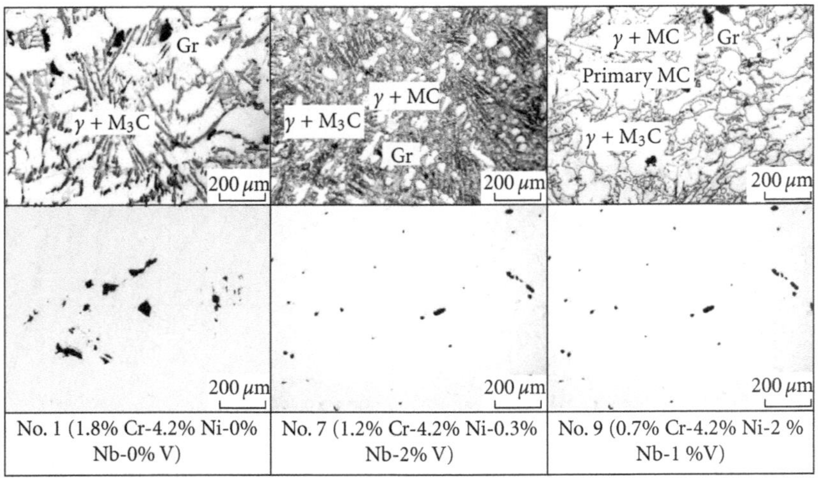

Figure 5: Microstructures of Ni-hard type cast irons.

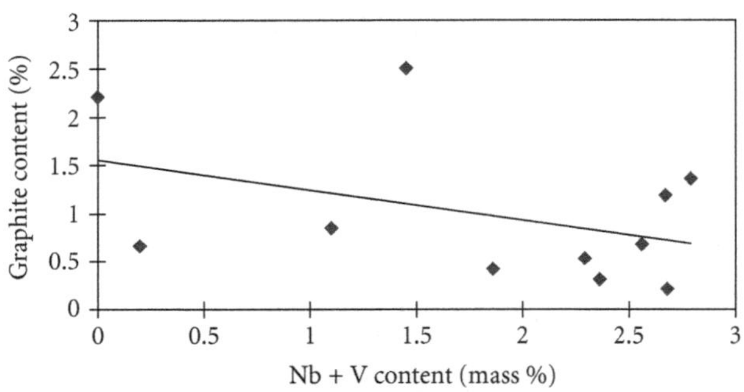

Figure 6: Effect of Nb and V content on the amount of graphite.

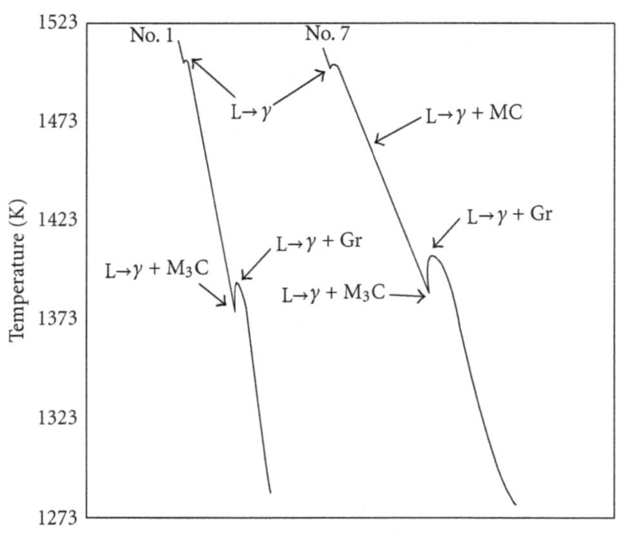

Figure 7: Thermal analysis curves of Ni-hard samples.

Since the partition coefficients of Nb and V to primary γ are less than unity, both elements are enriched in the residual liquid during the growth of primary γ, and the liquid composition reaches the γ + MC eutectic composition at 1448 K. Moreover, the growth of γ + MC eutectic diminishes

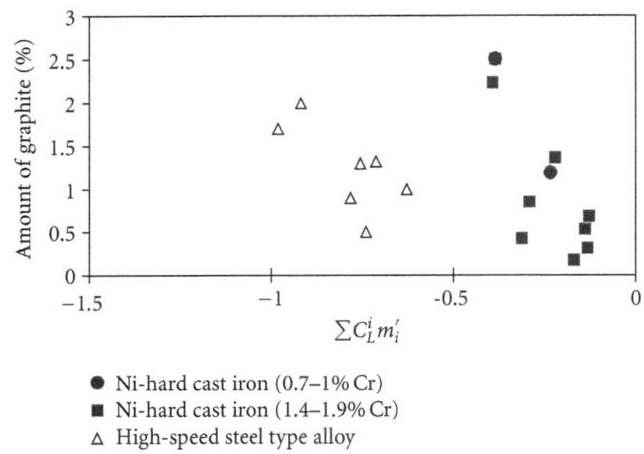

Figure 8: Relation between graphite's fraction and $\Sigma C_L^i m_i'$.

the Nb and V contents and increases the C content of residual liquid, and then $\gamma + M_3C$ eutectic starts crystallizing.

Moreover, the growth of $\gamma + M_3C$ eutectic decreases the carbide formers content and increases the Si and Ni contents on the residual liquid, and γ + graphite crystallizes in both lines at the final stage of solidification.

The relation between the amount of graphite and $\Sigma C_L^i m_i'$ values for all tested specimens was calculated as follows (2), and the results are show in Figure 8:

$$\Sigma C_L^i m_i' = 0.07[\%\text{Cr}] + 0.14[\%\text{V}] + 0.07[\%\text{Nb}] - 0.06[\%\text{Ni}] - 0.31[\%\text{Si}] + 0.02[\%\text{Mo}], \quad (2)$$

where ΣC_L^i and m_i' have the same meaning as (1).

The amount of graphite of Ni-hard type cast iron increases with the $\Sigma C_L^i m_i'$ value decreasing in a similar way as the case of high-speed steel type cast iron. A linear relationship is recognized for Ni-hard type cast irons and high-speed steel type cast irons except low-Cr Ni-hard type irons.

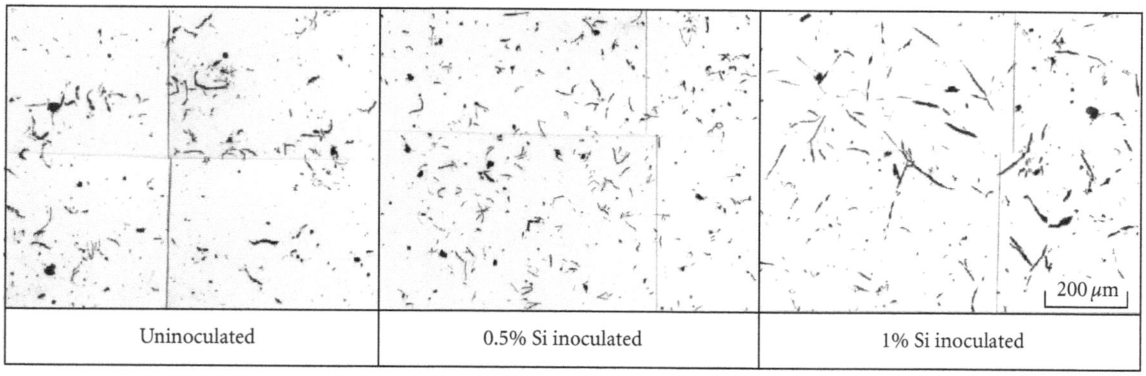

FIGURE 9: Inoculation effect on high-speed steel type cast iron.

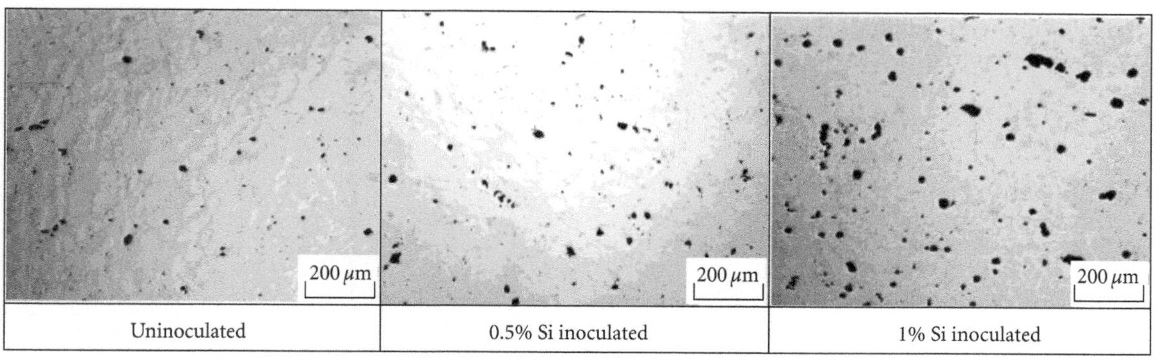

FIGURE 10: Effect of inoculation on Ni-hard type cast iron.

3.3. Inoculation Test

3.3.1. Inoculation Test in HS Type Cast Iron. The graphite distribution on alloy no. 7 specimens inoculated with Fe-75% Si alloy (0.5 and 1.0% Si added) is shown in Figure 9. The increasing of Si added as inoculant promotes the formation of graphite flakes, and the specimen with 1.0% Si added shows a random distribution of well-defined graphite flakes.

3.3.2. Inoculation Test in Ni-Hard Type Cast Iron. The specimen's composition no. 8 was used to study the effect of inoculation by using Fe-75% Si, and the results are shown in both Figures 10 and 11. On Figure 8, the photographs show that the amount and size of graphite increase when the amount of inoculant added increases. Moreover, the inoculation gives more uniform distribution of graphite flakes. Figure 11 shows that when 0.2% Si is added, the amount of graphite increases almost three times compared with not inoculated specimen and more than four times when 1.0% Si is added to the alloy.

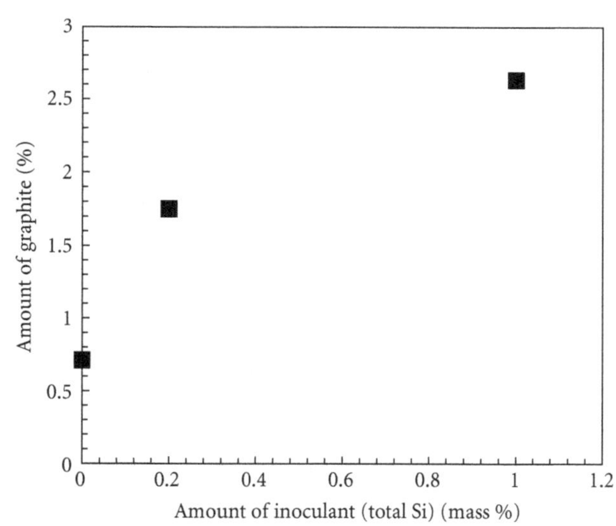

FIGURE 11: Relation between the amount of graphite and the amount of inoculant.

4. Conclusions

The effects of alloy elements and inoculation on graphite formation were investigated for high-speed steel type cast iron and Ni-hard type cast iron. The following conclusions were obtained.

(1) In case of high-speed steel type cast iron, the crystallization of hypereutectic alloy proceeds in the order of primary MC, γ + MC, $\gamma + M_6C$, $\gamma + M_7C_3$, and γ + graphite eutectic, while hypoeutectic alloys solidify in the order of primary γ, γ + MC, γ + graphite, $\gamma + M_6C$, and $\gamma + M_7C_3$ eutectic. The γ + graphite eutectic solidifies with the decrease of carbide forming elements V, Nb, and Cr and

the increase of Si and C contents in residual liquid during solidification.

(2) In case of Ni-hard type cast iron, by the addition of Nb and V, γ + MC eutectic reaction appears between the primary γ and γ + M_3C eutectic. γ + graphite eutectic crystallizes at the final stage of solidification. The solidification sequence is interpreted based on the behaviors of alloy elements and the change in the composition of residual liquid.

(3) The amount of graphite increases almost linearly with decreasing of solubility parameter $\Sigma C_L^i m_i'$ in both high-speed steel type and Ni-hard type cast iron.

(4) In both high-speed steel type cast iron and NI-hard type cast iron, the inoculation with Fe-75% Si alloy effectively increases the amount of graphite, and higher amount of inoculant results in more uniform distribution of larger flakes and particles graphite.

References

[1] S. K. Chang, D. G. Kim, and J. W. Choi, "Effects of alloying elements and austenite destabilization heat treatment on graphitization of high chromium cast iron," *ISIJ International*, vol. 32, no. 11, pp. 1163–1165, 1992.

[2] T. Himemiya, T. Umeda, G. Matsunoshita, and Y. Kimura, "Solidification behavior of high-nickel grain roll materials," *ISIJ International*, vol. 32, no. 11, pp. 1141–1149, 1992.

[3] A. Sawamoto, K. Ogi, and K. Matsuda, "Solidification structures of Fe–C–Cr–(V–Nb–W) alloys," *AFS Transactions*, vol. 94, pp. 403–416, 1986.

[4] Y. Ono, R. Shimizu, N. Murai, and K. Ogi, "Crystallization of graphite in eutectic solidification of high chromium cast iron," *Journal of Forest Science (JFS)*, vol. 63, pp. 340–346, 1991.

[5] F. Neumann, H. Schenck, and W. Patterson, Zeitschrift für das gesamte Giessereiwesen, 6 pp. 47, Heft 2, Jahrgan, January 1960.

11

Study of Tracer Diffusion Mechanism in Amorphous Metal

P. H. Kien,[1] H. V. Hue,[2] and P. K. Hung[3]

[1] *Department of Physics, Thai Nguyen University of Education, Luong Ngoc Quyen, Thai Nguyen, Vietnam*
[2] *Faculty of General Science, Ho Chi Minh City University of Food Industry, 140 Le Trong Tan, Ho Chi Minh, Vietnam*
[3] *Department of Computational Physics, Ha Noi University of Science and Technology, 1 Dai Co Viet, Ha Noi, Vietnam*

Correspondence should be addressed to P. H. Kien, huukienpham@yahoo.com

Academic Editor: E. Mittemeijer

The statistical relaxation (SR) simulation has been conducted to study the behavior of simplexes and bubbles (BB) in amorphous Co metal containing 2×10^5 atoms. The simulation reveals that the fraction of 4-simplex increases and n-simplex ($n > 4$) decreases depending upon relaxation degree. The simulation found that a large number of BB vary upon relaxation degree, which could play a role of diffusion vehicle for Co atoms in amorphous matrix. The idea of the diffusion mechanism in amorphous metal is described as follows: the elemental atomic movement includes a jump of neighboring atom into the BB and then a collective displacement of a large number of atoms around BB.

1. Introduction

The diffusion behavior in amorphous metals (AMs) has been intensively studied by both experiment and simulation for long times [1–8]. It is found that there are many specific properties of diffusion in AM compared to crystal matters. For example, the tracer diffusivity in a well-relaxed sample is much slower than the one in an as-quenched sample [9, 10]. This relaxation effect is interpreted by the reduction of vacancies in supersaturation until the relaxation is over. In the well-relaxed sample, conversely, the tracer atoms diffuse via collective movement of a group of neighboring atoms. The experimental studies in [11–13] on isotope effect, pressure dependence, and irradiation-enhanced diffusivity are sometimes in contradiction with the diffusion described above. Furthermore, there is not a clear definition of vacancy. Computer simulation, on the other hand, reveals unstableness of vacancies in amorphous matrix. Several works found a continuous spectrum of spherical voids in amorphous matter, but their size is less than atomic radius [14–16]. The free volume and two-level state theories are also employed to interpret the diffusion behavior of amorphous matter, but they cannot properly describe the diffusivity in some amorphous matters such as $Ti_{60}N_{40}$ and $Fe_{40}Ni_{40}B_{20}$ which show the cooperative activated movement more like diffusivity in solid state than in liquid [2, 17–20]. In [21], Sietsma and Thijsse analyzed different types of holes in amorphous matters and found that the number of holes surrounded by ten or more atoms decreases strongly in a well-relaxed sample. Furthermore, they argue the importance of big holes for atomic diffusivity.

Our previous study shows that amorphous matters suffer from a number of spherical voids, the size of which is closed to atomic size. Because the concentration of those voids weakly depends on temperature, it is very sensitive to the relaxation degree. We call them "native vacancy," which fully disappears until the amorphous matters transform into crystalline solid. However, the analysis of this study is based only on the geometrical consideration. Therefore, the account of potential barrier could provide more accurate elucidation of the voids for diffusion in amorphous matters. The aim of the present work is the study of simplexes and their role of them in diffusion in Co amorphous metals by using SR simulation method as a way to clarify diffusion mechanism in amorphous matters.

2. Calculation Method

The simulation has been conducted for the sample consisting of 2×10^5 atoms in a cubic box with periodic boundary conditions. We use the Pak-Doyama potential [16], and

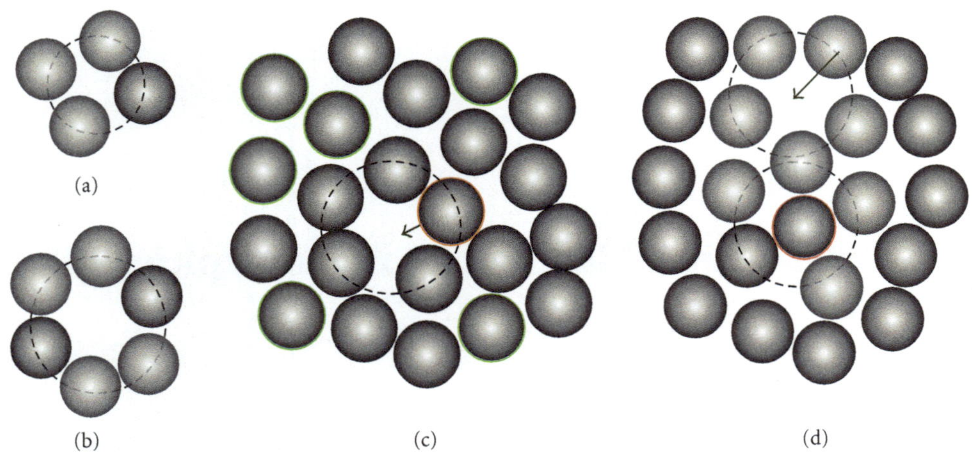

FIGURE 1: The schematic illustration of simplexes: (a) 4-simplex; (b) 6-simplex; (c) 5-simplex and neighboring atoms; (d) the new 5-simplex and neighboring atoms after RDA. The dash circle presents the CSP of four atoms, and the arrow is a path along which the MA jumps into CSP.

the density is taken from real amorphous metal. The interatomic potential has the following form:

$$\varphi(r_{ij}) = -0.12812(r_{ij}-1.82709)^4 + 1.15421(r_{ij}-2.50849)^2 - 0.13448, \quad 0 \le r_{ij} \le 3.44\,\text{Å}, \tag{1}$$

where r_{ij} is the distance between i and j atoms. Molecular dynamic (MD) simulation could reproduce the structure of AM; however, this method is time consuming. Thus, we employ the SR method, which provides the structure of AM similar to that of the low-temperature MD model. Accordingly, each atom moves by a length dr in the direction of the force acting on it from all remaining atoms. The length dr is equal to 0.01–0.4 Å. This movement is repeated many times until the system reaches an equilibrium state. More details about the SR method can be found in [16, 21].

Initial configuration is generated by randomly placing all atoms in a simulation box. Then, the sample is relaxed over several thousand steps until the system attains the equilibrium, for example, the energy of system fluctuates around a constant value and the pressure is equal to zero.The model A is constructed by relaxing with $dr = 0.4$ Å over 200 SR steps and then is treated with $dr = 0.01$ Å by 10^6 SR steps. To investigate the relaxation effect, two additional samples (samples B and C) are prepared with the same density as sample A, but their potential energy is lower (more stable state). Note that the lower the energy of the system the bigger its relaxation degree. More stable samples B and C can be constructed many times with relaxing the model A that has $dr = 0.4$ Å, which likes the shaking of atomic structure. Then, they are again relaxed with $dr = 0.01$ Å until the system reaches a new equilibrium.

Four neighboring atoms form a tetrahedron and have a circumsphere (CSP) whose surface passes through the vertices of this tetrahedron. Consider only such tetrahedron with CSP that does not contain any atom inside. For example, those atoms are the nearest neighboring. Let R_{CSP} and n be the radius of CSP and the number of atoms on the surface of CSP, respectively. The atom on surface of CSP is determined as one that is located from the center of CSP at a distance in range of $R_{\text{CSP}} \pm 0.1$ Å. hereafter we call it n-simplex (see Figures 1(a) and 1(b)). If n and R_{CSP} is large enough, then n atoms form an atomic cage like a bubble (BB), for example, a large group of atoms gathered around a large void. The BB is unstable and it may break up leading to diffusion. Therefore, it is interesting to clarify which one among n-simplexes is BB and how BB breaks up. Hence, the next calculation is performed as follows: first for every simplex, we test all n atoms to perform jump inside the CSP. Second, the atom selected to move is inserted inside CSP and the system is relaxed until it reaches equilibrium.

3. Results and Discussion

3.1. The Sample. Figure 2 shows the radial distribution function (RDF) $g(r)$ of the amorphous samples. As shown in Figure 2, although the energy per atom of the considered samples changes from −0.9336 to −0.9534 eV, the RDF $g(r)$ for the considered samples is identical. This indicates the nonsensitivity of the function $g(r)$ to the change in local microstructure of AMs. This result further indicates that the relaxation degree almost does not affect RDF, but it is reflected by the concentration of simplex. To test the validity of the constructed samples, we have compared our obtained RDF with the experimental data. As shown from Figure 3 the structural characteristics of our samples are in accordance with the experiment data [22]. In addition, the function $g(r)$ has a splitting second peak, which is often thought to be related to the icosahedrons in amorphous systems [17, 18, 20]. The main characteristics of the samples are listed in Table 1. The comparison with experiment shows agreement in the position and height of the first peak of RDF, which indicates the reality of the constructed sample.

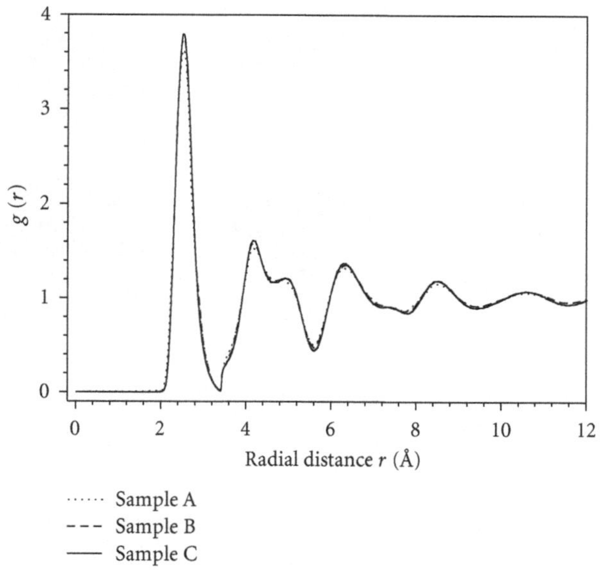

FIGURE 2: The radial distribution function of the obtained samples.

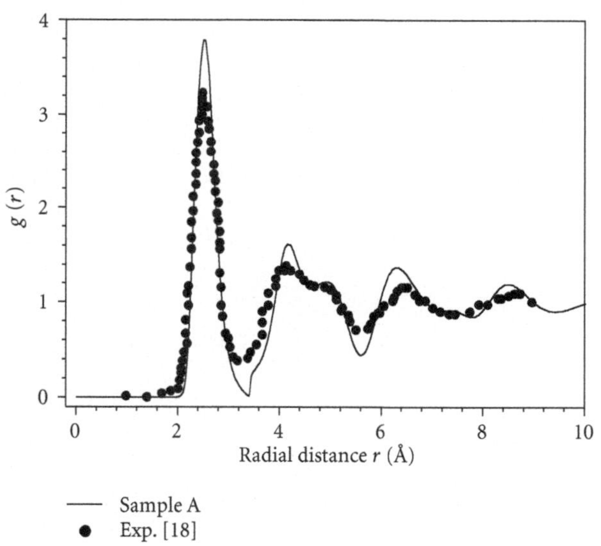

FIGURE 3: The radial distribution function of sample A and the experimental data in [18].

TABLE 1: Structure characteristics of obtained samples. Here, r_i is the position of ith peak and $g(r)$ is the height of first peak of the RDF.

Systems	r_1	r_2/r_1	r_3/r_1	r_4/r_1	r_5/r_1	$g(r)$
Sample A	2.51	1.67	1.99	2.52	3.39	3.64
Sample B	2.52	1.67	1.99	2.51	3.38	3.75
Sample C	2.50	1.66	2.01	2.54	3.44	3.78
[a]Co	2.49	1.69	1.93	2.49	3.35	3.20
[b]Co	—	1.65	1.90	2.57	3.46	—

[a,b]Experimental data in [22–24].

3.2. Simplex and Role of Simplex. Table 2 presents the energy per atom and fraction of simplex found in the obtained samples. One can see that the energy per atom decreases

TABLE 2: The energy per atom (ε) and the fraction of n-simplexes in the obtained samples. Here, n is number of atom in one simplex.

Systems	ε(eV)	4 ($\times 10^0$)	5 ($\times 10^{-1}$)	6 ($\times 10^{-2}$)	≥ 7 ($\times 10^{-3}$)
Sample A	−0.9336	0.961	0. 374	0.173	0.02
Sample B	−0.9462	0.963	0. 358	0.167	0.00
Sample C	−0.9534	0.964	0.339	0.164	0.00

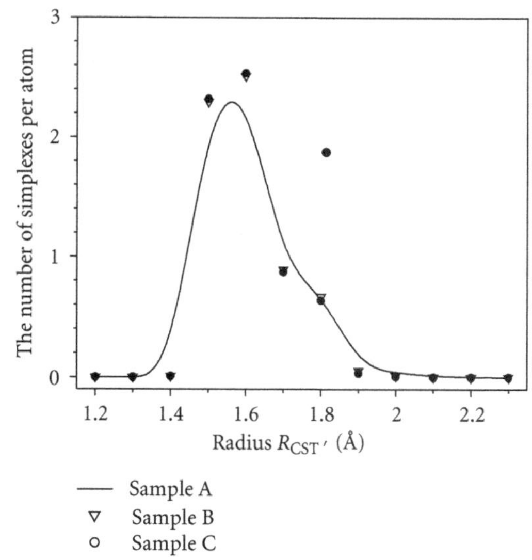

FIGURE 4: The number of simplexes per atom as function of radius R_{CST}.

from −0.9336 to −0.9534 eV, upon relaxation degree. The fraction of 4-simplex increases as the sample potential energy decreases, that is, the more stable state is, bigger the number of 4-simplexes is. For other kinds of simplex, we observe the opposite trend. Thus, it means that due to fast quenching from liquid the parking of atoms in AM is not efficient. Therefore, there is always an amount of structural defects like large void (free volume). The relaxation is accompanied by annihilation of those defects. As mentioned above, the bigger number n is, the bigger the size of simplex and void inside it is. Therefore, Table 2 specifies that the monotony decrease in the number of n-simplex with $n > 4$ from less relaxed (sample A) to well relaxed state (sample C) indicates the annihilation of structural defects in amorphous matrix. This result also can be seen in the radius distribution of simplex as in Figure 4. From Figure 4, one can see clearly the significant decrease in the number of large simplex ($R_{CSP} > 1.8$ Å) in the sample A as compared to the sample C. We proposed that the large simplex ($n = 6, 7, 8$) is broken up into 4-simplex under the relaxation degree. So as the concentration of 4-simplex increases, the large simplex decreases.

We now turn to discuss the role of large simplex for diffusivity in AM. For each simplex, we examine the potential energy profile (PEP) for neighboring atoms moving into CSP of the simplex. The PEP is determined as follows: each of n atoms of the simplex is taken and then we move this atom step by step towards the center of CSP. The step length is set to be 0.02 Å. At each step, the potential

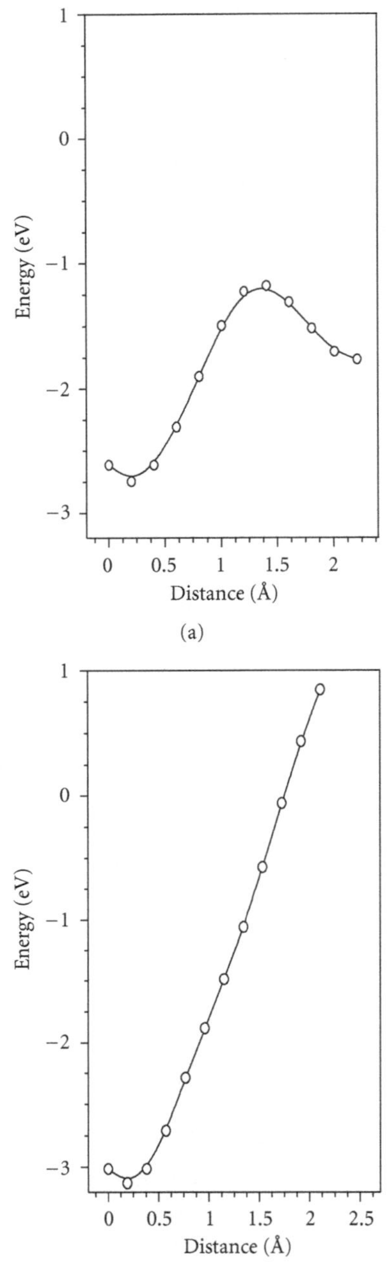

FIGURE 5: Potential energy profiles for atom moving into simplex.

energy of the moving atom (MA) is recorded. Finally, we obtain the PEP for MA. Figure 5 displays the typical PEPs for several simplexes detected in our samples. Curve (*b*) (Figure 5(b)) shows the monotonous increase in the MA energy indicating that the MA could not jump into the CSP due to very high potential barrier. Curve (*a*) (Figure 5(a)), in converse, have a pronounced maximum. Here, the barrier height is determined by the maximum and initial points in PEP. They are like the PEP for tracer atom jumping in vacancy incrystalline lattice. Therefore, the MA can jump into CSP, and the simplex plays a role of the diffusion vehicle assisting tracer atom to move in amorphous matrix. The

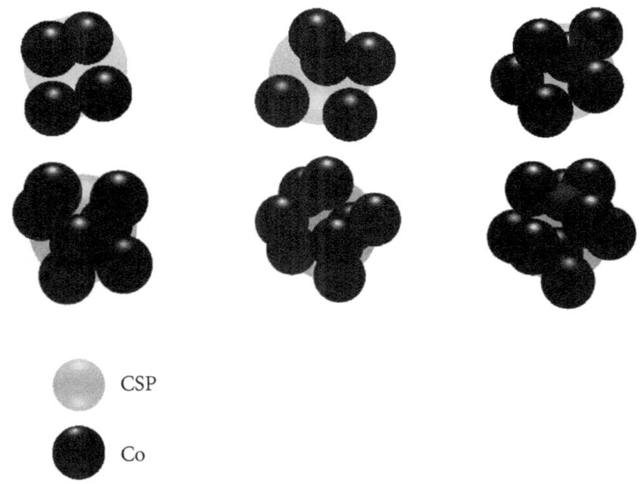

FIGURE 6: The snapshot of typical bubbles detected in the obtained sample.

simplex having PEP like curves *a*, with corresponding MA, is called the bubbles (BB) and the diffusing atom (DA), respectively. Figure 6 shows some typical *n*-bubbles in the obtained samples. The number of BB (n_{BB}) is found in Table 3. As shown in Table 3, the number of BB decreases depending upon relaxation degree. The dependence of the number of DA on their corresponding barrier height is shown in Figure 7. Most frequent barrier height lies in the interval of 0.5–1.9 eV. Because the diffusion coefficient in general is proportional to n_{BB}, it is given as follows:

$$D = \frac{1}{6} f \nu_0 \langle d^2 \rangle \exp\left(\frac{\Delta s_m}{k_B}\right) n_{\mathrm{BB}} \exp\left(-\frac{E_m}{k_B T}\right). \quad (2)$$

Here, f is the correlation factor, ν_0 is the attempt frequency, d is the jump length, $\Delta s_m, E_m$ are the effective migration entropy and energy for diffusion in disordered media, respectively, k_B is the Boltzmann constant. Note that the parameter $\langle d^2 \rangle$ represents the mean square displacement of different jump lengths. Because the SR model is in fact the molecular dynamic (MD) model in the limit of zero temperature, the number of simplexes found in the samples is independent of temperature and it varies only with the relaxation degree. As a result, the activation energy is the migration energy (E_m), and the preexponential factor D_0 is given as

$$D_0 = \frac{1}{6} f \nu_0 \langle d^2 \rangle \exp\left(\frac{\Delta s_m}{k_B}\right) n_{BB}. \quad (3)$$

Furthermore, the decrease in diffusion coefficient under relaxation is ascribed by the annihilation of the bubbles. Therefore, the decrease in Co diffusivity is found to be 27 times for samples A and C.

Several experimental findings shows that (i) sudden temperature change during diffusion annealing results in instantaneous change of diffusion coefficient (D_R) and (ii) The self-diffusion enthalpy H of transition metal for certain amorphous alloy seems to be the migration enthalpy, but

TABLE 3: The number of BB (n_{BB}); the pre-exponential factor (D_0) at 560 K in obtained samples.

Systems	n_{BB} ($\times 10^{-3}$)	D_0 (m^2/s)
Samples A	8.39	9.3×10^{-6}
Samples B	1.34	6.1×10^{-7}
Samples C	0.31	1.1×10^{-7}
[c]$Co_{89}Zr_{11}$	—	8.0×10^{-7}
[d]$Co_{79}Nb_{14}B_7$	—	3.3×10^{-6}

[c,d]Experimental data in [25, 26].

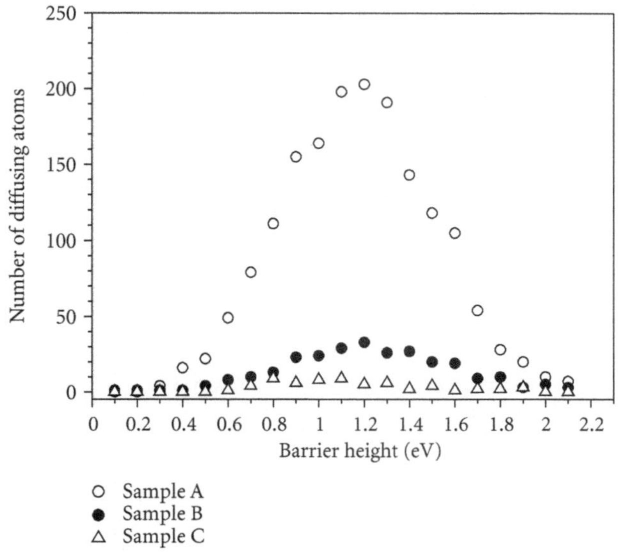

FIGURE 7: The distribution of barrier height of diffusing atom.

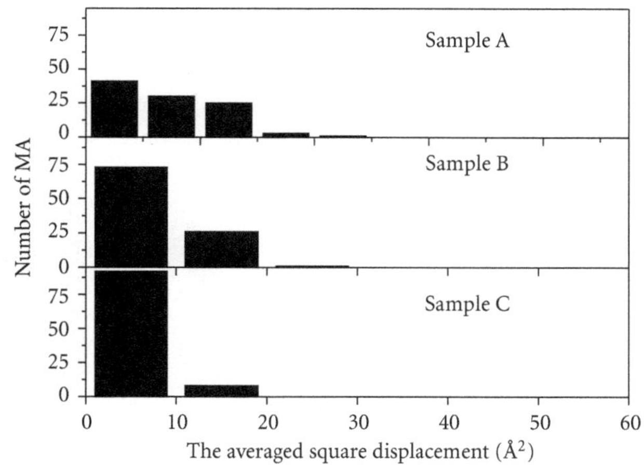

FIGURE 8: The distribution of averaged square displacement of moving atoms.

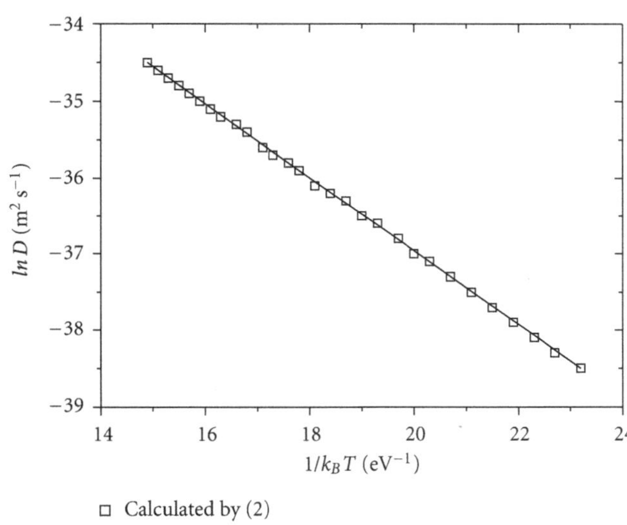

FIGURE 9: The temperature dependence of diffusion coefficient; a line is drawn to show the Arrhenius behavior.

not a sum of migration and formation enthalpy of diffusion vehicles as it is the case of self-diffusion in crystal [3]. These experimental observations are interpreted as a result of direct diffusion mechanism occurring in relaxed amorphous sample. Our simulation result shows possible diffusivity via the bubble. It is consistent with the experiment because the concentration of bubbles is independent of temperature, and they vary only with relaxation degree. So, the activation energy for diffusion via the bubble is equal to the migration energy (E_m).

To estimate the parameter $\langle d^2 \rangle$ in (2), we replace DA into CSP and then relax the system until a new equilibrium is attained. This procedure for convenience is called the remove diffusion atom (RDA). The distribution of mean square displacements for 100 RDAs is displayed in Figure 8. For sample C, the parameter $\langle d^2 \rangle$ of most RDAs is less than $10\,\text{Å}^2$ (92 atoms). Because the jump length of DA lies in the interval of 2.0–3.0 Å, the contribution of DA to $\langle d^2 \rangle$ will be essential and the RDA locates only in the small region nearby the bubble. In the case of sample A, we observe a very large value of $\langle d^2 \rangle$ (more than $10\,\text{Å}^2$), which represents a collective movement of a large number of atoms. Obviously, the atomic movement is spread over a large volume inside the amorphous matrix as shown in Figure 1(d). Therefore, one may see two distinct diffusion mechanisms occurring in the as-quenched (sample A) and well-relaxed metastable states (sample C). The first one is like the hoping mechanism via bubble. The second one relates to the collective mechanism involving a large number of atoms in each elemental diffusion movement. Assume that $f = 1$, $\exp(\Delta s_m/k_B) \approx 1$, $\nu_0 = 10^{12}\,s^{-1}$, $\langle d^2 \rangle \approx 10\,\text{Å}^2$ for sample C and $100\,\text{Å}^2$ for sample A. The diffusion coefficient of Co atom can be estimated on the base of (2). Although the barrier heights for different jumps vary from 0.5 to 1.9 eV, the D_{Co} diffusion coefficient is found to obey the Arrehenius behavior as shown in Figure 9. According to (3), the pre-exponential factor is calculated as shown in Table 3. The calculation data is consistent with the experimental data for the amorphous samples [25, 26].

4. Conclusions

A comprehensive study on both the number of simplexes and a bubbles diffusion mechanism in AM has been carried out using SR simulation with the pair Pak-Doyama inter-atomic potential upon the change of relaxation degree. Conclusions are as followed:

(i) The microstructure of the cobalt samples containing 2×10^5 atoms is in agreement with the experimental data. The fraction of 4-simplex increases and n-simplex ($n > 4$) decreases depending upon relaxation degree. The simulation reveals that a large number of bubbles vary with relaxation degree in AM.

(ii) A bubbles diffusion mechanism in AM is proposed for the elemental atomic movement, which includes a jump of neighboring atom into the bubbles and then collective displacement of a number of atoms. It may be seen two distinct diffusion mechanisms occurring in the as-quenched and well-relaxed samples. The decrease in pre-exponential factor upon relaxation is ascribed to the partial annihilation of bubbles. The obtained pre-exponential factor of Co atom is found to be in accordance with the experiment data.

References

[1] J. Horváth, J. Ott, K. Pfahler, and W. Ulfert, "Tracer diffusion in amorphous alloys," *Materials Science and Engineering*, vol. 97, pp. 409–413, 1988.

[2] J. Pavlovsky, W. Ulfert, and W. Frank, "Self-diffusion of ^{58}Co in amorphous $Co_{79}Nb_{14}B_7$ during isothermal crystallization," *Materials Chemistry and Physics*, vol. 36, no. 3-4, pp. 383–388, 1994.

[3] W. Frank, A. Horner, P. Scharwaechter, and H. Kronmiiler, "Diffusion mechanisms in amorphous alloys," *Materials Science and Engineering*, vol. 97, pp. 415–418, 1988.

[4] A. K. Tyagi, M. P. Macht, and V. Naundorf, "Diffusion coefficients of ^{63}Ni in $Fe_{40}Ni_{40}B_{20}$ metallic glass," *Acta Metallurgica et Materialia*, vol. 39, no. 4, pp. 609–617, 1991.

[5] T. Schuler, J. Pavlovskýa, P. Scharwaechter, W. Ulfert, and W. Frank, "Change of the self-diffusion of ^{58}Co in $Co_{79}Nb_{14}B_7$ during transition from the amorphous to the nanocrystalline phase," *Nanostructured Materials*, vol. 6, no. 5–8, pp. 863–867, 1995.

[6] S. Flege, U. Fecher, and H. Hahn, "Diffusion in amorphous NiZrAl alloys," *Journal of Non-Crystalline Solids*, vol. 270, no. 1–3, pp. 123–128, 2000.

[7] A. Grieschea, T. Zumkleya, M.-P. Machta, S. Suzukib, and G. Frohbergb, "Diffusion in PdCuNiP alloys from the amorphous to the liquid state," *Materials Science and Engineering A*, vol. 375–377, pp. 285–287, 2004.

[8] S. Chakravarty, M. Gupta, A. Gupta et al., "Fe and N self-diffusion in amorphous FeN: a SIMS and neutron reflectivity study," *Acta Materialia*, vol. 57, no. 4, pp. 1263–1271, 2009.

[9] S. K. Sharma, S. Banerjee, Kuldeep, and A. K. Jain, "Diffusion measurements in the $Fe_{82}B_{18}$ amorphous alloy by rutherford backscattering spectrometry," *Acta Metallurgica*, vol. 36, no. 7, pp. 1683–1690, 1988.

[10] W. D. Roos, J. D. Plessis, and G. N. van Wyk, "Diffusion of silicon in Fe-based amorphous and crystalline alloys," *Applied Surface Science*, vol. 40, no. 4, pp. 303–307, 1990.

[11] Y. Limoge, "Role of energetic disorder on diffusion in amorphous alloys," *Journal of Non-Crystalline Solids*, vol. 117-118, no. 2, pp. 605–608, 1990.

[12] G. Ruitenberg, P. de Hey, F. Sommerb, and J. Sietsma, "Pressure dependence of the free volume in amorphous $Pd_{40}Ni_{40}P_{20}$ and its implications for the diffusion process," *Materials Science and Engineering A*, vol. 226-228, pp. 397–400, 1997.

[13] Y. Limoge, "Activation volume for diffusion in a metallic glass," *Acta Metallurgica et Materialia*, vol. 38, no. 9, pp. 1733–1742, 1990.

[14] R. S. Averback, "Defects and diffusion in amorphous alloys," *MRS Bulletin*, vol. 16, pp. 47–52, 1991.

[15] Y. Limoge, "Microscopic and macroscopic properties of diffusion in metallic glasses," *Materials Science and Engineering A*, vol. 226-228, pp. 228–236, 1997.

[16] P. K. Hung, H. V. Hue, and L. T. Vinh, "Simulation study of pores and pore clusters in amorphous alloys $Co_{100-x}B_x$ and $Fe_{100-y}P_y$," *Journal of Non-Crystalline Solids*, vol. 352, no. 30-31, pp. 3332–3338, 2006.

[17] A. Zhu, G. J. Shiflet, and S. J. Poon, "Diffusion in metallic glasses: analysis from the atomic bond defect perspective," *Acta Materialia*, vol. 56, no. 14, pp. 3550–3557, 2008.

[18] P. K. Leung and J. G. Wright, "Structural investigations of amorphous transition element films: II. Chromium, iron, manganese and nickel," *Philosophical Magazine*, vol. 30, no. 5, pp. 995–1008, 1974.

[19] V. Naundorf, M. P. Macht, A. S. Bakai, and N. Lazarev, "The pre-factor, D_0, of the diffusion coefficient in amorphous alloys and in grain boundaries," *Journal of Non-Crystalline Solids*, vol. 224, no. 2, pp. 122–134, 1998.

[20] P. K. Hung, P. H. Kien, and L. T. Vinh, "Evidence of 'microscopic bubbles' and a new diffusion mechanism for amorphous alloys," *Journal of Physics*, vol. 22, Article ID 035401, pp. 1–5, 2010.

[21] J. Sietsma and B. J. Thijsse, "Characterization of free volume in atomic models of metallic glasses," *Physical Review B*, vol. 52, no. 5, pp. 3248–3255, 1995.

[22] Y. Waseda and S. Tamaki, "Structures of 3D-transition metals in the liquid state," *Philosophical Magazine*, vol. 32, no. 2, pp. 273–281, 1975.

[23] P. K. Leung and J. G. Wright, "Structural investigations of amorphous transition element films I. Scanning electron diffraction study of cobalt," *Philosophical Magazine*, vol. 30, no. 1, pp. 185–194, 1974.

[24] P. K. Leung and J. G. Wright, "Structural investigations of amorphous transition element films—2. Chromium, iron, manganese and nickel," *Philosophical Magazine*, vol. 30, no. 5, pp. 995–1108, 1974.

[25] V. Naundorf, M.-P. Machta, A. S. Bakaib, and N. Lazarevb, "The pre-factor, D_0, of the diffusion coefficient in amorphous alloys and in grain boundaries," *Journal of Non-Crystalline Solids*, vol. 224, no. 2, pp. 122–134, 1998.

[26] J. Pavlovsky and J. Cermak, "Diffusion of ^{59}Fe in amorphous $Co_{79}Nb_{14}B_7$ alloy," *Scripta Metallurgica et Materialia*, vol. 30, no. 10, pp. 1317–1320, 1994.

12

Electrical Properties of Rapidly Annealed Ir and Ir/Au Schottky Contacts on n-Type InGaN

V. Rajagopal Reddy, B. Prasanna Lakshmi, and R. Padma

Department of Physics, Sri Venkateswara University, Tirupati 517 502, India

Correspondence should be addressed to V. Rajagopal Reddy, reddy_vrg@rediffmail.com

Academic Editor: Stefano Gialanella

The effect of annealing temperature on electrical characteristics of iridium (Ir) and iridium/gold (Ir/Au) Schottky contacts to n-type InGaN have been investigated by means of current-voltage (I-V) and capacitance-voltage (C-V) techniques. It is observed that the barrier height of Ir/n-InGaN and Au/Ir/n-InGaN Schottky diodes increases after annealing at 300°C for 1 min in N_2 ambient compared to the as-deposited. However, the barrier heights are found to be decreased somewhat after annealing at 500°C for the both Ir and Ir/Au Schottky contacts. From the above observations, it is clear that the optimum annealing temperature for both Ir and Ir/Au Schottky contacts is 300°C. Moreover, the barrier height (ϕ_b), ideality factor (n) and series resistance (R_S) are determined using Cheung's and Norde methods. Besides, the energy distribution of interface state densities are determined from the forward bias I-V characteristics by taking into account the bias dependence of the effective barrier height. Based on the above results, it is clear that both Ir and Ir/Au Schottky contacts exhibit a kind of thermal stability during annealing.

1. Introduction

III-V nitride semiconductors, particularly gallium nitride (GaN) and related alloys, have received much attention because of the (or their) wide range of practical applications. Among the ternary compounds of group III nitride materials, InGaN has drawn a great deal of attention. GaN and InGaN have been considered as the most important and crucial materials used for the fabrication of light emitters which were active in entire visible and part of the near UV spectral regions. The development of high-quality Schottky contacts to InGaN is difficult because in obtaining high-quality InGaN layers with high composition of In, their low carrier mobility (compared to GaN), and a large amount of surface defects in InGaN layers [1]. Hence, the fabrication of Schottky contacts to InGaN with a high barrier height and good thermal stability is still a challenge.

Many research groups have extensively investigated the electrical properties of n-GaN Schottky diodes using various metal schemes such as W, WSi_x [2], Ru, Ru/Au [3], Rh, Rh/Au [4], Ni/Au [5], and Re [6]. Kim et al. [2] investigated the thermal stability of W and WSi_x Schottky contacts on n-type GaN and reported that the WSi_x contacts showed improved thermal stability compared to pure W contacts. They also reported that the WSi_x contact was stable up to temperatures >600°C with the Schottky barrier height (SBH) of ~0.5 eV. Rajagopal Reddy et al. [3] studied the influence of thermal annealing temperature on the electrical properties of Ru and Ru/Au Schottky contacts to n-type GaN. They found that the Ru Schottky contacts exhibit a kind of thermal stability during annealing compared to the Ru/Au contacts. Rajagopal Reddy et al. [4] investigated the effect of annealing temperature on electrical properties of Rh and Rh/Au Schottky contacts to n-type GaN. They reported that there was a significant improvement in the electrical characteristics of Rh/Au Schottky contacts upon annealing compared to the Rh Schottky contacts. Miura et al. [5] investigated the thermal annealing effects on Ni/Au-based Schottky contacts on n-GaN and AlGaN/GaN after inserting high work function metals such as Pt, Ir, Pd, or Mo. Venugopalan and Mohney [6] showed that the Re Schottky contact was thermally stable on n-GaN and reported a barrier height of 0.82 eV and 1.06 eV after annealing at 500°C by I-V and C-V methods. Jang et al. [7] investigated

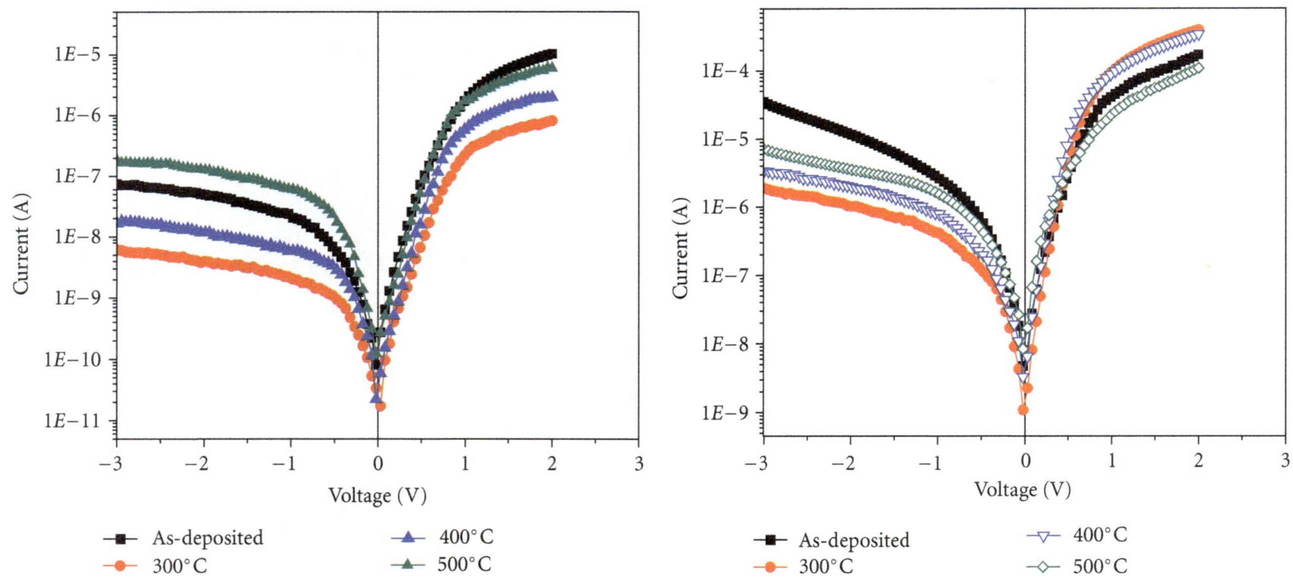

(a) Ir/n-InGaN Schottky diode.

(b) Au/Ir/n-InGaN Schottky diode.

FIGURE 1: (a) The typical current-voltage characteristics for the Ir Schottky contacts on n-type InGaN. (b) The typical current-voltage characteristics for the Ir/Au Schottky contacts on n-type InGaN.

the Schottky barrier characteristics of Pt contacts on n-InGaN by the *I-V* and *C-V* methods. They found that there was a large difference in SBHs obtained by the TE and TFE modes using the *I-V* data. Xue et al. [8] fabricated large-area Au/Pt/$In_{0.2}$ $G_{0.8}$N Schottky contacts for photovoltaic devices and investigated the current transport mechanisms of the Schottky diodes with different background carrier concentrations.

There are, however, only limited works on metal-InGaN ($N_d = 6 \times 10^{17}$ cm^{-3}) Schottky barriers. Hence, the main aim of the present work is to fabricate and characterize the Ir-based Schottky contacts on InGaN. To the best of our knowledge, Ir metal scheme has not been explored as Schottky contacts on InGaN. In this work, Ir is selected because it has a high work function of 5.46 eV. From a device point of view, it is important to know what happens to metal contacts when they are annealed. Therefore, our main focus in the present work is to investigate the electrical characteristics of Ir and Ir/Au Schottky contacts on n-InGaN by means of current-voltage (*I-V*) and capacitance-voltage (*C-V*) measurements as a function of annealing temperature.

2. Experimental Details

2 μm-thick unintentionally doped GaN layer on a 40 nm-thick nucleation layer/(0001) sapphire substrate was grown by metal organic chemical vapor deposition (MOCVD). This was followed by the growth of ~0.25 μm-thick n-InGaN: Si ($N_d = 6 \times 10^{17}$ cm^{-3}) with 10% of indium (In) composition. Prior to the metal deposition, the n-type InGaN layer was first ultrasonically degreased with warm trichloroethylene followed by acetone, methanol, and ethanol for 5 min in each step and then rinsed with deionized (DI) water. This degreased layer was then dipped into boiling aquarezia [HNO_3 : HCl = 1 : 3] for 10 min to remove the native oxide. For ohmic contacts, bilayer of Ti (25 nm)/Al (100 nm) was deposited on InGaN and then annealed at 650°C for 1 min in flowing N_2 atmosphere using rapid thermal annealing (RTA) system. Ir (50 nm) and Ir/Au (20 nm/30 nm) Schottky contacts were fabricated on n-InGaN with a circular diameter of 0.7 mm using electron beam evaporation system with a vacuum pressure of 5×10^{-6} mbar. In order to investigate the thermal annealing effect on Schottky diodes, the Ir and Ir/Au contacts were annealed at 300°C, 400°C, and 500°C for 1 min in N_2 ambient. The electrical characteristics of Schottky contacts were determined by means of current-voltage (*I-V*) and capacitance-voltage (*C-V*) characteristics using Keithley source measure unit (model no. 2400) and deep level transient spectrometer (DLS-83 D), respectively.

3. Results and Discussion

3.1. Current-Voltage (I-V) Characteristics. Figures 1(a) and 1(b) show the semilogarithmic plots of forward and reverse *I-V* characteristics of Ir and Ir/Au Schottky contacts measured as a function of annealing temperature. It is noted that the properties of Ir and Ir/Au Schottky diodes are uniform over different temperatures. It is observed that the leakage current of as-deposited Ir Schottky diode is 2.38×10^{-8} A at −1 V. For the contact annealed at 300°C, the leakage current decreases to 2.21×10^{-9} A, whereas for the contacts annealed at 400°C and 500°C, the leakage current values are 6.0×10^{-9} A and 6.49×10^{-8} A at −1 V, respectively. In the case of as-deposited Ir/Au Schottky diode, the leakage current at −1 V is 2.78×10^{-6} A. For the contacts annealed at 300°C, 400°C, and 500°C, the obtained leakage current values are

TABLE 1: The various parameters obtained from the *I-V* and *C-V* characteristics of Ir and Ir/Au Schottky contacts on n-InGaN as a function of annealing temperature.

Sample	Schottky barrier height (SBH) ϕ_b (eV)			From *I-V*	Cheung's functions			
					$dV/d\ln(I)$ versus I		$H(I)$ versus I	
	I-V	*C-V*	Norde	n	R_s (Ω)	n	ϕ_b (eV)	R_s(Ω)
Ir								
As-dep	0.79	1.17	0.81	1.60	188	2.15	0.80	226
300°C	0.85	1.21	0.88	1.10	62	1.38	0.86	74
400°C	0.84	1.11	0.86	1.25	83	1.61	0.85	102
500°C	0.80	1.07	0.82	1.50	122	1.93	0.81	133
Ir/Au								
As-dep	0.70	1.00	0.73	1.51	171	1.93	0.71	179
300°C	0.74	1.03	0.76	1.23	90	1.46	0.75	109
400°C	0.71	0.96	0.74	1.43	109	1.89	0.72	128
500°C	0.67	0.93	0.71	1.77	231	2.04	0.68	260

4.22×10^{-7} A, 7.87×10^{-7} A, and 1.66×10^{-6} A at −1 V, respectively. The smallest leakage current values are obtained for both Ir and Ir/Au Schottky contacts after annealing at 300°C, accompanied by the corresponding increase in the barrier height. This indicates that the electrical characteristics of both the Ir and Ir/Au contacts are improved upon annealing at 300°C compared to the as-deposited contacts. It also is noted that the reverse leakage current decreased for Ir/n-InGaN Schottky diode compared to the Au/Ir/n-InGaN Schottky diode upon annealing temperature.

The barrier height and the ideality factor of a metal-semiconductor contact are determined using thermionic emission theory and are given by [9]

$$I = I_S \exp\left[\left(\frac{q(V - IR)}{nKT}\right) - 1\right], \quad (1)$$

with

$$I_S = AA^{**}T^2 \exp\left(\frac{-q\phi_b}{kT}\right). \quad (2)$$

Here, I_S is the saturation current, q is the electron charge, V is the applied voltage, R is the series resistance, T is the absolute temperature, n is the ideality factor, k is the Boltzmann's constant, A is the contact area, and A^{**} is the effective Richardson constant. The theoretical value of A^{**} was 23 A cm^{-2} K^{-2} based on the effective mass ($m^* = 0.19m_o$) of the n-InGaN [7] and was used to deduce ϕ_b. In general by using a linear curve fit to the forward characteristics of $\ln(I)$ against V data, barrier height (ϕ_b) and ideality factor (n) can be determined from the y-intercept and the slope of the fitted curve, respectively. Calculations yield that the Schottky barrier height (SBH) of as-deposited Ir/n-InGaN Schottky diode is 0.79 eV. For the contacts annealed at 300°C, 400°C, and 500°C, the calculated barrier heights are 0.85 eV, 0.84 eV, and 0.80 eV. The Schottky barrier height estimated in the case of Ir/Au contact is 0.70 eV (as-deposited), 0.74 eV after annealing (at 300°C), 0.71 eV (400°C), and 0.67 eV (500°C).

The ideality factor is determined from *I-V* plot, specifically at small forward currents, where the effect of series resistance is considered to be small. In case of Ir and Ir/Au Schottky diodes, the typical values of ideality factor are found to be 1.60 and 1.51 (as-deposited), respectively. However, it is observed that the ideality factors are improved to 1.10 for Ir and 1.23 for Ir/Au Schottky contacts upon annealing at 300°C. The ideality factor calculated from the forward bias *I-V* plot is found to be greater than unity. This may be due to a potential drop in the interface layer and the presence of excess current and the recombination current through the interfacial state between the semiconductor/insulator layers [10]. Also, when the bias is applied, part of the voltage is dropped across the interface layers. Thus, the barrier height becomes dependent on the bias voltage, leading to an ideality factor greater than one. Moreover, the larger values of ideality factors are attributed to secondary mechanisms that include interface dipoles due to interface doping or specific interface structure as well as fabricate-induced defects at the interface [11, 12]. The high value of ideality factor can also be attributed to the presence of a wide distribution of low-SBH patches caused by laterally inhomogeneous according to Tung [13]. Another possibility may be the image-force effect, recombination generation, and tunneling mechanisms that could be lead to the ideality factor value greater than unity [14]. The barrier heights and the ideality factors of Ir and Ir/Au Schottky diodes as a function of annealing temperature are given in Table 1.

Generally, at low voltages, the linear forward bias current-voltage (*I-V*) characteristics in the semilogarithmic scale deviate considerably from linearity, which is due to the effect of parameters such as R_S and N_{SS} when the applied voltage is sufficiently large. The series resistance values are evaluated using the method developed by S. cheung and N. cheung [15]. Cheung's functions given as

$$\frac{dV}{d(\ln I)} = IR_S + n\left(\frac{kT}{q}\right), \quad (3)$$

$$H(I) = V - n\left(\frac{kT}{q}\right)\ln\left(\frac{I}{AA^{**}T^2}\right), \quad (4)$$

$$H(I) = IR_S + n\phi_b \quad (5)$$

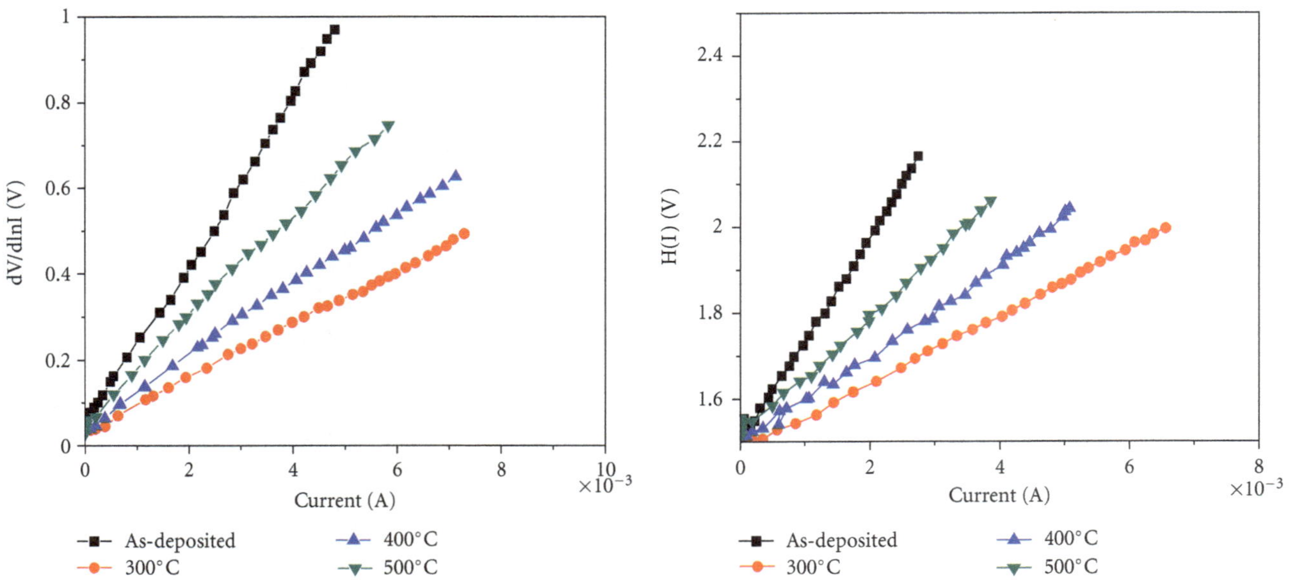

(a) Ir/n-InGaN Schottky diode.

(b) Ir/n-InGaN Schottky diode.

Figure 2: (a) Plot of $dV/d\ln I$ versus I for the Ir/n-InGaN Schottky contacts. (b) Plot of $H(I)$ versus I for the Ir/n-InGaN Schottky contacts.

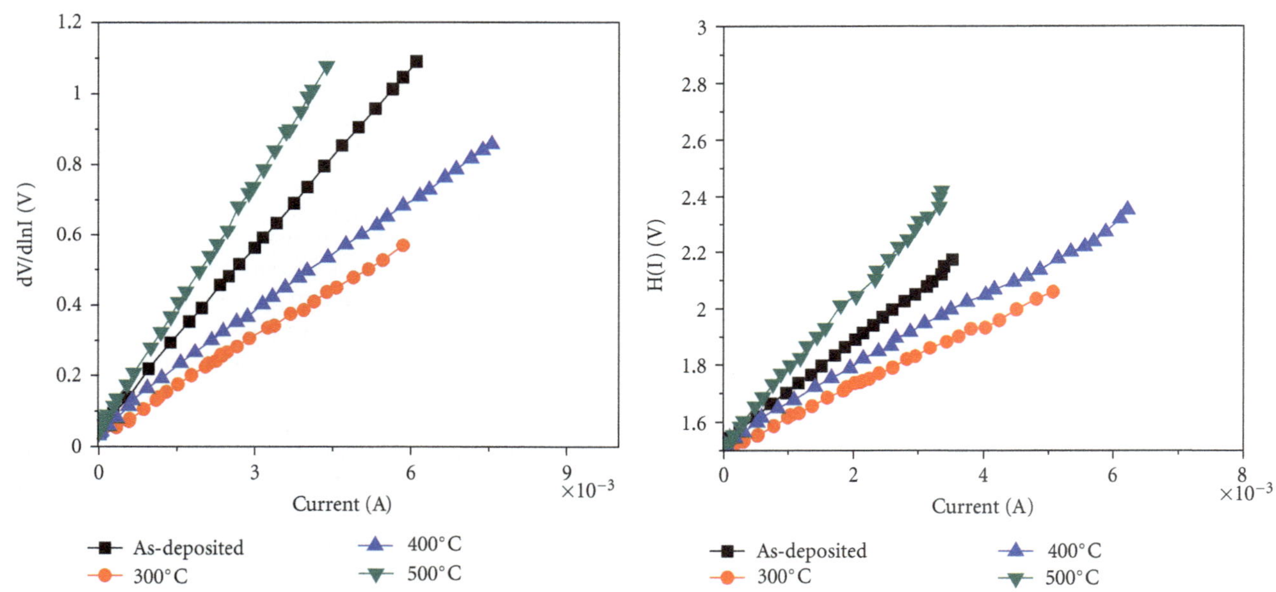

(a) Au/Ir/n-InGaN Schottky diode.

(b) Au/Ir/n-InGaN Schottky diode.

Figure 3: (a) Plot of $dV/d\ln I$ versus I for the Ir/Au/n-InGaN Schottky contacts. (b) Plot of $H(I)$ versus I for the Ir/Au/n-InGaN Schottky contacts.

should give a straight line for the data of downward curvature region in the forward bias I-V characteristics. The term IR_S is the current drop across the series resistance of Schottky diodes. Figures 2(a) and 3(a) show the experimental $dV/d\ln I$ versus I plot for the Ir and Ir/Au Schottky diodes, respectively. The values of R_S and nkT/q are calculated from the slope and y-axis intercepts of the $dV/d\ln I$ versus I plot, respectively. From Figures 2(b) and 3(b), it is evident that the plot of $H(I)$ versus I gives a straight line with the y-axis intercept equal to $n\phi_b$, and the slope parameter can be used to check the consistency of Cheung's method. From (3), the values of ideality factor and series resistance for Ir/InGaN and Au/Ir/InGaN Schottky diodes are calculated. The values of barrier height and series resistance for Ir/InGaN and Au/Ir/InGaN Schottky diodes are calculated from (5) which is given in Table 1. Thus, it can be clearly seen that there is relatively a difference between the values of ideality factor calculated from the downward curvature regions of forward bias I-V plots and from the linear regions of the same characteristics. This can be attributed to the existence of

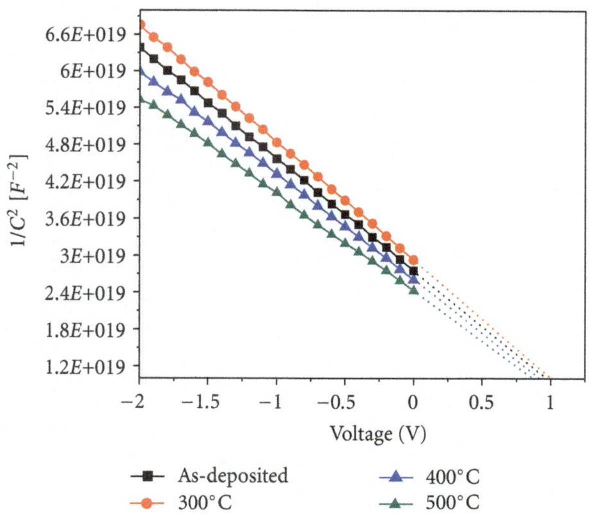

(a) Ir/n-InGaN Schottky diode.

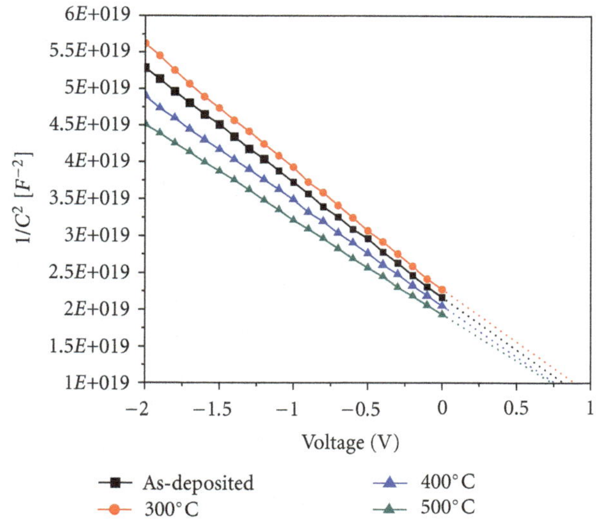

(b) Au/Ir/n-InGaN Schottky diode.

FIGURE 4: (a) Plot of $1/C^2$ versus V for the Ir Schottky contacts annealed at different temperatures. (b) Plot of $1/C^2$ versus V for the Ir/Au Schottky contacts annealed at different temperatures.

effects such as series resistance and the bias dependence of Schottky barrier height according to the voltage drop across the interfacial layer and change of the interface states with the bias in the low-voltage region of the current-voltage plot.

The accurate evaluation of the barrier height from the standard ln (I)-V plot can be effected by the high series resistance. To include this effect, Norde method is also used to compare the SBH of Ir and Ir/Au Schottky contacts. The Norde function $F(V)$, being plotted against $V \cdot F(V)$, is given by [16, 17]

$$F(V) = \frac{V}{2} - \frac{kT}{q} \ln\left[\frac{I(V)}{AA^{**}T^2}\right]. \quad (6)$$

The Schottky barrier height ϕ_b is given by

$$\phi_b = F(V_{\min}) + \frac{V_{\min}}{2} - \frac{kT}{q}, \quad (7)$$

where $F(V_{\min})$ is the minimum value of $F(V)$, and $V_{\min}$ is the corresponding voltage. Measurements showed that the SBHs are 0.81 eV for as-deposited, 0.88 eV for 300°C, 0.86 eV for 400°C, and 0.82 eV for 500°C for Ir Schottky contacts. In the case of Ir/Au Schottky contacts, the calculated barrier heights are 0.73 eV for as-deposited, 0.76 eV for 300°C, 0.74 eV for 400°C, and 0.71 eV for 500°C, respectively. It is observed that these values are in good agreement with those calculated by I-V method.

3.2. Capacitance-Voltage (C-V) Characteristics. Capacitance-voltage (C-V) characteristics of Ir and Ir/Au Schottky contacts are measured at a frequency of 1 MHz with an ac modulation of 100 mV in the dark. Figures 4(a) and 4(b) show plots of $1/C^2$ as a function of bias voltage for as-deposited and annealed contacts of Ir and Ir/Au Schottky contacts, respectively. The C-V relationship for Schottky diode is given by [9]

$$\frac{1}{C^2} = \frac{2[V_{bi} - (kT/q) - V]}{A^2 q N_d \varepsilon_S}, \quad (8)$$

where V_{bi} is the flat band voltage, N_d is the donor concentration, A is the area of the Schottky contact, and ε_s is the permittivity of the semiconductor ($\varepsilon_s = 8.4\varepsilon_o$). The x-intercept of the plot of $(1/C^2)$ versus V, V_o is related to the built in potential V_{bi} by equation $V_{bi} = V_o + kT/q$, where T is the absolute temperature. The barrier height (ϕ_{CV}) is given by the equation $\phi_{CV} = V_{bi} + V_n$ where $V_n = (kT/q)\ln(N_C/N_d)$. The density of states in the conduction band edge is given by $N_c = 2(2\pi m^* kT/h^2)^{3/2}$, $m^* = 0.19m_o$, and its value was $7 \times 10^{17}\,\text{cm}^{-3}$ for InGaN at room temperature. Measurements showed that the barrier heights of Ir Schottky contacts are 1.17 eV for as-deposited, 1.21 eV for 300°C, 1.11 eV for 400°C, and 1.07 eV for 500°C annealed contacts. For Ir/Au Schottky contacts, the estimated barrier heights are 1.00 eV (as-deposited), 1.03 eV after annealing the contacts (at 300°C), 0.96 eV (400°C), and 0.93 eV (500°C) by C-V method.

Figure 5(a) shows a plot of barrier height of Ir/n-InGaN as a function of annealing temperature. It can be seen clearly from Figure 5(a) that the barrier height of Ir/n-InGaN Schottky diode increases upon annealing at 300°C for 1 min in the N_2 ambient compared to the as-deposited contact. However, there is a slight decrease in the Schottky barrier height for the contacts annealed at 400°C and 500°C. The interface states and chemical reactions between metals and semiconductors at the interface may be the reason for the variation in the barrier heights [18]. From Figure 5(a), it is also noted that the ideality factor is improved and reaches near unity value (1.10) after annealing at 300°C. However, the values of

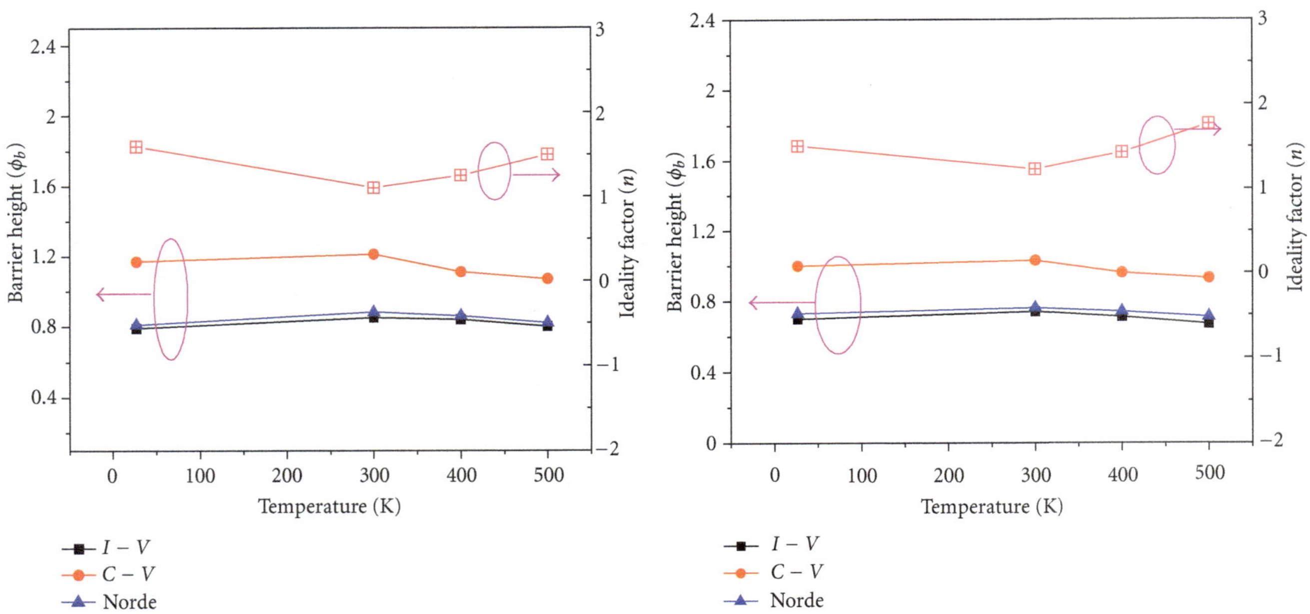

(a) Ir/n-InGaN Schottky diode.

(b) Au/Ir/n-InGaN Schottky diode.

FIGURE 5: (a) The barrier height and ideality factors of the Ir Schottky contact as a function of annealing. (b) The barrier height and ideality factors of the Ir/Au Schottky contact as a function of annealing.

ideality factor slightly increase upon annealing at 400°C and 500°C compared to the contact annealed at 300°C.

The plot of barrier heights and ideality factors of Ir/Au Schottky contacts as a function of annealing temperature is shown in Figure 5(b). It is noted that the barrier height calculated for the as-deposited Ir/Au contacts is less than annealed contacts (300°C and 400°C). When the contact is annealed at 300°C, it is found that the Schottky barrier height increases for Ir/Au contact. Rajagopal Reddy et al. [3] had observed similar behavior in Au/Ru/n-GaN Schottky diode. Also, Miura et al. [5] had observed similar behavior in Ni/Pt/Au, Ni/Au, and Ni/Mo/Au Schottky contacts on n-GaN. Upon annealing temperature, the increase in the barrier height may be due to the reduction in the nonstoichiometric defects in the metallurgical interface [19, 20]. Also, it is observed that the increase in the barrier height is accompanied by the corresponding decrease in the reverse leakage current. Duboz et al. [21] had reported that the metal/n-GaN interface remains unaffected even after annealing. They also found that the nature of the metal does not play a major role in the improvement of barrier height on annealing temperature. Many researchers had observed the pinning of Fermi level at metal/n-GaN interface because of the interface defects. Further, a modification in the defect density on annealing could change the pinning, which leads to the change of barrier height. Significant improvement in the electrical characteristics in the case of Ir/n-InGaN Schottky diode is observed upon annealing temperature compared to the Au/Ir/n-InGaN Schottky diode. It is also observed that the electrical characteristics of Ir Schottky contacts are fairly more thermally stable than that of Ir/Au Schottky contacts. Also, Figure 5(b) shows that the ideality factor value decreases upon annealing at 300°C. With a further increase in the annealing temperatures up to 500°C, the ideality factor is slightly increased.

It can also be seen from Figures 5(a) and 5(b) that the barrier heights obtained from *I-V* measurements are lower than those obtained from *C-V* measurements. This difference in the values of $\phi_{I\text{-}V}$ and $\phi_{C\text{-}V}$ may be due to the presence of native oxide layer at the metal/InGaN interface [22]. Lowering the barrier height by an image force due to current flow across the barrier can also have some effect [9]. However, with high-temperature annealing, the metal perhaps comes into intimate contact with the semiconductor. Furthermore, the differences in the barrier height calculated from *I-V* and *C-V* methods may also cause the Schottky barrier height inhomogeneities that prevail at the metal semiconductor interface [23]. Another possibility is that the surface damage at the metal-semiconductor interface affects *I-V* measurements because defects may act as recombination centers for trap-assisted tunneling currents. *C-V* measurements are normally less prone to interface states. Hence, the calculated barrier heights are considered more reliable, though the depletion width can be altered by interface defects if they are deeper into the space charge region [24]. It may also be the transport mechanism in these diodes which is not purely due to thermionic emission. To these diodes, ϕ_b obtained from *I-V* method is voltage or electric field sensitive unlike the ϕ_b obtained from *C-V*.

3.3. Interface State Density (N_{SS}) from I-V and C-V Measurements. It is noted from Figures 6(a) and 6(b) that the interface state density N_{SS} has an exponential increase with the bias from the midgap towards the top of the conduction

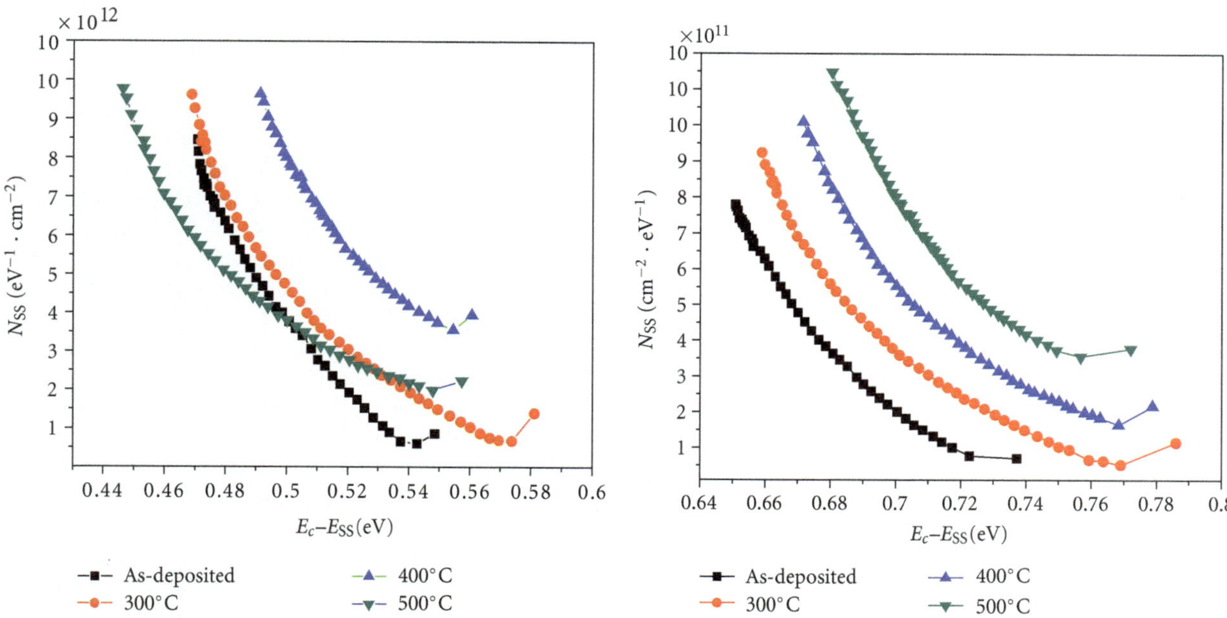

(a) Ir/n-InGaN Schottky diode.

(b) Au/Ir/n-InGaN Schottky diode.

FIGURE 6: (a) Interface state density distribution profiles as a function of $E_{CC} - E_{SS}$ for Ir Schottky contacts. (b) Interface state density distribution profiles as a function of $E_{CC} - E_{SS}$ for Ir/Au Schottky contacts.

band. According to Card and Rhoderick, for a real Schottky diode having interface states in equilibrium with the semiconductor, the ideality factor n becomes greater than unity and is given by [9, 25]

$$n(V) = 1 + \frac{\delta}{\varepsilon_i}\left[\frac{\varepsilon_S}{W_d} + q \cdot N_{ss}(V)\right], \tag{9}$$

where N_{SS} is the density of interface states, W_d is the width of the semiconductor, and ε_S and ε_i are the permittivity of the semiconductor and the interfacial layer, respectively. The voltage-dependent ideality factor $n(V)$ can be expressed as $n(V) = (qV)/(kT\ln(I/I_s))$ [26]. The expression for the interface state density can be given as

$$N_{SS} = \frac{1}{q}\left[\frac{\varepsilon_i}{\delta}(n(V) - 1) - \frac{\varepsilon_S}{W_d}\right]. \tag{10}$$

In the n-type semiconductor, the energy of the interface states E_{SS} with respect to the bottom of the conduction band at the surface of the semiconductor is given by [25, 27, 28]

$$E_C - E_{SS} = q(\phi_e - V), \tag{11}$$

where ϕ_e is the effective barrier height. The voltage dependence of the effective barrier height ϕ_e is contained in the ideality factor n through the relation

$$\frac{d\phi_e}{dV} = \beta = 1 - \frac{1}{n(V)}, \tag{12}$$

where β is the voltage coefficient of ϕ_e. The effective barrier height ϕ_e is given by [29]

$$\phi_e = \phi_b + \beta(V). \tag{13}$$

From Figures 6(a) and 6(b), it can be seen that the N_{SS} value decreases with the increase in the $E_C - E_{SS}$ value. The increase in N_{SS} values forms from midgap towards the bottom of the conduction band which is probably due to the increase in the series resistance (R_S). The change in the charge of the interface states and thus the interface state energy distribution due to the potential drop across the interfacial layer varies with the bias. It alters the diffusion potential and therefore the depletion capacitance [9, 30–33].

4. Conclusions

We have investigated the electrical properties of Ir and Ir/Au Schottky contacts on n-type InGaN ($N_d = 6 \times 10^{17}\,\text{cm}^{-3}$) by *I-V* and *C-V* measurements as a function of annealing temperature. Measurements showed that the barrier height of as-deposited Ir Schottky contact is 0.79 eV (*I-V*) and 1.17 eV (*C-V*). Upon annealing at 300°C for 1 min in nitrogen ambient, the Schottky barrier height increases to 0.85 eV (*I-V*) and 1.12 eV (*C-V*). After annealing at 400°C and 500°C, the Schottky barrier heights decrease to 0.84 eV (*I-V*), 1.11 eV (*C-V*) and 0.80 eV (*I-V*), 1.07 (*C-V*). In the case of as-deposited Ir/Au Schottky contacts, the estimated barrier height is 0.70 eV (*I-V*) and 1.00 eV (*C-V*). However, the Schottky barrier height increases upon annealing temperature at 300°C, and the corresponding value is 0.74 eV (*I-V*) and 1.03 eV (*C-V*). Further, with an increase in annealing temperature up to 500°C, the barrier height decreases to 0.67 eV (*I-V*) and 0.93 eV (*C-V*), respectively. Based on the above results, the optimum barrier heights are obtained after annealing at 300°C for both Ir and Ir/Au Schottky contacts. Therefore, the optimum annealing temperature for both Ir and Ir/Au Schottky contacts is 300°C. The values of diode parameters such as ideality factor (n), series resistance (R_S), and barrier height (ϕ_b) are also calculated by Cheung's and Norde's methods. The interface state density has an exponential raise with bias from the midgap towards the top of the conduction band. From the above observations, it is clear that the electrical properties of Ir and Ir/Au Schottky contacts are improved with an increase in annealing temperature compared to the as-deposited one. Moreover, both Ir and Ir/Au Schottky contacts showed relatively thermal stability during annealing temperature.

Acknowledgment

The authors thank the University Grant Commission (UGC), New Delhi, for providing the financial assistance (Grant no. 39-530/2010 (SR)).

References

[1] S. X. Li, K. M. Yu, J. Wu et al., "Fermi-level stabilization energy in group III nitrides," *Physical Review B*, vol. 71, no. 16, pp. 1–4, 2005.

[2] J. Kim, F. Ren, A. G. Baca, and S. J. Pearton, "Thermal stability of WSix and W Schottky contacts on n-GaN," *Applied Physics Letters*, vol. 82, no. 19, pp. 3263–3265, 2003.

[3] V. Rajagopal Reddy, C. K. Ramesh, and K. S. R. Koteswara Rao, "Effect of annealing temperature on electrical characteristics of ruthenium-based Schottky contacts on n-type GaN," *Journal of Materials Science*, vol. 17, no. 12, pp. 999–1004, 2006.

[4] V. Rajagopal Reddy and N. Ramesha Reddy, "Influence of thermal annealing temperature on electrical properties of Rh and Rh/Au Schottky contacts to n-type GaN," *Journal of Optoelectronics and Advanced Materials*, vol. 9, no. 12, pp. 3871–3876, 2007.

[5] N. Miura, T. Nanjo, M. Suita et al., "Thermal annealing effects on Ni/Au based Schottky contacts on n-GaN and AlGaN/GaN with insertion of high work function metal," *Solid-State Electronics*, vol. 48, no. 5, pp. 689–695, 2004.

[6] H. S. Venugopalan and S. E. Mohney, "Thermally stable rhenium Schottky contacts to n-GaN," *Applied Physics Letters*, vol. 73, no. 9, pp. 1242–1244, 1998.

[7] J. S. Jang, D. Kim, and T. Y. Seong, "Schottky barrier characteristics of Pt contacts to n-type InGaN," *Journal of Applied Physics*, vol. 99, no. 7, Article ID 073704, 2006.

[8] J. J. Xue, D. J. Chen, B. Liu et al., "Au/Pt/InGaN/GaN heterostructure schottky prototype solar cell," *Chinese Physics Letters*, vol. 26, no. 9, Article ID 098102, 2009.

[9] E. H. Rhoderick and T. H. Williams, *Metal-Semiconductor Contacts*, Oxford Science, Oxford, UK, 1988.

[10] D. T. Quan and H. Hbib, "High barrier height Au/n-type InP Schottky contacts with a POxNyHz interfacial layer," *Solid-State Electronics*, vol. 36, no. 3, pp. 339–344, 1993.

[11] C. R. Crowell, "The Richardson constant for thermionic emission in Schottky barrier diodes," *Solid State Electronics*, vol. 8, no. 4, pp. 395–399, 1965.

[12] H. Çetin and E. Ayyildiz, "Electrical characteristics of Au, Al, Cu/n-InP Schottky contacts formed on chemically cleaned and air-exposed n-InP surface," *Physica B*, vol. 394, no. 1, pp. 93–99, 2007.

[13] R. T. Tung, "Recent advances in Schottky barrier concepts," *Materials Science and Engineering: R*, vol. 35, no. 1–3, pp. 1–13, 2001.

[14] L. Messick, "A D.C. to 16 GHz indium phosphide MISFET," *Solid State Electronics*, vol. 23, no. 6, pp. 551–555, 1980.

[15] S. K. Cheung and N. W. Cheung, "Extraction of Schottky diode parameters from forward current-voltage characteristics," *Applied Physics Letters*, vol. 49, no. 2, pp. 85–87, 1986.

[16] H. Norde, "A modified forward I-V plot for Schottky diodes with high series resistance," *Journal of Applied Physics*, vol. 50, no. 7, pp. 5052–5053, 1979.

[17] J. S. Jang, S. J. Park, and T. Y. Seong, "Formation of low resistance Pt ohmic contacts to p-type GaN using two-step surface treatment," *Journal of Vacuum Science and Technology B*, vol. 17, no. 6, pp. 2667–2670, 1999.

[18] S. Yuunping, X. M. Shen, J. Wang et al., "Thermal annealing behaviour of Ni/Au on n-GaN Schottky contacts," *Journal of Physics D*, vol. 35, no. 20, pp. 2648–2651, 2002.

[19] W. E. Spicer, I. Lindau, P. Skeath, C. Y. Su, and P. Chye, "Unified mechanism for schottky-barrier formation and III-V oxide interface states," *Physical Review Letters*, vol. 44, no. 6, pp. 420–423, 1980.

[20] R. H. Williams, V. Montgomery, and R. R. Varma, "Chemical effects in Schottky barrier formation," *Journal of Physics C*, vol. 11, no. 17, pp. L735–L738, 1978.

[21] J. Y. Duboz, F. Binet, N. Laurent et al., "Influence of surface defects on the characteristics of GaN Schottky diodes," *Material Research Society Symposium Proceedings*, vol. 449, pp. 1085–1090, 2009.

[22] V. Rajagopal Reddy, M. Ravinandan, P. Koteswara Rao, and C. J. Choi, "Effects of thermal annealing on the electrical and structural properties of Pt/Mo Schottky contacts on n-type GaN," *Journal of Materials Science*, vol. 20, no. 10, pp. 1018–1025, 2009.

[23] J. H. Werner and H. H. Güttler, "Barrier inhomogeneities at Schottky contacts," *Journal of Applied Physics*, vol. 69, no. 3, pp. 1522–1533, 1991.

[24] C. Fontaine, T. Okumura, and K. N. Tu, "Interfacial reaction and Schottky barrier between Pt and GaAs," *Journal of Applied Physics*, vol. 54, no. 3, pp. 1404–1412, 1983.

[25] H. C. Card and E. H. Rhoderick, "Studies of tunnel MOS diodes I. Interface effects in silicon Schottky diodes," *Journal of Physics D*, vol. 4, no. 10, pp. 1589–1601, 1971.

[26] O. Güllü and A. Türüt, "Electrical analysis of organic interlayer based metal/interlayer/ semiconductor diode structures," *Journal of Applied Physics*, vol. 106, no. 10, Article ID 103717, 2009.

[27] A. Singh, K. C. Reinhardt, and W. A. Anderson, "Temperature dependence of the electrical characteristics of Yb/p-InP tunnel metal-insulator-semiconductor junctions," *Journal of Applied Physics*, vol. 68, no. 7, pp. 3475–3483, 1990.

[28] M. K. Hudait and S. B. Krupanidhi, "Interface states density distribution in Au/n-GaAs Schottky diodes on n-Ge and n-GaAs substrates," *Materials Science and Engineering B*, vol. 87, no. 2, pp. 141–147, 2001.

[29] A. Turut, M. Saglam, H. Efeoglu, N. Yalcin, M. Yildirim, and B. Abay, "Interpreting the nonideal reverse bias C-V characteristics and importance of the dependence of Schottky barrier height on applied voltage," *Physica B*, vol. 205, no. 1, pp. 41–50, 1995.

[30] S. M. Sze, *Physics of Semiconductor Structures*, John Wiley & Sons, New York, NY, USA, 2nd edition, 1981.

[31] S. Karatas, S. Altindal, and M. Cakar, "Current transport in Zn/p-Si(1 0 0) Schottky barrier diodes at high temperatures," *Physica B*, vol. 357, no. 3-4, pp. 386–397, 2005.

[32] S. Karatas and A. Turut, "The determination of electronic and interface state density distributions of Au/n-type GaAs Schottky barrier diodes," *Physica B*, vol. 381, no. 1-2, pp. 199–203, 2006.

[33] A. Turut and M. Saglam, "Determination of the density of Si-metal interface states and excess capacitance caused by them," *Physica B*, vol. 179, no. 4, pp. 285–294, 1992.

13

Developing 1D MM of Axisymmetric Transient Quenched Chromium Steel to Determine LHP

Abdlmanam S. A. Elmaryami and Badrul Omar

Mechanical Engineering Department, Universiti Tun Hussein Onn Malaysia, Batu Pahat, Johor, 86400 Parit Raja, Malaysia

Correspondence should be addressed to Abdlmanam S. A. Elmaryami, damer604@yahoo.com

Academic Editor: Elena V. Pereloma

The modelling of an axisymmetric industrial quenched chromium steel bar AISI-SAE 8650H based on finite element method has been produced to investigate the impact of process history on metallurgical and material properties. Mathematical modelling of 1-dimensional line (radius) element axisymmetric model has been adopted to predict temperature history and consequently the hardness of the quenched steel bar at any point (node). The lowest hardness point (LHP) is determined. In this paper hardness in specimen points was calculated by the conversion of calculated characteristic cooling time for phase transformation $t_{8/5}$ to hardness. The model can be employed as a guideline to design cooling approach to achieve desired microstructure and mechanical properties such as hardness. The developed mathematical model is converted to a computer program. This program can be used independently or incorporated into a temperature history calculator to continuously calculate and display temperature history of the industrial quenched steel bar and thereby calculate LHP. The developed program from the mathematical model has been verified and validated by comparing its hardness results with commercial finite element software results.

1. Introduction

Quenching is a heat treatment usually employed in industrial processes in order to control mechanical properties of steels such as hardness [1]. The process consists of raising the steel temperature above a certain critical value, holding it at that temperature for a specified time and then rapidly cooling it in a suitable medium to room temperature [2]. The resulting microstructures formed from quenching (ferrite, cementite, pearlite, upper bainite, lower bainite, and martensite) depend on cooling rate and on chemical composition of the steel [3].

Quenching of steels is a multiphysics process involving a complicated pattern of couplings among heat transfer, because of the complexity, coupled (thermal-mechanical-metallurgical) theory, and nonlinear nature of the problem, no analytical solution exists; however, numerical solution is possible by finite difference method, finite volume method, and the most popular one-finite element method (FEM) [4]. During the quenching process of the steel bar, the heat transfer is in an unsteady state as there is a variation of temperature with time [5]. The heat transfer analysis in this paper will be carried out in 3-dimensions. The three dimensional analysis will be reduced into a 1-dimensional axisymmetric analysis to save cost and computer time [4, 6, 16]. This is achievable because in axisymmetric conditions, there is no temperature variation in the theta ($\boldsymbol{\theta}$) direction and in ($\mathbf{z}$) direction, the temperature deviations are only in ($\mathbf{r}$). The Galerkin weighted residual technique is used to derive the mathematical model. In this paper, 1-dimensional line (radius) element will be developed.

2. Mathematical Model

The temperature history of the quenched cylindrical steel bar at any point would be calculated; three-dimensional heat transfer can be analyzed using one dimensional axisymmetric elements as shown in Figures 1, 2, and 3.

The linear temperature distribution for an element (radius) line, T is given by:

$$T^{(R)} = a_1 + a_2 R, \tag{1}$$

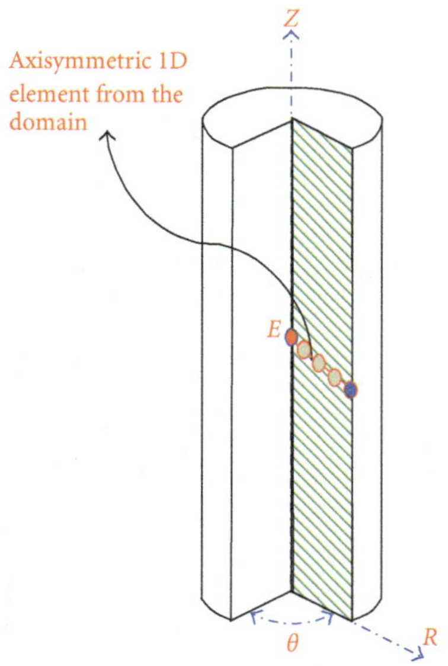

FIGURE 1: The axisymmetric one-dimensional line (radius) element from the domain, on the cylindrical chromium steel bar which had been heated and then submerged in sea water.

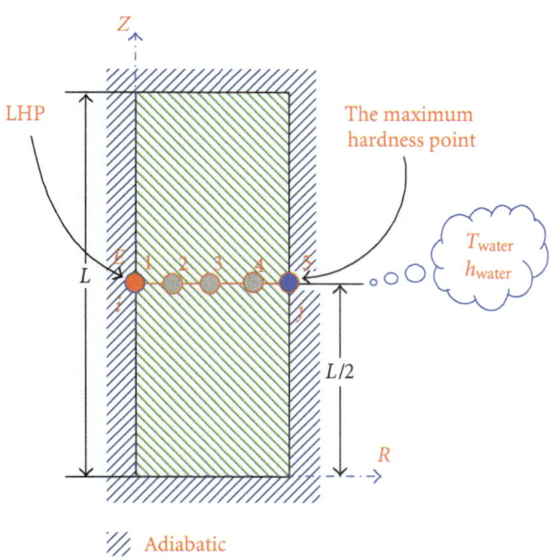

FIGURE 2: The axisymmetric one-dimensional line (radius) element from the domain on the axisymmetric rectangular cross section.

where $T(R)$ = nodal temperature as the function of R, a_1 and a_2 are constants, R is any point on the (radius) line element.

2.1. Shape Function of the Axisymmetric Triangular Element. The shape functions were to represent the variation of the field variable over the element. The shape function of axisymmetric 1-dimensional line (radius) element expressed in terms of the r coordinate and its coordinate as shown

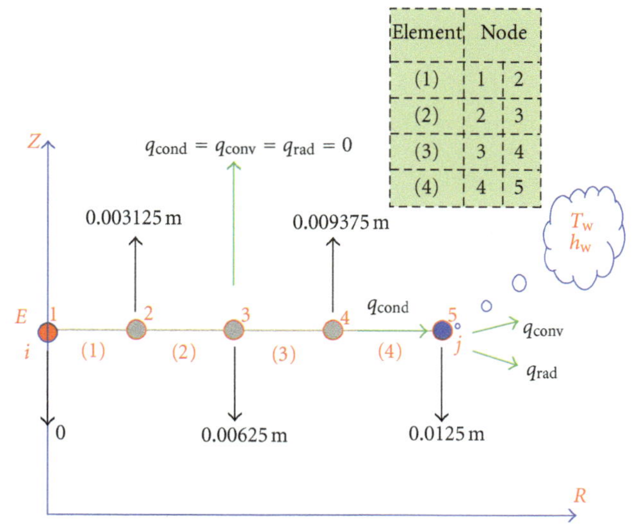

FIGURE 3: The axisymmetric one dimensional line (radius) element from the domain, the selected 4 elements with 5 nodes and the boundary at node $j(5)$ for an element 4.

in Figure 4; which are derived to obtain the following shape functions as shown:

$$S_i = \left(\frac{R_j - R}{R_j - R_i}\right) = \left(\frac{R_j - R}{L}\right),$$
$$S_j = \left(\frac{R - R_i}{R_j - R_i}\right) = \left(\frac{R - R_i}{L}\right). \quad (2)$$

Thus the temperature distribution of 1D radius for an element in terms of the shape function can be written as:

$$T^{(R)} = S_i T_i + S_j T_j = \left[S^{(r)}\right]\{T\}, \quad (3)$$

where $[S^{(r)}] = [\, S_i \; S_j \,]$ is a row vector matrix and

$$\{T\} = \begin{Bmatrix} T_i \\ T_j \end{Bmatrix} \quad (4)$$

is a column vector of nodal temperature of the element.

Equation (3) can also be expressed in matrix form as:

$$T(R) = \left[S_i \;\; S_j\right]\begin{Bmatrix} T_i \\ T_j \end{Bmatrix}. \quad (5)$$

Thus for 1-dimensional element we can write in general:

$$\Psi^{(e)} = \left[S_i \;\; S_j\right]\begin{Bmatrix} \psi_i \\ \psi_j \end{Bmatrix}, \quad (6)$$

where Ψ_i and Ψ_j represent the nodal values of the unknown variable such as in our case temperature also the unknown can be deflection, velocity, and so forth.

2.2. Natural Area Coordinate. Using the natural length coordinates and their relationship with the shape function by simplification of the integral of Galerkin solution.

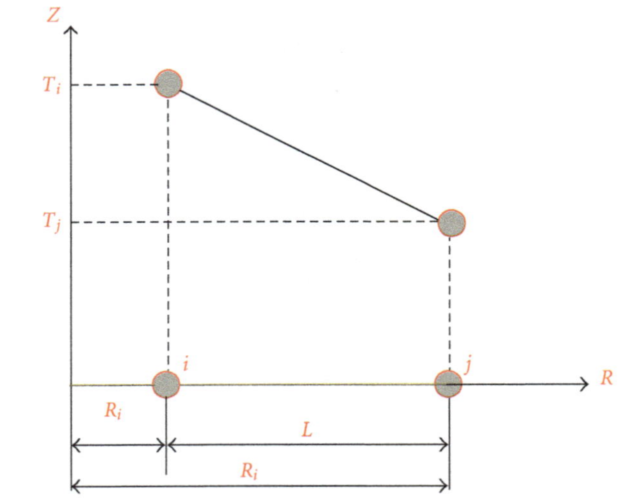

FIGURE 4: One-dimensional linear temperature distribution for an element (radius) line in global coordinate system.

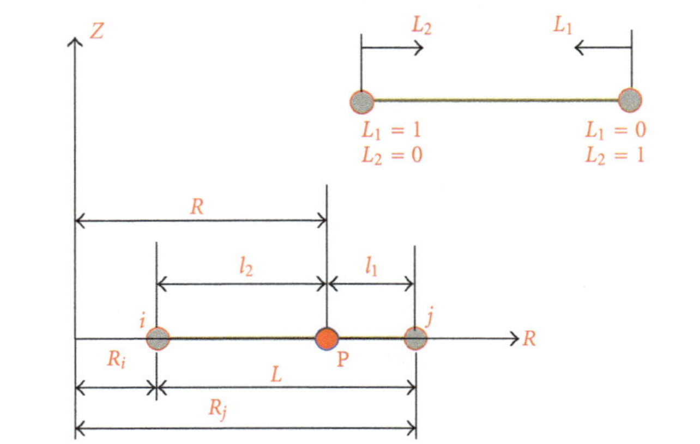

FIGURE 5: Two-node line element showing interior point P and the two naturals coordinates L_1 and L_2.

The two length natural coordinates L_1 and L_2 at any point P inside the element are shown in Figure 5 from which we can write:

$$L_1 = \frac{R_j - R}{R_j - R_i} = \frac{l_1}{L},$$
$$L_2 = \frac{R - R_i}{R_j - R_i} = \frac{l_2}{L}. \tag{7}$$

Since it is a one-dimensional element, there should be only one independent coordinate to define any point P. This is true even with natural coordinates as the two natural coordinates L_1 and L_2 are not independent, but are related as:

$$L_1 + L_2 = 1 \quad \text{or} \quad L_1 + L_2 = \frac{l_1}{L} + \frac{l_2}{L} = 1. \tag{8}$$

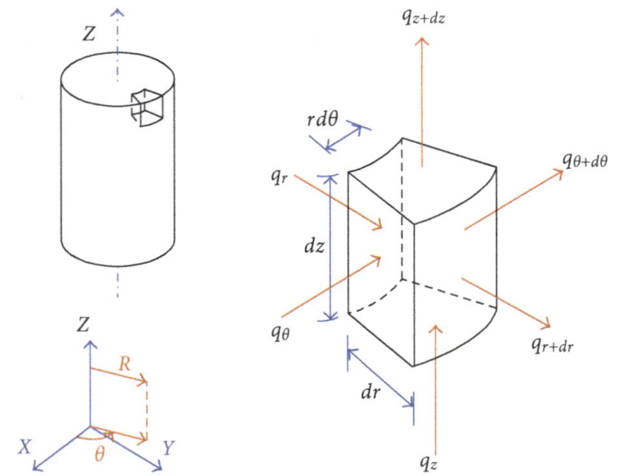

FIGURE 6: Axisymmetric element from an axisymmetric body.

The natural coordinates L_1 and L_2 are also the shape functions for the line element, thus:

$$
\begin{gathered}
S_i = \left(\frac{R_j - R}{R_j - R_i}\right) = \left(\frac{R_j - R}{L}\right) = L_1,\\
S_j = \left(\frac{R - R_i}{R_j - R_i}\right) = \left(\frac{R - R_i}{L}\right) = L_2,\\
S_i = L_1, \qquad S_j = L_2,\\
R = R_j L_2 + R_i L_1 = R_i S_i + R_j S_j,\\
\frac{\partial [S]_i}{\partial r} = \frac{\partial L_1}{\partial r} = \frac{-1}{R_j - R_i} = -\frac{1}{L},\\
\frac{\partial [S]_j}{\partial r} = \frac{\partial L_2}{\partial r} = \frac{1}{R_j - R_i} = \frac{1}{L}.
\end{gathered}
\tag{9}
$$

2.3. Develop Equation for All Elements of the Domain. Derivation of equation of heat transfer in axisymmetric one-dimensional line (radius) elements by pplying the conservation of energy to a differential volume cylindrical segment as shown in Figure 6,

$$E_{\text{in}} - E_{\text{out}} + E_{\text{generated}} = E_{\text{stored}}. \tag{10}$$

The transient heat transfer within the component during quenching can mathematically be described by simplifying the differential volume term; the heat conduction equation is derived and given by

$$\frac{1}{r}\frac{d}{dr}\left(K_r r \frac{dT}{dr}\right) + \frac{1}{r^2}\frac{d}{d\theta}\left(K_\theta \frac{dT}{d\theta}\right) + \frac{d}{dz}\left(K_z \frac{dT}{dz}\right) + q = \rho c \frac{dT}{dt}, \tag{11}$$

k_r = heat conductivity coefficient in r-direction, W/m·°C, k_θ = heat conductivity coefficient in θ-direction, W/m·°C, k_z = heat conductivity coefficient in z-direction, W/m·°C, T = temperature, °C, q = heat generation, W/m^3, ρ = mass density, kg/m^3, c = specific heat of the medium, J/kg·K, t = time, s.

The assumptions made in this problem were:

(i) for axisymmetric situations one dimensional line (radius) element, there is no variation of temperature in the **Z**-direction as shown in Figures 1, 2, and 3. Because we already assumed that in steel quenching and cooling process of the steel bar is insulated from convection at the cross section of the front and back.

It means that we have convection and radiation at one node only which is on the surface (node 5), in our research we focus to calculate LHP which is at (node 1), where it is the last point will be cooled, this give the maximum advantage to make our assumption more safe, because it is the last point which will effect by convection and radiation from the front and back cross-section of the steel bar therefore we can write, $(\partial T/\partial z = 0)$,

(ii) for axisymmetric situations, there is no variation of temperature in the $\boldsymbol{\theta}$-direction, because it is clear from Figures 1, 2, and 3 that the temperature distribution along the radius will be the same if the radius move with angle θ, 360° therefore, $(\partial T/\partial \theta = 0)$,

(iii) the thermal energy generation rate $\dot{q}$ represent the rate of the conversion of energy from electrical, chemical, nuclear, or electromagnetic forms to thermal energy within the volume of the system. Such as conversion is the electric field which will be studied with details in the 2nd part of our research, however in this manuscript no heat generation therefore,

$$\dot{q} = 0. \tag{12}$$

After simplifying, (11) becomes

$$\frac{k}{r}\frac{\partial}{\partial r}\left(r\frac{\partial T}{\partial r}\right) = \rho c\frac{\partial T}{\partial t}, \tag{13}$$

and also known as residual or partial differential equation

$$\{\mathfrak{R}\} = \frac{k}{r}\frac{\partial}{\partial r}\left(r\frac{\partial T}{\partial r}\right) - \rho c\frac{\partial T}{\partial t} = 0. \tag{14}$$

2.4. Galerkin Weighted Residual Method Formulation. From the derived heat conduction equation, the Galerkin residual for 1-dimensional line (radius) element in an unsteady state heat transfer by integration the shape functions times the residual which minimize the residual to zero becomes

$$\int_V [S]^T\{\mathfrak{R}\}^{(e)}dv = 0, \tag{15}$$

where $[S]^T$ = the transpose of the shape function matrix, $\{\mathfrak{R}\}^{(e)}$ = the residual contributed by element (e) to the final system of equations

$$\frac{k}{r}\int [S]^T\frac{\partial}{\partial r}\left(r\frac{\partial T}{\partial r}\right)dv - \int [S]^T\rho c\frac{\partial T}{\partial t}dv = 0. \tag{16}$$

2.5. Chain Rule. The terms 1 and 2 of (16) can be rearranged using the chain rule which states that;

$$(fg)^{-} = fg^{-} + gf^{-}. \tag{17}$$

Therefore, $fg^{-} = (fg)^{-} - f^{-}g$,

$$\frac{\partial}{\partial r}\left([S]^T r\frac{\partial T}{\partial r}\right) = [S]^T\left\{\frac{\partial}{\partial r}\left(r\frac{\partial T}{\partial r}\right)\right\} + r\frac{\partial T}{\partial r}\frac{\partial [S]^T}{\partial r}. \tag{18}$$

Term 1 of (16) is rearranged, thus

$$[S]^T\left\{\frac{\partial}{\partial r}\left(r\frac{\partial T}{\partial r}\right)\right\} = \frac{\partial}{\partial r}\left([S]^T r\frac{\partial T}{\partial r}\right) - r\frac{\partial T}{\partial r}\frac{\partial [S]^T}{\partial r}. \tag{19}$$

By substituting (19) in to (16), get

$$\begin{aligned} &= \overbrace{\frac{k}{r}\int\left\{\frac{\partial}{\partial r}\left([S]^T r\frac{\partial T}{\partial r}\right)\right\}dv}^{A} - \overbrace{\frac{k}{r}\int\left\{\frac{\partial [S]^T}{\partial r}r\frac{\partial T}{\partial r}\right\}dv}^{B} \\ &\quad - \underbrace{\int [S]^T\left\{\rho c\frac{\partial T}{\partial t}\right\}dv.}_{C} \end{aligned} \tag{20}$$

Term A is the heat convection terms and contributes to the conductance and thermal load matrix. Term B is the heat conduction terms and contributes to the conductance matrix. Term C is the transient equation and contributes to the capacitance matrix, where

$$\begin{aligned} &\overbrace{\frac{k}{r}\int\frac{\partial}{\partial r}\left([S]^T r\frac{\partial T}{\partial r}\right)dV}^{A} \\ &\quad = \underbrace{-2\pi hrz[S]^T[S]\{T\}^e}_{1} + \underbrace{2\pi hrz[S]^T T_f}_{2} \\ &\quad - \underbrace{2\pi rz\varepsilon_s\sigma[S]^T[S]\{T\}^e([S]\{T\}^e)^3}_{3} + \underbrace{2\pi rz\varepsilon_s\sigma[S]^T T_f^4.}_{4} \end{aligned} \tag{21}$$

Note that term (1) and term (3) contributed to the conductance matrix since they contains the unknown temperature $\{T\}$. Terms (2) and (4) contributed to the thermal load matrix as T_f is the known fluid temperature. Term (3) and term (4) heat radiation are very important if our heat treatment is annealing (cooling in the furnace) or normalizing (cooling in air or jet air), but can be ignore (neglect) if the cooling is quenching in water as in our paper.

From earlier explanations, derivation and after simplifying we can formulate the conductance matrix in the r-direction for Term B finally we get:

Term B (the conduction term) contributes to the conductance matrix:

$$\underbrace{\overbrace{\frac{k}{L}(R_j + R_i)\begin{bmatrix} 1 & -1 \\ -1 & 1 \end{bmatrix}\begin{Bmatrix} T_i \\ T_j \end{Bmatrix}}}_{k_c}. \tag{22}$$

Similarly, Term C is the unsteady state (transient) which contributes to the Capacitance Matrix,

Term C (heat stored) contributes to the Capacitance Matrix:

$$\underbrace{\frac{L\rho c}{6}\begin{bmatrix}(3R_j+R_i) & (R_i+R_j)\\ (R_i+R_j) & (R_i+3R_j)\end{bmatrix}\begin{Bmatrix}T_i\\ T_j\end{Bmatrix}}_{C}. \quad (23)$$

Term A (heat convection):

(i) *Term* A_1: contributes to conductance matrix,

Term A_1 (the convection term) contributes to the conductance matrix:

$$\underbrace{-2hR_j\begin{bmatrix}0 & 0\\ 0 & 1\end{bmatrix}\begin{Bmatrix}T_i\\ T_j\end{Bmatrix}}_{k_h}, \quad (24)$$

(ii) *Term* A_2: contributes to thermal load Matrix:

Term A_2 (the convection term) contributes to thermal load matrix

$$\underbrace{-2hR_jT_w\begin{bmatrix}0\\ 1\end{bmatrix}}_{f_h}. \quad (25)$$

2.6. Construct the Element Matrices to the Global Matrix. Assemble the global, conductance, capacitance, and thermal load matrices and the global of the unknown temperature matrix for all the elements in the domain, that is, the element's conductance, capacitance and thermal load matrices have been derived. Assembling these elements is necessary in all finite element analysis.

Constructing these elements will result into the following finite element equation:

$$[\mathbf{K}]^{(G)}\{\mathbf{T}\}^{(G)} + [\mathbf{C}]^{(G)}\{\dot{\mathbf{T}}\}^{(G)} = \{\mathbf{F}\}^{(G)}, \quad (26)$$

with $[\mathbf{K}] = [\mathbf{K_c}] + [\mathbf{K_h}]$. Conductance matrix due to conduction (elements 1 to 4) and heat loss through convection at the element's boundary (element 4 node 5) as shown in Figures 1 and 2:

$\{T\}$: Temperature value at each node, °C,

$[C]$: Capacitance matrix, due to transient equation (heat stored),

$\{\dot{T}\}$: Temperature rate for each node, °C/s,

$\{F\} = \{F_h\} + \{F_{\dot{q}}\}$: heat load due to heat loss through convection at the element's boundary (element 4 node 5) and internal heat generation (element 4 node 5).

2.7. Euler's Method. Two point recurrence formulas will allow us to compute the nodal temperatures as a function of time. In this paper, Euler's method which known as the backward difference scheme (BDS) will be used to determine the rate of change in temperature, the temperature history at any point (node) of the steel bar [3].

If the derivative of T with respect to time t is written in the backward direction and if the time step is not equal to zero ($\Delta t \neq 0$), we have that,

$$\{\dot{T}\} \approx \left\{\frac{T(t) - T(t-\Delta t)}{\Delta t}\right\}, \quad (27)$$

with $\dot{T}$ = temperature rate (°C/s); $T(t)$ = temperature at t s(°C); $T(t-\Delta t)$ = temperature at $(t-\Delta t)$ s, (°C) Δt = selected time step (s) and t = time (s) (at starting time, $t = 0$).

By substituting the value of $\{\dot{T}\}$ into the finite element global equation, we have that

$$[K]^{(G)}\{T(t)\}^{(G)} + [C]^{(G)}\left\{\frac{T(t) - T(t-\Delta t)}{\Delta t}\right\}^{(G)} = \{F(t)\}^{(G)}. \quad (28)$$

Finally, the matrices become

$$\left[[K]^{(G)}\Delta t + [C]^{(G)}\right]\{T\}_{i+1}^{(G)} = [C]^{(G)}\{T\}_{i}^{(G)} + \{F\}_{i+1}^{(G)}\Delta t. \quad (29)$$

From (29) all the right hand side is completely known at time t, including $t = 0$ for which the initial condition apply.

Therefore, the nodal temperature can be obtained for a subsequent time given the temperature for the preceding time.

Once the temperature history is known the important mechanical properties of the steel bar can be obtained such as hardness and strength.

3. Application

3.1. Calculation the Temperature History. The present developed mathematical model is programmed using MATLAB to simulate the results of the temperature distribution with respect to time in transient state heat transfer of the industrial quenched steel bar. The cylindrical chromium steel bar has been heated to 850°C. Then being quenched in water with $T_{\text{water}} = 32$°C, the water convection heat transfer coefficient, $h_{\text{water}} = 5000\ \text{W/m}^2 \cdot$ °C.

The temperature history at any point (node) of the cylindrical steel bar after quenched is being identified in Figures 6 and 7. The cylindrical bar was made from chromium steel 8650H, with properties as mentioned below.

Thermal capacity, ρc (J/m^3 · °C):

$0 \leq T \leq 650$°C, $\rho c = (0.004T + 3.3) \times 10^6$,

$650 < T \leq 725$°C, $\rho c = (0.068T - 38.3) \times 10^6$,

$725 < T \leq 800$°C, $\rho c = (-0.086T + 73.55) \times 10^6$,

$T > 800$°C, $\rho c = 7.55 \times 10^6$.

Thermal conductivity, k (W/m·°C)

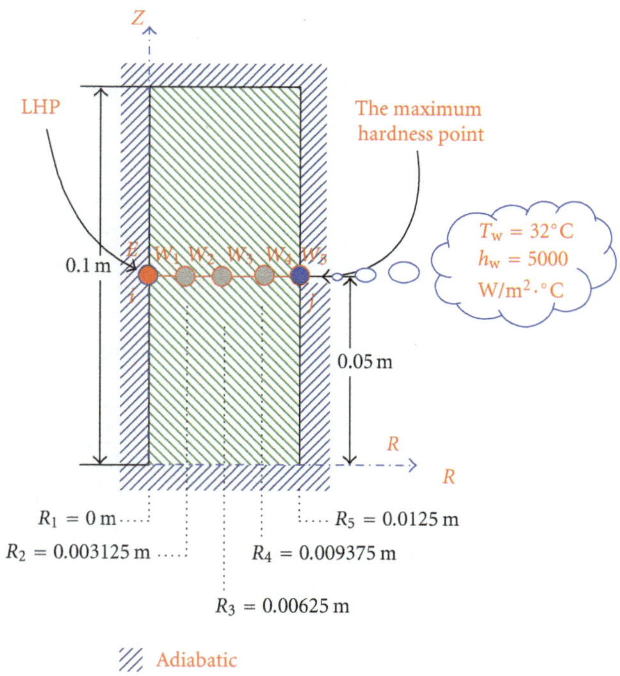

FIGURE 7: The axisymmetric one dimensional line (radius) element from the domain, the selected 4 elements with 5 nodes and the boundary at node $j(5)$ for an element 4.

$$0 \leq T \leq 900°\text{C},\ k = -0.022T + 48,$$
$$T > 900°\text{C},\ k = 28.2,$$

where in our case the global conductance matrix $[K]^{(G)}$, the global capacitance matrix $[C]^{(G)}$ and the global thermal load matrix $\{F\}^{(G)}$ can be computed easily as the following:

$$[K]^{(G)} = [K_c]^{(1)} + [K_c]^{(2)} + [K_c]^{(3)} + [K_c]^{(4)} + [K_h]^{(4)}, \tag{30a}$$

where $[K_c]^{(1)}$, $[K_c]^{(2)}$, $[K_c]^{(3)}$, $[K_c]^{(4)}$ are the conductance matrices due to conduction in 1-dimensional element for the 1st element, the 2nd, the 3rd, and the 4th element respectively while $[K_h]^{(4)}$ because we note that there is convection in element 4 at node $j(5)$ only as shown clearly in Figures 3 and 7:

$$[C]^{(G)} = [C]^{(1)} + [C]^{(2)} + [C]^{(3)} + [C]^{(4)}, \tag{30b}$$

where $[C]^{(1)}$, $[C]^{(2)}$, $[C]^{(3)}$, $[C]^{(4)}$ are the capacitance matrices due to transient [unsteady state] in 1-dimensional line (radius) element:

$$\{F\}^{(G)} = \{F_h\}^{(4)}. \tag{30c}$$

We have convection in element 4 at node $j(5)$ only as shown clearly in Figures 3 and 7.

With the input data and boundary conditions provided, a sensitivity analysis is carried out with the developed program to obtain the temperature distribution at any point (node) of the quenched steel bar, as example, is the transient state temperature distribution results of the selected five nodes from the center $[W_1]$ to the surface $[W_5]$ of the quenched steed bar which were computed as shown in Figures 7 and 8.

TABLE 1: Cooling time and rate of cooling for each node of 1-dimensional line (radius) element.

Node	t_c(s)	ROC (°C/s)
W_1	8.245	36.38568
W_2	8.138	36.864094
W_3	7.406	40.507696
W_4	5.311	56.486537
W_5	1.93	155.44041

3.2. LHP Calculation

3.2.1. Calculating the Cooling Time Required. In this study, we choose to calculate the cooling time between 800°C and 500°C [3, 7–9], where the characteristic cooling time, relevant for phase transformation in most structural steels is the time of cooling from 800 to 500°C (time $t_{8/5}$) [10–17]:

$$t_c = t_{800} - t_{500}. \tag{31}$$

From Figure 7 we can determine the time taken for node W_1 to reach 800°C,

$$t_{800°\text{C}} = 4.830 \text{ sec}. \tag{32}$$

By the same way the time taken for node W_1 to reach 500°C is $t_{500°\text{C}} = 13.075$ sec.

So the Cooling time t_c for node W_1;

$$t_c = t_{500°\text{C}} - t_{800°\text{C}} = 13.075 - 4.830 = 8.245 \text{ sec}. \tag{33}$$

For nodes W_2 to W_5, the cooling time t_c calculated by the same way, the final results shown in Table 1.

Table 1 shows the cooling time t_c and the rate of cooling ROC.

3.2.2. Calculating the Jominy Distance from Standard Jominy Distance versus Cooling Time Curve. Cooling time, t_c, obtained will now be substituted into the Jominy distance versus cooling time curve to get the correspondent Jominy distance. Jominy distance can also be calculated by using polynomial expressions via polynomial regression via Microsoft Excel.

The standard Table (cooling rate at each Jominy distance (Chandler, 1999)) can be used too [18].

Then Jominy distance of nodes W_1 to W_5 will be calculated, the final results shown in Table 2, where the rate of cooling, ROC, was defined by;

$$\text{ROC} = \frac{800\,°\text{C} - 500\,°\text{C}}{t_c} = \frac{800\,°\text{C} - 500\,°\text{C}}{t_{500\,°\text{C}} - t_{800\,°\text{C}}}\ (°\text{C/s}). \tag{34}$$

3.2.3. Predict the Hardness of the Quenched Steel Bar. The HRC of AISI-SAE 8650H can be calculated by using the relation between the J-Distance and the HRC from the Practical

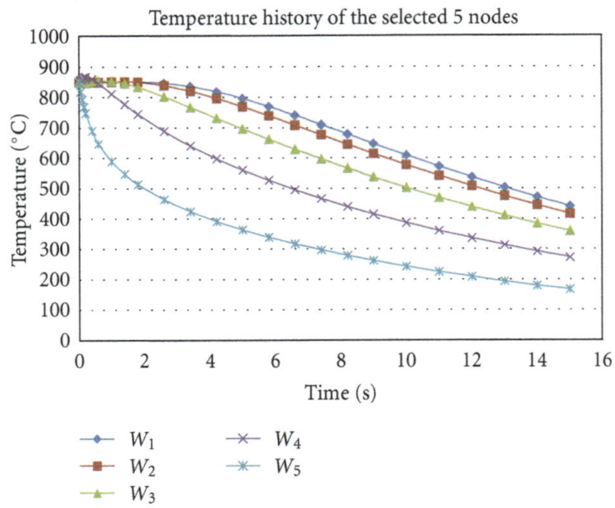

Figure 8: Graph of temperature history along WW cross-section from MATLAB program.

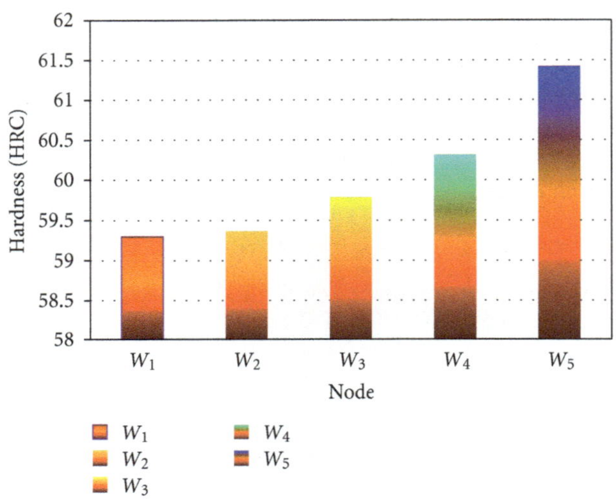

Figure 9: Hardness distribution along WW cross-section for the nodes W_1 to W_5 from the centre to the surface, respectively, at half the length at the centre of the quenched steel bar.

date Handbook, the Timken Company 1835 Duebex Avenue SW Canton, Ohio 44706-2798 1-800-223, the final results are shown in Figure 9 and Table 3.

4. Mathematical Model Verification

The same data input for the steel properties and boundary condition used in the mathematical model is applied to the ANSYS software to verify the temperature simulation results. The temperature distribution from the ANSYS analysis is depicted figuratively as shown in Figures 10(a) and 10(b).

The temperature time graph from the ANSYS analysis is depicted as shown in Figure 11.

From the graphs shown in Figure 8 by mathematical model and Figure 11 by ANSYS, it can be clearly seen that the temperature history of the quenched steel bar have

Table 2: Cooling time, Cooling rate, and Jominy distance for the nodes W_1 to W_5.

Node	t_c(s)	ROC (°C/s)	Jominy distance (mm)
W_1	8.245	36.38568	8.685
W_2	8.138	36.864094	8.621
W_3	7.406	40.507696	8.179
W_4	5.311	56.486537	6.935
W_5	1.93	155.44041	3.431

Table 3: Cooling time, cooling rate, Jominy distance, and HRC for the nodes W_1 to W_5, water cooled.

Node	t_c(s)	ROC (°C/s)	Jominy distance (mm)	Hardness (HRC)
W_1	8.245	36.38568	8.685	59.297
W_2	8.138	36.864094	8.621	59.359
W_3	7.406	40.507696	8.179	59.775
W_4	5.311	56.486537	6.935	60.316
W_5	1.93	155.44041	3.431	61.421

Table 4: Cooling time, Cooling rate, Jominy distance, and HRC for the nodes A_1 to A_5, water cooled by ANSYS.

Node	Cooling time	Cooling rate	J distance (mm)	HRC
A_1	9.7	30.87574	9.556	58.492
A_2	9.589	31.284	9.472	58.518
A_3	9.055	33.130	9.174	58.613
A_4	5.905	50.797	7.286	60.208
A_5	*2.180*	*137.580*	*3.828*	*61.295*

the same pattern. The heat transfer across the steel bar is uniform. From Figure 11 the cooling time, Jominy-distance, and consequently the hardness of the quenched chromium steel bar at any point (node), even the lowest hardness point (LHP) is determined by ANSYS too, the final results shown in Table 4 and Figure 12.

From our results we found that in the mathematical model for the 1st node with W_1 in the center, we found that HRC = 59.297. While in ANSYS for the same node A_1, we found that HRC = 58.492.

And for the nodes on the surfaces W_5 and A_5, it was found that HRC = 61.421 and 61.295 for the mathematical model and ANSYS, respectively. From the above, it can be seen that there is a strong agreement between both results. The difference between all the results of the mathematical model and the ANSYS simulations can be accounted due to the fact that the ANSYS software is commercial purpose, and thereby has some automated input data. But the developed mathematical model is precisely for a circular steel bar axisymmetric cross section. However, there is strong agreement between both results and thereby the result is validated, where the comparison indicated reliability of the proposed model.

Also the results showed that the node on the surface will be the 1st which completely cooled after quenching because

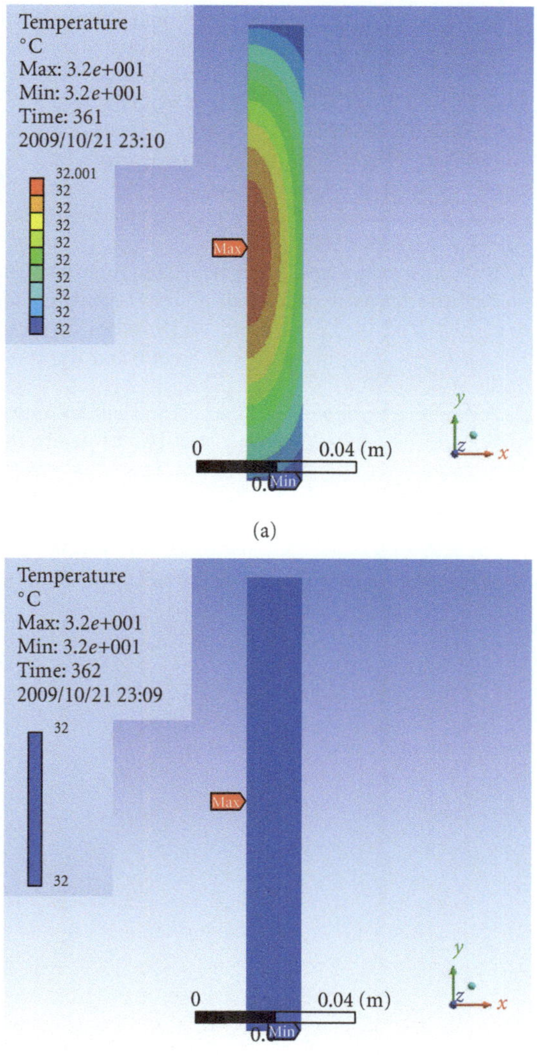

Figure 10: (a) shows the temperature distribution just before the steel bar becomes completely cooled and (b) shows the temperature distribution at the moment that the entire steel bar becomes completely cooled after 362 s.

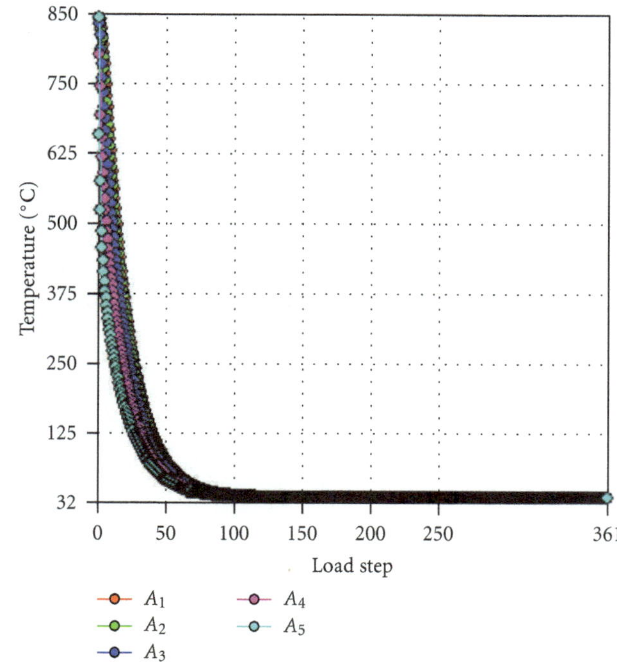

Figure 11: Temperature-time graph from ANSYS.

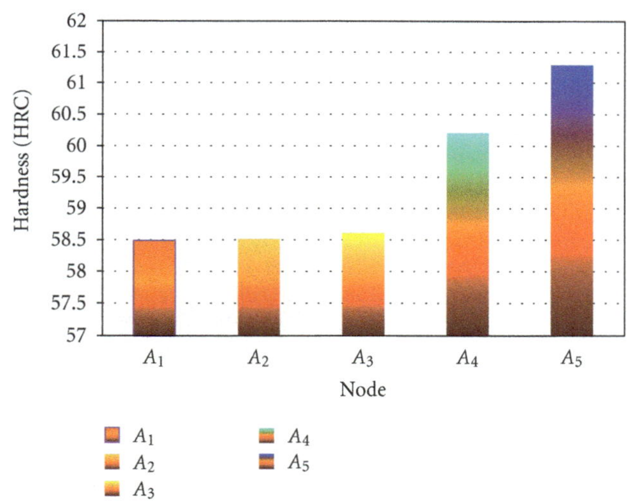

Figure 12: Hardness distribution by ANSYS along AA cross-section for the nodes A_1 to A_5 from the centre to the surface, respectively, at half the length at the centre of the quenched steel bar.

it is in the contact with the cooling medium then the other nodes on the radial axis to the centre, respectively, and the last point will be completely cooled after quenching will be at half the length at the centre. Hence LHP will be at half the length at the centre of the quenched industrial chromium steel bar. It will be more important to know LHP once the radius of the quenched steel bar is large because LHP will be low, in other words, it will be lower than the hardness on the surface, that means increasing the radius of the bar inversely proportional with LHP.

LHP calculation experimentally is an almost impossible task using manual calculation techniques also the earlier methods only used hardness calculated at the surface, which is higher than LHP, which has negative consequence that can result to the deformation and failure of the component.

5. Conclusion

A mathematical model of steel quenching has been developed to compute LHP of the quenched chromium steel bar at any point (node) in a specimen with cylindrical geometry. The model is based on the finite element Galerkin residual method. The numerical simulation of quenching consisted of numerical simulation of temperature transient field of cooling process. This mathematical model was verified and validated by comparing the hardness results with ANSYS software simulations. From the mathematical model and ANSYS results, it is clear that the nodes on the surface (W_5

and A_5), respectively, cools faster than the nodes on the center (W_1 and A_1) because $t_{cW_5} < t_{cW_1}$ and $t_{cA_5} < t_{cA_1}$, this means that the mechanical properties will be different such as hardness where the hardness on the surface nodes (W_5 and A_5) will be higher than the hardness on the center nodes (W_1 and A_1).

Acknowledgments

The authors would like to thank the University Tun Hussein Onn Malaysia for supporting this research under the Science Fund Grant. The corresponding author grateful to the Postgraduate Centre of UTHM for their support of this research, where they accepted him under the university Ph.D. scholarship.

References

[1] K. N. Robert, *Quenching and Tempering of Welded Steel Tubular*, The Fabricator Articles, July 2001.
[2] B. Omar, M. Elshayb, and A. S. A. Elmaryami, "Unsteady state thermal Behavior of industrial quenched steel bar," in *Proceedings of the 18th World IMACS Congress and International Congress on Modelling and Simulation (MODSIM '09)*, Cairns, Australia, 2009.
[3] E. Mohamed and Y. K. Bing, *Application of Finite Difference and Finite Element Methods*, University Malaysia Sabah Kota Kinabalu, Sabah, Malaysia, 2000.
[4] A. S. A. Elmaryami and B. Omar, "Developing of unsteady state axi-symmetric FEMM to predict the temperatureof industrial quenched steel," *Journal of Metals Science and Heat Treatment*. In press.
[5] K. G. Budinski, *Engineering Material: Properties and Selection*, Prentice-Hall, Upper Saddle River, NJ, USA, 4th edition, 1992.
[6] A. S. A. Elmaryami, *Heat treatment of steel by developing finite element mathematical model and by simulation*, M.S. thesis, University Tun Hussein Onn, Johor, Malaysia, 2010.
[7] A. S. A. Elmaryami, S. B. Haji Hasan, B. Omar, and M. Elshayeb, "Customer reviews unsteady state hardness prediction of industrial quenched steel bar [One and Two Dimensional]," in *Proceedings of the Materials Science & Technology Conference and Exhibition*, Pittsburgh, Pa, USA, 2009.
[8] S. Moaveni, *Finite Element Analysis; Theory and Application with ANSYS*, Pearson Education International, 2008.
[9] A. Rose et al., *Atlas zur Wärmebe- Handlung der Stähle I*, Verlag Stahleisen, Düsseldorf, Germany, 1958.
[10] B. Smoljan, "Mathematical modelling of austenite decomposition during the quenching," in *Proceedings of the 13th International Scientific Conference Achievements in Mechanical and Materials Engineering*, The Polish Academy of Science, 2005.
[11] B. Smoljan, "Prediction of mechanical properties and microstructure distribution of quenched and tempered steel shaft," *Journal of Materials Processing Technology*, vol. 175, no. 1–3, pp. 393–397, 2006.
[12] B. Donnay, J. C. Herman, V. Leroy, U. Lotter, R. Grossterlinden, and H. Pircher, *Microstructure Evolution of C-Mn Steel in the Hot Deformation Process: The Stripcam Model.*
[13] B. Liščić, "System for process analysis and hardness prediction when quenching axially-symmetrical workpieces of any shape in liquid quenchants," *Materials Science Forum*, vol. 638-642, pp. 3966–3974, 2010.
[14] B. Smoljan, D. Iljkić, and S. Smokvina Hanza, "Computer simulation of working stress of heat treated steel specimen," *Journal of Achievements in Materials and Manufacturing Engineering*, vol. 34, no. 2, pp. 152–156, 2009.
[15] R. I. Hsieh, H. Y. Liou, and Y. T. Pan, "Effects of cooling time and alloying elements on the microstructure of the Gleeble-simulated heat-affected zone of 22% Cr duplex stainless steels," *Journal of Materials Engineering and Performance*, vol. 10, no. 5, pp. 526–536, 2001.
[16] A. S. A. Elmaryami and B. Omar, "Computer simulation to predict the hardness of transient axi-symmetric industrial quenched steel bar at different radial axises," *International Journal of Emerging Technology in Science and Engineering*. In press.
[17] A. S. A. Elmaryami and B. Omar, "Transient computer simulation of industrial quenched steel bar to determine LHP of molybdenum and boron steel at 850°C as austenitizing temperature quenched in different medium," *International Journal of Material Science*. In press.
[18] H. Chandler, *Hardness Testing Applications*, Hardness Testing, ASM International, Indianapolis, Ind, USA, 2nd edition, 1999.

14

Mould Filling Ability Characterisation of Cast Aluminium Alloys Using Design of Experiments

**Samavedam Santhi,[1] S. B. Sakri,[2]
Dharwada Hanumantha Rao,[3] and Srinivasan Sundarrajan[4]**

[1] *Department of Metallurgical and Materials Engineering, MGIT, Hyderabad 500075, India*
[2] *Special Fabrication Division, DRDL, Hyderabad 500058, India*
[3] *Mechanical Engineering Department, M.V.S.R. Engineering College, Hyderabad 501510, India*
[4] *National Institute of Technology, Trichy 620015, India*

Correspondence should be addressed to Samavedam Santhi, santhi_samave@yahoo.com

Academic Editor: Brij Kumar Dhindaw

Mould filling ability characteristic has been influencing the liquid metal flow, heat transfer, and solidification there by affecting the quality of aerospace castings. Mould filling is a critical parameter in the production of sound and quality castings, especially in the case of complex-shaped castings where section thickness is varying considerably. The mould filling ability of aluminium alloys LM6 and LM25 has been studied in the present investigation. Experimentation has been carried out using orthogonal array experimental layout. The process parameters studied in the present investigation are alloy composition, sand fineness, mould coating, and pouring temperature. The mould filling ability of selected aluminium alloy has been studied using pin test piece with cylindrical cores. The results from experimentation are analyzed to find the influence of the process parameters on mould filling ability. Based on the above, LM6 alloy has been found to have better mould filling ability characteristics and the analysis of variance has also revealed the same optimum factor combination.

1. Introduction

Metal casting is unique among metal forming processes for a variety of reasons [1]. Aluminum castings have played an integral role in the growth of the aluminum industry since its inception in the late 19th century. Popular aluminum casting alloys contain, in addition to strengthening elements, sufficient amounts of eutectic-forming elements (usually silicon) in order to have adequate fluidity to make the metal flow through the cavities. The aluminum castings and applications rapidly expanded to address the requirements of a wide range of engineering specifications and continuing market requirement to reduce vehicle weight and to increase fuel efficiency. Complicated physics involved in the casting process is fluid mechanics with phase change, shrinkage, porosity, macrosegregation in alloys, heat transfer between metal and mould, and thermal stress in the solidifying ingot [2]. Thin-wall aluminium alloy casting techniques can be used to produce clean castings of lower weight and with superior mechanical and physical properties compared to those produced by more complex traditional methods and are ideal for automotive applications [3]. The filling of the mould is an essential part of the entire casting cycle. Mould-filling ability is the bottleneck of manufacturing thin-walled castings [4].

The solidification of the melt during the mould filling continuously changes the properties of liquid and solid phases. The filling conditions play a significant role on the casting quality. A good mould-filling metal has the ability to fill out the mould pattern and reproduce fine details of the mould [5, 6]. Many defects in a casting generally originate at the filling stage [6].

The gating design is driven by the ideal mould filling time, which depends on the cast metal, casting weight, and minimum wall thickness [7]. Fast filling leads to turbulence-related defects (such as mould erosion, air aspiration, and

Table 1: Chemical composition (% wt).

Alloy	Si	Fe	Cu	Mn	Mg	Ni	Zn	Pb	Sn	Ti	Each	Total	Al
LM6	10.5–13.5	0.65	0.15	0.55	0.1	0.1	0.15	0.1	—	0.2	0.05	0.15	Rem
LM25	6.5-7.5	0.55	0.2	0.35	0.2–0.65	0.15	0.15	0.15	0.05	0.05–0.25	0.05	0.15	Rem

inclusions). On the other hand, slow filling may cause defects related to premature solidification (such as cold shuts and misruns).

The integrity of a casting will be degraded substantially if high levels of free surface turbulence are experienced by the liquid metal during mould filling [8, 9]. This is due to the entrapment and breakup of surface oxide film formed on the surface of liquid metal. This oxide film becomes trapped in the casting and behaves like a crack. This generally appears when large flat castings filled horizontally [10].

Mould filling is strongly dependent on the surface tension and viscosity of the metal. It is affected by the mould characteristic, melt composition, and melt cleanliness. The mould filling process that closely related to molten metal flow, heat transfer, and mass transfer process is a flow process at changeable temperature accompanied with heat loss and solidification [4]. Factors affecting mould filling generally are alloy composition, mould, sand fineness, wall thickness, mould coat [7], gating system [7, 11], pouring issue (pouring temperature), and filters [12].

Taguchi's methods belong to the class of approaches that attempt to ensure quality through design. The design of experiments using Taguchi's approach can economically satisfy the needs of problem solving and product/process design optimization projects in the manufacturing industry. Multiple process variables can be studied at the same time with Taguchi's orthogonal array-efficient designs, instead of in a hit-and-miss approach, providing very reproducible results. Due to the statistical balance of the designs, more potential combinations of process variables (at different levels) can be evaluated for the best overall combination, in small number of experiments.

Orthogonal array technique allowed many factors to be analysed with limited number of experiments. Orthogonal array of design of experiments makes the design to be balanced so that factor levels are weighted equally [5]. Because of this, each factor can be evaluated independently of all the other factors, so the effect of one factor does not influence the estimation of another factor. MINITAB software is used for experimental design [13, 14]. Orthogonal arrays provide a set of well-balanced (minimum) experiments.

2. Experimental Plan

Designing of experiments is an efficient approach for improving a process. Changing more than one factor at a time to quickly obtain meaningful results and draw conclusions about how factors interact is the primary objective of the design of experiments. A full scale of experimentation requires 16 experiments which have been reduced to 8 experiments using orthogonal array design of experiments.

Table 2: Factors and their levels.

	Factor 1 Alloy	Factor 2 Sand fineness (AFS No.)	Factor 3 Mould coat	Factor 4 Pouring temperature (°C)
Level 1	LM25	25	Graphite	T
Level 2	LM6	40	No coating	$T + 20°C$

Table 3: L8 Orthogonal array.

Exp. run order	Alloy	Sand fineness (AFS no.)	Mould coat	Pouring temperature (°C)
1	LM25	25	Graphite (GC)	T
2	LM25	25	No coating (NC)	$T + 20$
3	LM25	40	Graphite (GC)	T
4	LM25	40	No coating (NC)	$T + 20$
5	LM6	25	No coating (NC)	T
6	LM6	25	Graphite (GC)	$T + 20$
7	LM6	40	Graphite (GC)	$T + 20$
8	LM6	40	No coating (NC)	T

For studying mould filling ability, the influencing parameters selected are alloy composition, sand fineness, mould coat, and pouring temperature.

Alloy composition affects the fluidity, filling characteristics, mechanical properties, and structure of the cast product. To study the influence of alloy composition, two alloys are taken for the present study; these are LM6 and LM25. Table 1 gives the chemical composition of the two alloys.

Another processing parameter is fineness of the sand which shows significant difference in the casting characteristics like surface finish and strength. AFS sand fineness numbers 25 and 40 are considered.

The third processing parameter is mould coat [5]. Generally mould coat provides smooth surface. Pouring temperature [12] influences fluidity, porosity, strength, and structure of the casting. Hence, pouring temperature with 20°C of super heat is considered as one of the process parameters for the present study.

Table 2 shows the details of the factors and their levels for the present study. Orthogonal array L8 (2**4) is used with four factors and 2 levels (for 8 runs) for experimental investigation as shown in Table 3.

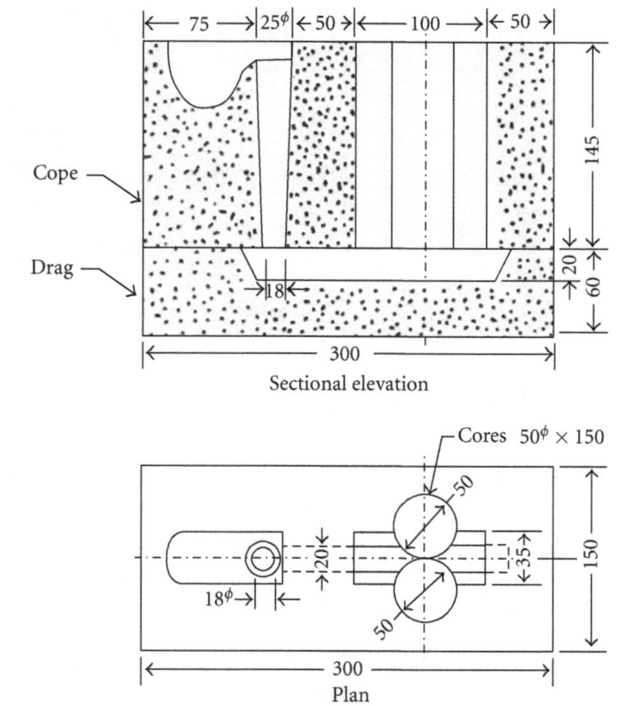

Figure 1: Schematic diagram of mould used for experimentation.

3. Experimental Details

The mould filling ability of these alloys has been studied using pin test piece with cylindrical cores designed by Engler and Ellerbrok as shown in Figure 1. 50 mm diameter cylindrical cores are considered for the present study. The test casting consists of two fins of metal at both sides. The inverse of the diameter of curvature of the edge tip of the fin gives the value of the mould filling ability. The mould filling ability pin test piece is based on the determination of the diameter of meniscus of liquid metal just at the time of solidification in the test casting. It is possible to quantitatively determine the values of mould filling ability at various pressure heads using the pin test piece. The mould filling ability as the percentage length of the spiral which has successfully reproduced the contour to the total length gives only empirical values.

3.1. Preparation of Green Sand Mould. Testing arrangement for the mould filling ability is shown in Figure 2. Tooling for experiments and assembled mould for this experiment are given in Figure 2. The mould filling pattern and cylindrical cores are provided in the cope box and it is shown in Figure 2(a). The cope box, drag, and cylindrical cores are given in Figure 2(b). Top view of the assembled mould for mould filling ability experiments is shown in Figure 2(c).

The moulds are provided with dowel pins for perfect matching of cope and drag. Moulds are prepared with slight ramming. The patterns have been stripped after 3 hours.

Moulds are prepared using green sand process. The composition consists of Bentonite 5-6% of sand weight and water is 5–8% of sand weight. In case of mould coatings the graphite paint is sprayed on to the mould and mould is dried immediately by lightening a flame on the painted surface.

3.2. Alloys. Alloys used in the present investigation are LM6 (Al-12Si) and LM25 (Al-7Si-0.4 Mg). Each alloy melting done with their respective same batch (belong to same heat) for all the 8 castings.

3.3. Melting and Pouring. The alloys are melted in an electric resistance furnace of capacity 20 Kg provided with mild steel crucible. Temperature is measured with the help of a thermocouple. The furnace is put off and the crucible is lifted and put in a tilting device. The metal is tapped into a smaller crucible for pouring into the mould.

3.4. Process Variables Studied. The variables studied are alloy composition, sand fineness, mould coat, and pouring temperature.

4. Analysis of Results

Metal filling is an important phase of the casting process. The ability to fill and the contours play a significant role in reproducing the intricate details of the casting. The mould filling ability depends on the metallostatic balance between the metal pressure head and the surface tension of the mould. H is the height of liquid metal column above the filling point. A higher metallostatic pressure gives higher velocity of molten metal and thereby higher fluidity. The casting fluidity and filling are driven by metallostatic pressure head.

The test casting obtained for experimental run order 7 is shown in Figure 3.

Using this test piece the mould filling ability value can be determined at various pressure heads (Hs) as shown in Figure 3. Pressure head is the height of liquid metal column above the filling point. The casting consists of two fins of metal at both sides.

The hot metal poured into the sprue fills the curved cavity between the two cylindrical cores having a line contact at the center. But it solidifies before it fills up the complete casting. The inverse of the diameter of curvature of the edge tip of the fin gives the value of the mould filling ability. The diameter at the tip of the fin gives the meniscus diameter of the liquid metal at the time of solidification as represented in Figure 4 for one particular pressure head.

It is difficult to measure the diameter of the tip of the edge and hence an indirect way of calculation has been used. As per Figure 5,

$$\begin{gathered} R^2 + (r + x)^2 = (r + R)^2, \\ \text{where } R = \text{radius of the sand core, mm,} \\ R = 25\text{ mm (radius of cylindrical core).} \\ r = \text{radius of the meniscus } (2r = d)\text{, mm,} \\ 2x = \text{distance between edges, mm,} \\ \text{so } \frac{1}{d} = \frac{(R - x)}{x^2}, \\ \frac{1}{d} = \text{mould filling ability, } \frac{1}{\text{mm}}, \end{gathered} \tag{1}$$

(a) Mould filling pattern and cylindrical cores (b) Cope box, drag, and cylindrical cores (c) Assembled mould for mould filling ability

FIGURE 2: Tooling for mould filling experiments and assembled mould for this experiment.

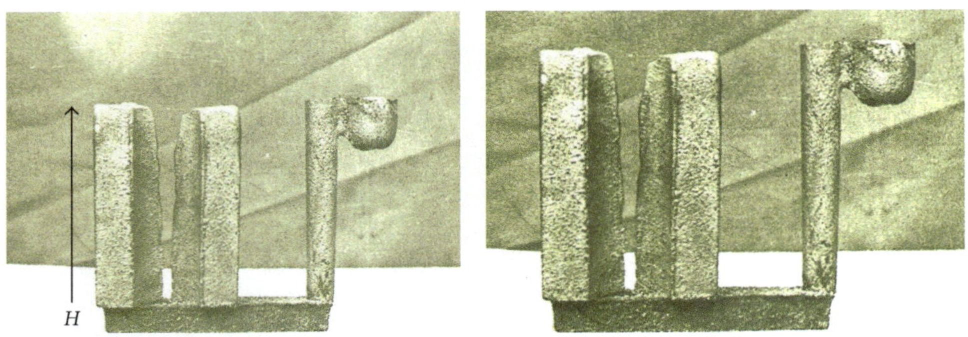

(a) (b)

FIGURE 3: Mould filling ability test casting.

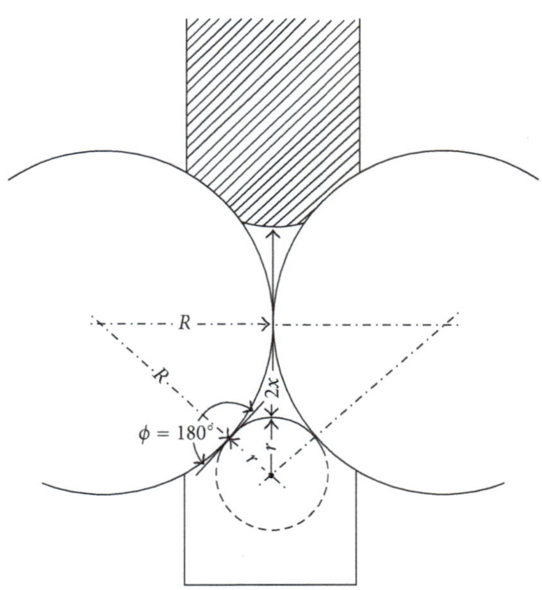

FIGURE 4: Measurement of mould filling ability.

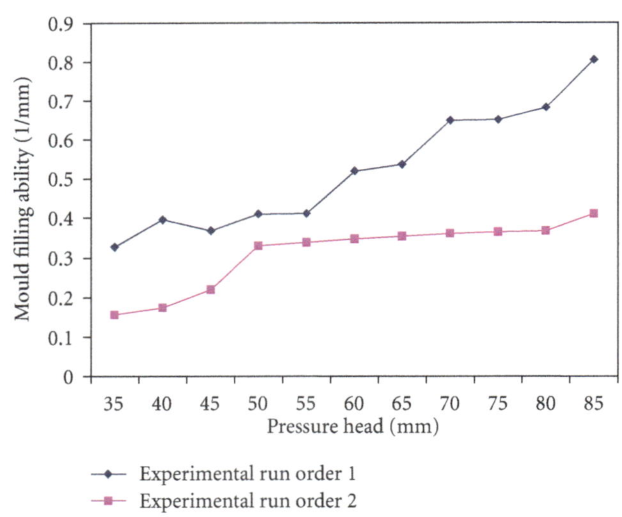

FIGURE 5: Influence of pressure head on mould filling ability.

Height gauge has been used to mark the height at every interval of 5 mm for each solidified casting. The distance between the fin edges, $2x$, is measured with vernier micrometer for each 5 mm interval. The $2x$ values for 8 experiments are given at Table 4. The x values are substituted in (1) and the mould filling ability calculated for all the 8 experiments. The mould-filling ability values have been calculated for every 5 mm increment by using (1).

The mould filling ability values at various pressure heads for experimental run order 8 are given in Table 5. The mould

TABLE 4: The distance between the fin edges, $2x$ in mm.

H, mm	Exp. run order 1	Exp. run order 2	Exp run order 3	Exp run order 4	Exp run order 5	Exp run order 6	Exp run order 7	Exp run order 8
35	5.4		7		9.6	9.759	10.6	7.5
40	5.9	3.8	7.3	3.5	9.8	9.79	11	8
45	5.7	4	7.4	4.5	10	9.988	11.1	8.1
50	6	4.5	7.5	4.6	10.1	10.252	11.2	8.2
55	6	5.5	7.8	5	10.2	10.35	11.8	8.3
60	6.7	5.5	7.9	5.5	10.5	10.45	12	8.5
65	6.8	5.55	8	6	10.55	10.65	12.5	8.5
70	7.6	5.6	8.5	6.5	10.6	10.75	13	8.5
75	7.3	5.65	8.5	6.8	10.7	10.85	13.5	8.6
80	7.6	5.68	9	6.9	10.9	10.98	13.6	8.7
85	8.2	5.7	10	7	11	11.5	14	9
90	7.3	6	11	7.1		11.55	14.1	10

TABLE 5: Mould filling ability calculations.

S. no.	H, mm	$2x$, mm	$1/d$ mm^{-1}, $[(R-x)/x^2]$
1	0	5	0.278
2	5	5.5	0.34
3	10	6	0.408
4	15	6.2	0.438
5	20	6.5	0.485
6	25	6.8	0.536
7	30	7	0.569
8	35	7.5	0.662
9	40	8	0.762
10	45	8.1	0.783
11	50	8.2	0.806
12	55	8.3	0.826
13	60	8.5	0.871
14	65	8.5	0.871
15	70	8.5	0.871
16	75	8.6	0.892
17	80	8.7	0.9165
18	85	9	0.99
19	90	10	1.25

filling ability for 8 experimental run orders has been given in Table 6.

4.1. Influence of Pressure Head on Mould Filling Ability of Aluminium Alloys. The mould filling ability is expressed as the reciprocal of the diameter of the meniscus of fin edge in the solidified test casting. It is expressed as a curve showing its variation with respect to pressure head (H) as measured on the test casting. So in present study the mould filling ability can be expressed as a curve showing its variation with pressure head. The mould filling ability values have been calculated for every 5 mm increment by using (1). The mould filling ability values at various pressure heads for experimental run orders 1 to 4 are calculated and plotted (Figure 5) showing that effect of pressure head on mould filling ability of LM25. This indicates that the mould filling ability increases with increase in pressure head.

The mould filling ability expressed as reciprocal of the diameter of the meniscus of fin edge in the solidified test casting. The mould filling ability is also plotted as a curve showing its variation with pressure head. Pressure head plays a very important role in the mould filling ability of metals. Experiments have been conducted to calculate mould filling ability of Aluminium alloys (LM6 and LM25). The mould filling ability increases with increase in pressure head. A higher-pressure head gives higher velocity of molten metal and better filling. Increase in pressure head leads to increase in the metallostatic force, allowing the liquid to enter the finer contours between the cylindrical cores thereby reducing the fin edge and increasing the mould filling ability values.

If the moulds were at the same temperature as of the molten metal during pouring, the mould filling ability follows the equation given below:

$$\frac{1}{d} = \frac{h.\rho.g}{2\sigma \cdot \cos\varphi},$$

$$\frac{1}{d} = \text{mould filling ability, 1/mm},$$

$$h = \text{pressure head, mm},$$

$$\rho = \text{density, gm/cc}, \tag{2}$$

$$g = \text{acceleration due to gravity, m/sec}^2,$$

$$\sigma = \text{surface tension, dynes/cm},$$

$$\varphi = \text{angle of contact}.$$

The factors which influence the mould filling ability in liquid region are surface tension and contact angle. The factors which influence the mould filling ability in solidifying region are cooling conditions. In the present studies the experiments have been conducted in solidifying region. So the mould filling ability depends on pressure head while the surface

Table 6: The mould-filling ability for 8 experimental run orders.

H, mm	Exp. run order 1	Exp. run order 2	Exp. run order 3	Exp. run order 4	Exp. run order 5	Exp. run order 6	Exp. run order 7	Exp. run order 8
35	0.326906	0.14784	0.569767	0.124	1.14	1.1833	1.425888	0.6617
40	0.394671	0.156277	0.624005	0.131	1.19	1.191794	1.551282	0.7619
45	0.366704	0.17391	0.642723	0.22	0.125125	1.2466	1.583676	0.7829
50	0.409091	0.22	0.661765	0.233	1.2783	1.322	1.616495	0.8043
55	0.41	0.33	0.720853	0.2777	1.307	1.35	1.822513	0.826
60	0.51836	0.338	0.741211	0.33988	1.39577	1.38	1.894737	0.86
65	0.535185	0.3464	0.761905	0.409	1.41	1.44	2.083333	0.87
70	0.648113	0.3531	0.870482	0.48565	1.42588	1.47	2.283784	0.88
75	0.65	0.3598	0.881815	0.535	1.4566	1.503	2.496575	0.89
80	0.681132	0.3639	0.987805	0.55	1.5193	1.544	2.540659	0.9136
85	0.8043	0.3666	1.25	0.5697	1.5512	1.7175	2.722222	0.98
90	0.826	0.4091	1.551282	0.587529	1.583676	1.735	2.768942	1.25

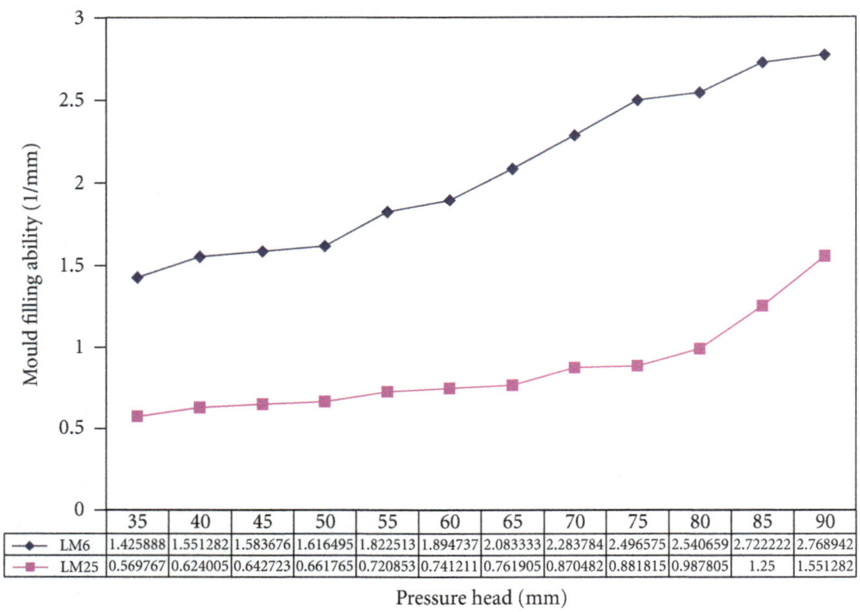

Figure 6: Influence of silicon on mould filling ability of LM6 and LM25.

tension and density are neglected and an angle of contact of 180° is assumed. Higher pressure heads lead to higher mould filling ability because the metal has to be in contact with larger area in order to penetrate into cavity which promotes faster solidification.

4.2. Influence of Silicon Content on Mould Filling Ability. The mould filling ability increases with increase in silicon content in the alloy. LM6 is the eutectic alloy and it contains more silicon content of 12 weight percentage. Higher the silicon content higher will be the fluidity. Silicon additions improve casting characteristics by improving fluidity, feeding, and hot tear resistance.

This has been shown in Figure 6. LM6 having better mould filling ability than LM25 alloy attributes mainly to the higher silicon content in LM6 alloy. The mould filling ability values corresponding to Figure 6 have been calculated using (1). The mould filling ability values at various pressure heads for LM6 and LM25 are plotted.

4.3. Influence of Pouring Temperature on Mould Filling Ability. Pouring temperature affects the mould filling ability. Mould filling ability value increases with increase in pouring temperature. From the present experiments, it has been shown in Figure 7. Increase in pouring temperature increases the heat content of the alloy, resulting in alloy being liquid for longer duration. This promotes the metallostatic pressure head to act on the liquid metal for a longer period of time, thus forcing the metal to easily enter the fine contours.

4.4. Influence of Sand Fineness on Mould Filling Ability. Sand fineness plays a very important role in the moulding process.

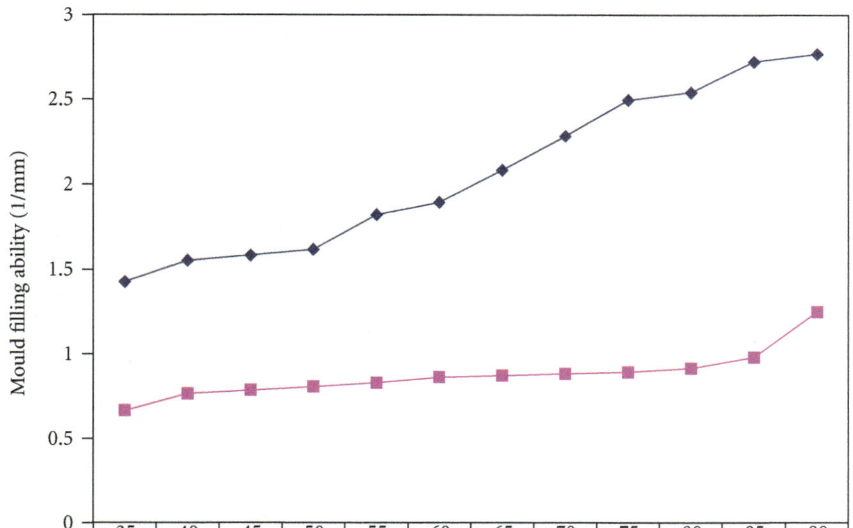

Pressure head (mm)	35	40	45	50	55	60	65	70	75	80	85	90
LM6 $T + 20$	1.425888	1.551282	1.583676	1.616495	1.822513	1.894737	2.083333	2.283784	2.496575	2.540659	2.722222	2.768942
LM6 T	0.6617	0.7619	0.7829	0.8043	0.826	0.86	0.87	0.88	0.89	0.9136	0.98	1.25

FIGURE 7: Influence of pouring temperature on mould filling ability of LM6.

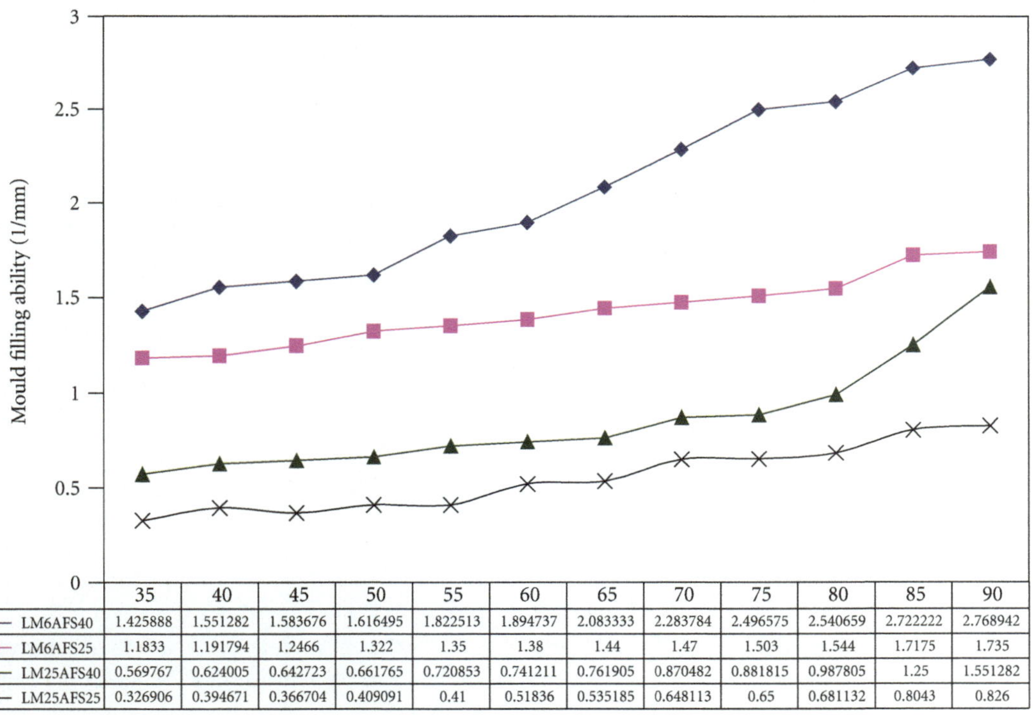

Pressure head (mm)	35	40	45	50	55	60	65	70	75	80	85	90
LM6AFS40	1.425888	1.551282	1.583676	1.616495	1.822513	1.894737	2.083333	2.283784	2.496575	2.540659	2.722222	2.768942
LM6AFS25	1.1833	1.191794	1.2466	1.322	1.35	1.38	1.44	1.47	1.503	1.544	1.7175	1.735
LM25AFS40	0.569767	0.624005	0.642723	0.661765	0.720853	0.741211	0.761905	0.870482	0.881815	0.987805	1.25	1.551282
LM25AFS25	0.326906	0.394671	0.366704	0.409091	0.41	0.51836	0.535185	0.648113	0.65	0.681132	0.8043	0.826

FIGURE 8: Influence of fineness of sand on mould filling ability of LM6 and LM25.

In the following experiments, the fineness of sand being used for cylindrical cores varies from fine sand and to coarse sand. Experimental results show that the mould filling ability increases with increase in sand fineness from Figure 8. The coarse sand is having more chilling power and solidifies the metal faster. This restricts the filling up of the cavity contours between the cylindrical sand cores and reduces the mould filling ability.

4.5. Influence of Mould Coating on Mould Filling Ability. Mould coating plays a significant role in the moulding process. Experiments have been conducted with graphite coating

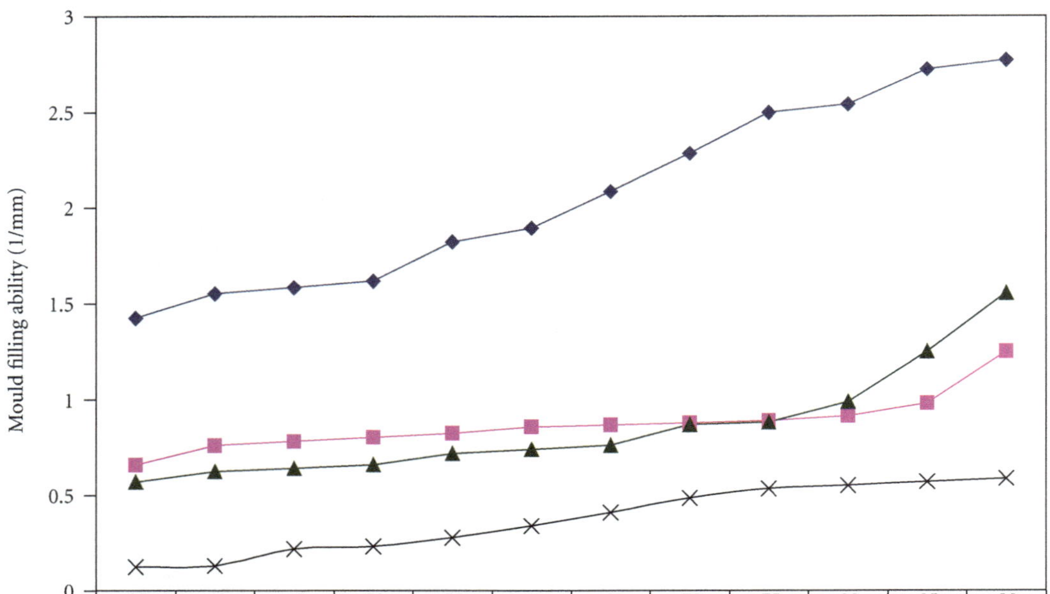

	35	40	45	50	55	60	65	70	75	80	85	90
LM6GC	1.425888	1.551282	1.583676	1.616495	1.822513	1.894737	2.083333	2.283784	2.496575	2.540659	2.722222	2.768942
LM6NC	0.6617	0.7619	0.7829	0.8043	0.826	0.86	0.87	0.88	0.89	0.9136	0.98	1.25
LM25GC	0.569767	0.624005	0.642723	0.661765	0.720853	0.741211	0.761905	0.870482	0.881815	0.987805	1.25	1.551282
LM25NC	0.124	0.131	0.22	0.233	0.2777	0.33988	0.409	0.48565	0.535	0.55	0.5697	0.587529

Pressure head (mm)

FIGURE 9: Influence of graphite coating on mould filling ability of LM6 and LM25.

TABLE 7: Response table.

Exp. run order	Alloy	Sand fineness (AFS no.)	Mould coat	Pouring temperature (°C)	1/d mm^{-1}
1	LM25	25	Graphite (GC)	T	0.624005
2	LM25	25	No coating (NC)	$T + 20$	0.40909
3	LM25	40	Graphite (GC)	T	1.55
4	LM25	40	No coating (NC)	$T + 20$	0.58753
5	LM6	25	No coating (NC)	T	1.55
6	LM6	25	Graphite (GC)	$T + 20$	1.73453
7	LM6	40	Graphite (GC)	$T + 20$	2.7689
8	LM6	40	No coating (NC)	T	1.25

on the mould. Mould coating improves the mould filling ability. This is by creating a smooth mould metal interface which provides less resistance to metal filling the contours. Results are shown in Figure 9.

4.6. Analysis of Variance (ANOVA). To study the characteristics of mould filling ability, it is necessary to study, all the process parameters together. In the present study Taguchi orthogonal array L8 (2**4) is used with four factors and 2 levels (for 8 runs) as indicated in Table 2 and response values are given in Table 7. The interactions selected for this study are

(i) alloy and sand fineness (AFS no.),

(ii) alloy and mould coat,

(iii) alloy and pouring temperature.

ANOVA values for mould filling ability are calculated using MINITAB and it is given at Table 8.

P is used to determine whether a factor is significant, typically to compare against an alpha (α) value of 0.1. Alpha is frequently referred to as the level of significance and is expressed as a probability ranging between 0 and 1. If the P value is lower than 0.1, then the factor is significant. Choosing a larger value for α is better for aerospace applications. The analysis of variance indicates that the alloy factor is considered more significant in influencing mould filling ability. It is seen that the alloy content and pouring temperature interaction contribute more to mould filling ability value.

Use of main effects' plot in conjunction with an analysis of variance (ANOVA) is useful when there have been several factors. The effect is the mean for that level minus the overall

Table 8: ANOVA.

Source	Degrees of freedom	Sum of squares	Mean squares	Factor	P
Alloy	1	2.13501	2.13501	8.33	0.045
AFS no.	1	0.42265	0.42265	1.02	0.370
Pouring temperature	1	0.03459	0.03459	0.14	0.732
Mould coat	1	1.03739	1.03739	4.05	0.114
Alloy * sand fineness	1	0.01712	0.01712	0.04	0.849
Alloy * mould coat	1	0.03459	0.03459	0.14	0.732
Alloy * pouring temperature	1	1.03739	1.03739	4.05	0.114
Error	4	1.02461	0.25615		
Total	11	5.74335			

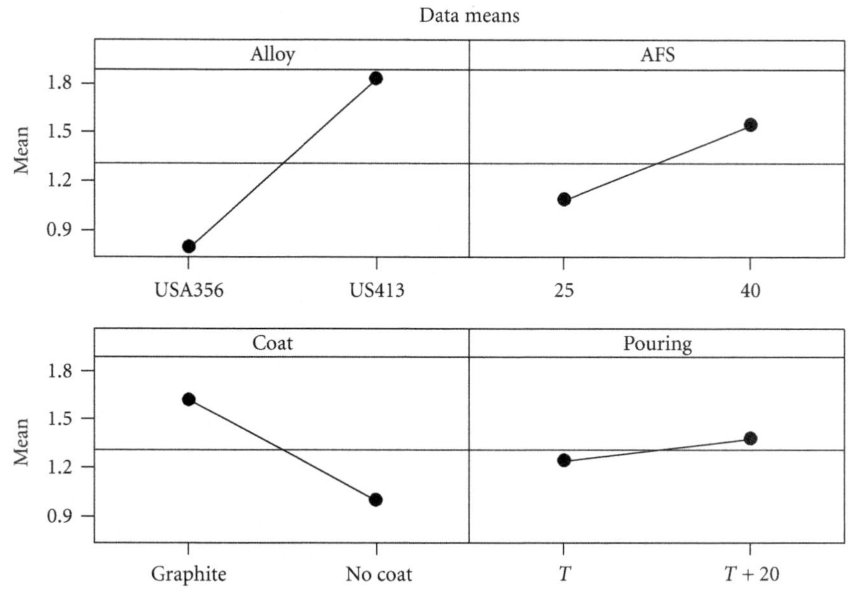

Figure 10: Main effects' plot for mould filling ability.

Table 9: The effects are summarized.

Factor	Mean	Effect (mean – overall mean)	
LM6	1.82586	1.82586 – 1.309	0.51686
LM25	0.79266	0.79266 – 1.309	−0.51634
Pouring temp T	1.24350	1.24350 – 1.309	−0.0655
Pouring temp $T + 20$	1.37501	1.37501 – 1.309	0.06601
GC	1.66936	1.66936 – 1.309	0.36036
NC	0.94915	0.94915 – 1.309	−0.35985
25	1.07941	1.07941 – 1.309	−0.22959
40	1.53911	1.53911 – 1.309	0.23011

mean for the factor. The effects are summarized at Table 9. The overall mean is 1.309 for the present study from the ANVOA calculations.

Main effects' plot for mould filling ability values isshown in Figure 10. The magnitude of an effect is related to its distance from the mean. A reference line is drawn for the representing the overall mean. Lines are not horizontal (parallel to the x-axis), for all the four factors of the present study, so there is a main effect present. The main effects plot indicates that the alloy, mould coat, sand fineness, and pouring temperature are influencing the mould filling ability. In case of alloy factor, the difference in the vertical position of the plotted points is more, hence magnitude of the main effect is greater.

The main effects' plot indicates an optimum factor combination of LM6, graphite coat, sand fineness no. 40, and pouring temperature $T + 20$ for getting better mould filling ability.

Correlation coefficient quantifies the strength of the linear association between two variables.

To ascertain the strength of association, correlation coefficient between the two alloys is calculated using the formula

$$\text{Correlation}(r) = \left[\frac{(N\Sigma XY - (\Sigma X)(\Sigma Y))}{\text{Sqrt}\left(\left[N\Sigma X^2 - (\Sigma X)^2\right]\left[N\Sigma Y^2 - (\Sigma Y)^2\right]\right)}\right], \quad (3)$$

Table 10: Correlation coefficient for mould filling ability values.

N	X (LM6)	Y (LM 25)	$X \times Y$	X^2	Y^2
1	1.55	0.624	0.967208	2.4025	0.389382
2	1.7345	0.4091	0.709579	3.008594	0.167355
3	2.7689	1.55	4.291795	7.666807	2.4025
4	1.25	0.5875	0.734413	1.5625	0.345192

where N = number of values, X = mould filling ability values for LM6, and Y = mould filling ability values for LM 25.

The results of mould filling ability values used to find the correlation coefficient value are given in Table 10.

The correlation coefficient always takes a value between -1 and 1, with $+1$ indicating perfect correlation.

The correlation coefficient value obtained is 0.899242 indicating a strong, positive relationship.

5. Conclusions

The mould filling ability characteristics of LM6 and LM25 alloys have been studied using pin test piece with cylindrical cores. Mould filling ability increases with increase in metallostatic/pressure head. Silicon content in the alloys improves the mould filling ability. LM6 is having better mould filling ability than LM25. The mould filling ability increases with increase in fineness of sand because coarse sand is having more chilling power and solidifies the metal faster. Mould coat improves mould filling ability by providing less resistance to liquid metal flow. Alloy factor has been considered more significant in influencing mould filling ability. The main effects' plot has been indicating an optimum factor combination of LM6, graphite coat, sand fineness no. 40, and pouring temperature $T + 20$ for getting better mould filling ability.

Acknowledgments

The authors extend their thanks to the Director, the Directorate of Engineering, and the Director DRDL for providing support and permission for carrying out this R&D work.

References

[1] *ASM Metals Handbook, Volume 15, Casting*, ASM INTERNATIONAL, The Materials Information Company.

[2] S. V. Shepel and S. Paolucci, "Numerical simulation of filling and solidification of permanent mold castings," *Applied Thermal Engineering*, vol. 22, no. 2, pp. 229–248, 2002.

[3] Prepared by Energetics, Incorporated: Aluminum industry roadmap for the automotive market: Enabling Technologies and Challenges for Body Structures and Closures, The Aluminum Association, Inc., with support from the U.S. Department of Energy, 1999.

[4] Y. Su, T. Zhang, J. Guo et al., "Physical simulation of mold-filling processing of thin-walled castings under traveling magnetic field," *Journal of Materials Science and Technology*, vol. 20, no. 1, pp. 27–30, 2004.

[5] S. Sundarrajan and H. M. Roshan, "Studies on mould filling ability characteristics of Mg-Al Alloys," *Transactions of the American Foundrymen's Society*, pp. 607–616, 1989.

[6] R. W. Lewis, A. S. Usmani, and J. T. Cross, "Efficient mould filling simulation in castings by an explicit finite element method," *International Journal for Numerical Methods in Fluids*, vol. 20, no. 6, pp. 493–506, 1995.

[7] M. C. Flemings, *Solidification Processing*, Mcgraw-Hill, London, UK, 1974.

[8] N. R. Green and J. Campbell, "Influence in oxide film filling defects on the strength of Al-7si-Mg Alloy Castings," *Transactions of the American Foundry Society*, vol. 114, pp. 341–347, 1994.

[9] C. Reilly, N. R. Green, M. R. Jolly, and J. C. Gebelin, "Using the calculated froude number for quality assessment of casting filling methods," in *Proceedings of the 12th International Conference on Modeling of Casting, Welding, and Advanced Solidification Processes*, pp. 419–426, June 2009.

[10] M. Jolly, "Prof. John Campbell's ten rules for making reliable castings," *JOM*, vol. 57, no. 5, pp. 19–28, 2005.

[11] C. J. Cooksey, V. Kondic, and J. Wilcock, "The casting fluidity of some foundry alloys," *The British Foundryman*, vol. 52, p. 381, 1959.

[12] B. Sirrell and J. Campbell, "Mechanism of filtration in reduction of casting defects due to surface turbulence during mould filling," *Transactions of the American Foundrymen's Society*, vol. 105, pp. 645–654, 1998.

[13] User Manual of MINITAB.

[14] D. Hanumantha Rao, G. R. N. Tagore, and G. Rangajanardhan, "Process simulation of aluminium alloy casting solidification for controlled porosity and cooling rate," *Indian Foundry Journal*, vol. 52, no. 5, pp. 26–34, 2006.

15

Preparation of Niobium Metal Powder by Two-Stage Magnesium Vapor Reduction of Niobium Pentoxide

T. Satish Kumar,[1] S. Rajesh Kumar,[1] M. Lakshmipathi Rao,[2] and T. L. Prakash[1]

[1] *Refractory Metals Division, Centre for Materials for Electronics Technology (C-MET), IDA Phase-III, HCL (PO), Cherlapally, Hyderabad 500051, India*
[2] *Department of Physics, Osmania University, Hyderabad 500007, India*

Correspondence should be addressed to S. Rajesh Kumar; s_k_rajesh@yahoo.com

Academic Editor: Herbert Ipser

Magnesium vapor reduction of niobium pentoxide was studied using a laboratory system. Niobium powder was prepared by the magnesium vapor reduction at 1123 K for 5 hours and it contained about 8 mass % oxygen. However, the oxygen concentration could be decreased to 0.65% when it was prepared by double-step reduction by magnesium vapor and a chemical treatment. Controlled and diluted supply of magnesium vapor to the reaction front has averted excess heat generation at the reaction front and thereby fine particles were produced. Effects of various factors on the vapor reduction process were studied and discussed.

1. Introduction

Extractive metallurgy of niobium has attained immense academic interest [1–14], mainly due to its application in the emerging niobium and its niobium oxide electrolytic capacitors and also in the superconducting radio frequency cavities for particle accelerators, also used in mint metal [1]. Reductants such as aluminum and calcium are used for the preparation of niobium metal from its oxide [2–5]. Niobium metal and its alloys such as ferro niobium and nickel niobium alloys are being commercially produced by aluminothermic reduction process. High purity niobium metal is prepared by electron beam melting and refining of the niobium metal which is obtained from the aluminothermic reduction of its oxide [2]. Carbothermic process can be considered as another process exploited for its commercial production [6]. Awasti prepared niobium by silicothermic reduction of niobium in ultrahigh vacuum [7]. These processes result in the formation of chunk lets and coarsened powders and hence may not be useful in the preparation of fine niobium metal powder. However, requirement of niobium metal powder with stringent purity specifications for its possible use in electrolytic capacitor has necessitated the development of new techniques. Niobium metal powder is generally prepared by hydriding, crushing of niobium hydride, and dehydriding [8]. It is reported that niobium metal powder is prepared from its oxide by calcium using electron-mediated reduction process [9]. Similar to tantalum, niobium fine powder is also prepared by liquid-liquid reduction of K_2NbF_7 with sodium [10, 11]. Niobium metal powders can be also prepared by using magnesiothermic reduction process [12–14]. Using a cyclone separator assembly, niobium is prepared by the reaction between the magnesium vapor and niobium oxide [12]. Okabe reported a preforms reduced process (PRP) for niobium metal using magnesiothermic reduction process [13].

There are a number of factors to be considered for the selection of a reductant in a metallothermic reduction. In order for a metal M^* to reduce the MO_x, the M^* must have a greater affinity with oxygen than M:

$$MO_x + M^* \Longleftrightarrow M^*O_x + M. \quad (1)$$

From the free energy in thermodynamic database [15] as listed in (Table 1), it is found that calcium, magnesium, and aluminium have more affinity to oxygen than the niobium. Calcium has high reactivity for the reduction of niobium oxide to niobium. However, magnesium has an advantage over calcium that it has a high vapor pressure (0.078 bar) even at moderate temperature of 1123 K. Due to this, adequate

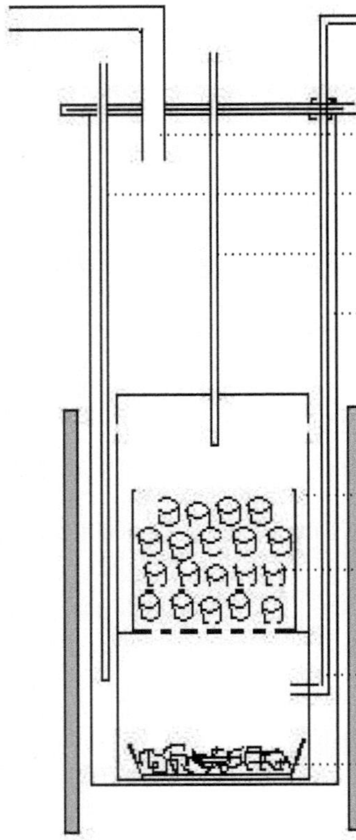

FIGURE 1: Schematic representation of magnesium vapor reduction system.

TABLE 1: Free energy of formation of some oxides at 298 K [15].

Metal oxides	ΔG°_{298} (kJ/g-atom of O)
MoO_3	−229
FeO	−245
WO_3	−254.1
Nb_2O_5	−353.6
Ta_2O_5	−390
MgO	−568.4
CaO	−597.7
Al_2O_3	−526.7

control over the reaction conditions is possible for the reduction of niobium oxide with magnesium vapor. Stefan had overcome these problems by controlling the feed and gas flow rates and obtained fine powders. However, niobium metal prepared by the cyclone reduction techniques has resulted in high concentration of oxygen (7.68%) [12].

There are not many studies reported in the literature on the preparation of phase pure niobium metal powder by magnesium vapor reduction of niobium oxide. In this work, an attempt has been made to prepare the niobium metal powder with lower oxygen content using the magnesium vapor reduction process. Influence of various factors in the magnesium vapor reduction of niobium oxide and deoxidation of niobium metal powder in a two-stage reduction process using a simple laboratory system are discussed.

2. Experimental Studies

2.1. Raw Materials. Magnesium metal turnings (99% purity, S.D. Fine chemicals), hydrochloric acid (GR grade, Merck), niobium pentoxide (>99% pure), and argon gas (UHP grade) are used for the reduction experiments. Niobium pentoxide is prepared from Indian columbite-tantalite ore (AMD, India) by solvent extraction and calcination process [16].

2.2. Magnesium Vapor Reduction Setup. A laboratory system was fabricated for the reduction experiment and the schematic sketch is shown in Figure 1. The reaction chamber is a SS430 cylindrical vessel with a lid, an inlet port, and distributor for argon gas. At the bottom of the reaction chamber, magnesium turnings are kept in a crucible. The niobium oxide compacts are filled in another crucible with perforated bottom and placed at the middle of the chamber. Both crucibles are also made up of SS430. The reaction chamber is further housed in a retort which is placed in a furnace.

Niobium pentoxide powder was compacted to pellets at a pressure of 2.94 bar with a green density of 1.2 g/cm^3. Batch size of 200 g niobium pentoxide was used for the experiments and a soaking time of 5 hours was set at 1123 K. Magnesium metal turnings are taken in the 1 : 1 ratio by weight with respect to niobium oxide. Reactor chamber is evacuated (10^{-3} bar) and filled with argon gas (0.98 bar) to remove the atmospheric air and to create the inert atmosphere in the chamber. This process is repeated for four times. Continuous flow of argon is adjusted to 1-2 liter/minute and maintained throughout the experiment. The temperature of the system is increased from room temperature to 1123 K. Magnesium vapor generated at the bottom crucible is diluted with the argon gas which acts as a carrier gas. Magnesium vapor with carrier gas enters into the top crucible through the perforated bottom and reacts with niobium oxide. As the by-product of the reaction, MgO, is not volatile, an increase in the weight of the product is observed. Based on this increase in weight, the percentage of reduction is calculated. After soaking at the required temperature and time, the reaction chamber is allowed to cool to room temperature while continuing the argon flow. Air is slowly introduced to the system to passivate the reduced mass. The reduced mass is transferred to a beaker containing water, and it is treated with 20% hydrochloric acid. The resultant powder is washed with distilled water, filtered, and dried at 373 K in vacuum oven (10^{-3} bar). For the second stage reduction, the powder obtained from first reduction is compacted and repeated the vapor reduction process with magnesium under the same inert atmosphere and followed by chemical treatment process. The powder thus obtained was soaked in dilute hydrofluoric acid (1% solution) for 15 minutes and washed with water, filtered, and dried under vacuum. Elemental analysis of niobium powder is carried out by Inductively Coupled Plasma Optical Emission Spectroscopy using JY2000. Scanning electron microscopy studies were performed by using Philips XL 30 microscope. The powder X-ray diffraction (XRD) patterns of samples were obtained using a Panalytical X'pert (Cu Kα = 0.1542 nm) X-ray diffract meter. Oxygen content of the niobium powder is analyzed by O–N analyzer using inert fusion technique using LECO TC-236. Particle size analysis is carried out by Microtrac laser diffraction particle size analyzer.

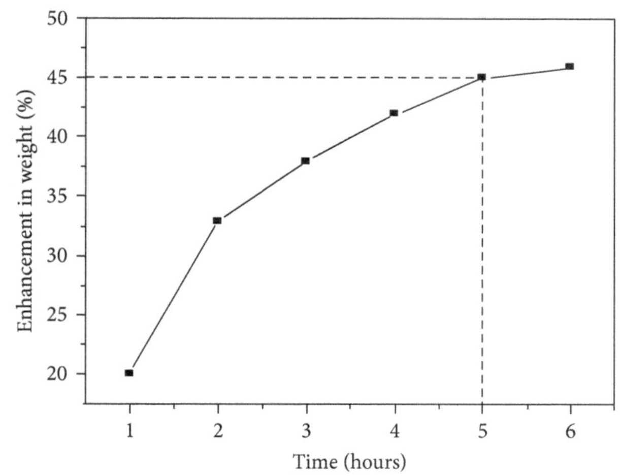

Figure 2: Effect of soaking time on the weight of the product at 1123 K.

Figure 3: Lateral cross-section of the niobium pellet (15 mm diameter) after reduced at 1123 K for 5 hours.

3. Results and Discussions

When magnesium vapor reduction experiments were conducted for niobium pentoxide pellets of uniform size (5 mm diameter and 5 mm thickness) at 1123 K, it is observed that, with the increase in the soaking time, the weight of the product increased. Magnesium vapor reacts with niobium pentoxide and the products are niobium metal and magnesium oxide. The relation between soaking time and weight gain is shown in Figure 2. There is a weight gain of 20% for the first hour of soaking which is further increased to 43%, 45%, and 46% when soaking time increased to 4, 5, and 6 hours respectively. This shows that magnesium vapor reduction of niobium oxide depends on the time of soaking. Magnesium vapor initiates the reaction at the exterior surface of the compacts and further percolates into the interior of the pellets. The rate of reaction of niobium pentoxide with magnesium is controlled by the rate of diffusion of magnesium vapors and depends on soaking time. In another experiment, pellets with different diameters were used for the reduction experiment, where it was observed that pellets with larger diameter are not fully converted to metal even after 5 hours of soaking at 1123 K. Therefore, the pellet size also plays an important role in the rate of reaction. The photograph of a lateral cross-section of a 15 mm thick niobium pellet after a reaction time of 5 hours is shown in Figure 3. The exterior portion of the pellet is black in color whereas the interior portion is still white in color. There is a bluish tinge at the interface. White color in the interior portion indicates unreacted niobium pentoxide indicating that the magnesium vapor has not diffused into the region. The black portion at the exterior portion indicates that the reaction starts at the outer surface.

Powder XRD pattern of niobium powder obtained after the first stage of magnesium vapor reduction of 5 mm pellets soaked at 1123 K for 5 hours is shown in Figure 4. It shows that, though niobium pentoxide is reduced to niobium metal, the reduction is not completed because there are significant peaks corresponding to niobium dioxide. It also shows the peaks corresponding to magnesium oxide present in the reduced mass, formed as a result of reduction of niobium pentoxide. The presence of niobium dioxide is further confirmed by

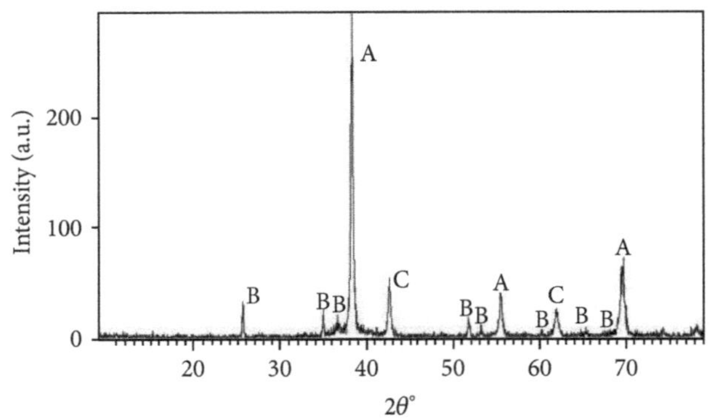

FIGURE 4: XRD pattern of niobium powder after the first stage of magnesium vapor reduction process.

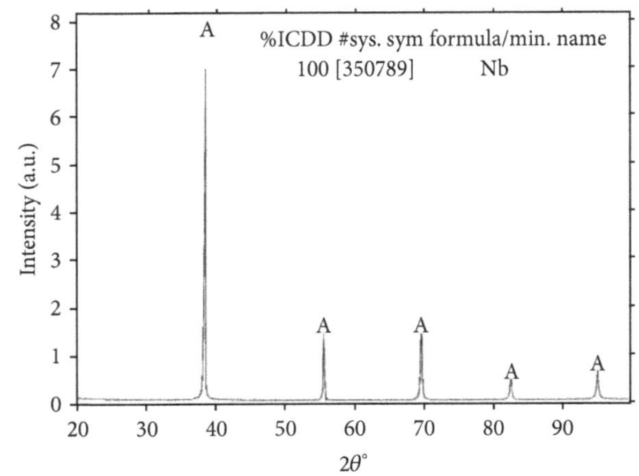

FIGURE 5: XRD pattern of niobium powder after double-stage reduction with magnesium vapor.

FIGURE 6: SEM micrograph of niobium powders obtained after two stage reduction.

TABLE 2: Chemical analysis of magnesium vapor reduced niobium powders (in ppm).

Element	Niobium pentoxide	1st stage niobium powder	After 2nd stage niobium powder
Ni	<20	35	35
W	420	550	555
Ta	925	1150	1160
Co	<20	<20	<20
Cr	<20	<20	<20
Mn	<10	<10	<5
Mg	<20	1200	925
O	—	8%	6500
Fe	<20	<20	<20

the oxygen analysis. Oxygen content of the reduced mass leached with HCl is as high as 8% (Table 2). There are no peaks corresponding to magnesium niobate as reported in earlier works [12].

Powder XRD pattern of the dried niobium metal powder obtained by the second reduction step followed by leaching with HCl and treatment with 1% HF is shown in Figure 5. This pattern is exactly matching to standard XRD pattern of niobium metal (JCPDS card no. 035-0789). Niobium dioxide peaks observed for single-stage reduced powder are completely absent in this pattern. It shows a complete reduction of oxide to metal. Further magnesium oxides peaks are also absent in the pattern as magnesium and magnesium oxide from the reduced mass are completely leached from the reduced mass. Lower concentration of oxygen is also achieved by the second-stage magnesium vapor reduction (Table 2). Fine metal powders are found to take up oxygen during powder-processing steps. Oxygen is chemisorbed into the fresh surface of metal powder, which further diffuses into the bulk during heat treatment steps [17]. In this work, a washing procedure was adopted, where niobium metal powder obtained by the second stage of reduction and leached with HCl is soaked with dilute hydrofluoric acid (1% solution) further washed with water, and dried under vacuum. Surface oxygen was removed during this process and hence resulted in low oxygen content in the metal powder, as low as 0.65%. Chemical analysis shows that there is no appreciable increase in the concentration of iron, chromium, and nickel contents in niobium metal powder (Table 2). Chances of getting contaminated from magnesium are also very little as magnesium vapor leaves behind the impurities associated with in the bottom crucible. Magnesium oxide and magnesium from the second-stage reduced mass are leached out with hydrochloric acid treatment, and magnesium content in the niobium powder is ~900 ppm.

SEM micrograph of the niobium powder is shown in Figure 6. Fine niobium powder of <1 μm particles is agglomerated from 5 to 10 μm. The absence of coarse microstructure reveals that there is no uncontrollable reaction. Magnesium vapor is diluted with the fact that carrier gas argon percolates the niobium pentoxide bed and undergoes the reaction. As the magnesium supply is very limited, heat generated in a given time is also limited. This combination of factors such as limited and dilute supply has resulted in fine and uniform powders.

The average particle size of the niobium pentoxide is 25 microns as measured by using laser beam technique

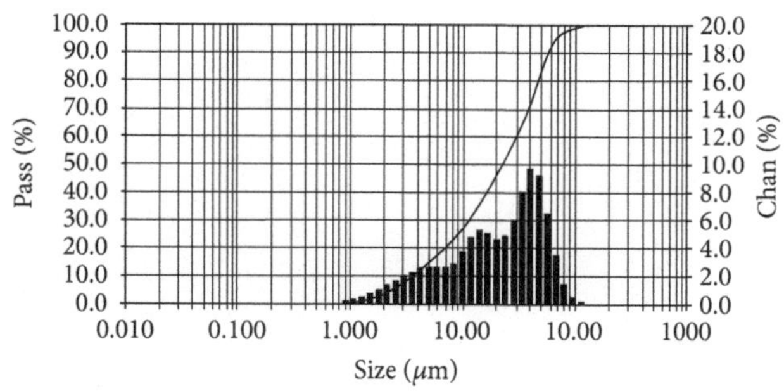

FIGURE 7: Particle size analysis of niobium pentoxide powder.

in a wet particle size analyzer as shown in Figure 7. The green density of the niobium oxide is varied from 1.3 to 1.5 g/cm^3. The pellets are broken into small pieces during the reaction. The reaction front proceeds via niobium pentoxide to niobium dioxide and to niobium metal. The reduction of oxide starts at the surface of the oxide pellet. Magnesium had to diffuse through the reduced mass consisting of niobium dioxide, niobium, and magnesium oxide to the inner surface of the porous bed of unreduced niobium pentoxide. The presence of unreacted magnesium also contributes to the weight of the reduced mass. This is reflected in the presence of 8% oxygen (Table 2) in the niobium metal prepared after first stage of reduction for 5 hours soaking even though weight enhancement is >45%, which is an index of complete reduction. Second-stage reduction has resulted in pure niobium metal, without any traces of oxides. Chemical treatment of the resulted niobium powder removed the surface oxygen chemisorbed into the surface and resulted in oxygen content of 0.65%. This work proves that, using simple laboratory system, niobium pentoxide can be reduced to niobium metal powder with low oxygen content using two-stage vapor reduction technique. From the chemical analysis, it is observed that impurities like Fe, Co, Cr, Ni, and Na did not increase even after the two stages of reduction process. Hence, the purity of the raw materials is an important factor for future applications of the product. After the reduction, the particle size of the product is observed in submicron range and ultimately the surface area increases. This can enhance the charge storage capacity per gram for electrolytic capacitors.

Thus, these results show that the reduction of niobium pentoxide to niobium metal is a two-step process. First step involves the reduction of niobium pentoxide to niobium metal and niobium dioxide and the second step involves the reduction of niobium dioxide also to niobium by the repeating the same process. Similar process may be utilized for the preparation of fine powders of various metals including tantalum. Powders thus produced by this technique can be utilized for metallurgical or electrical applications.

4. Conclusion

A simple laboratory system is fabricated for magnesium vapor reduction of niobium pentoxide to niobium metal powder by using the two-stage reduction by magnesium vapor. This system is successfully used in the preparation of high-quality niobium metal powder having fine and uniform microstructure with oxygen content as low as 0.65%. Powders thus produced by this technique can be utilized for metallurgical or electrical applications.

Acknowledgments

The authors are indebted to the Department of Science and Technology, Government of India for the financial assistance. They are also thankful to Dr. MRP Reddy for chemical analysis and Dr. Tanay Seth for SEM.

References

[1] G. Aggarwal, S. J. Park, and I. Smid, "Development of niobium powder injection molding: part I. Feedstock and injection molding," *International Journal of Refractory Metals and Hard Materials*, vol. 24, no. 3, pp. 253–262, 2006.

[2] G. R. Kamat and C. K. Gupta, "Open aluminothermic reduction of columbium (Nb) pentoxide and purification of the reduced metal," *Metallurgical Transactions*, vol. 2, no. 10, pp. 2817–2823, 1971.

[3] T. K. Mukaharjee, G. R. Kamat, and C. K. Guptha, "Preparation of columbium metal by calcium hydride reduction of columbium pentoxides," *Journal of Metals*, vol. 22, no. 2, pp. 50–53, 1970.

[4] C. K. Guptha and P. K. Jena, "calciothermic reduction of niobium (columbium) Pentoxide," *Transactions of Metallurgical Society of AIME*, vol. 230, pp. 1433–1438, 1964.

[5] J. C. Sehra, D. K. Bose, and P. K. Jena, "Preparation of niobium and tantalum powders by calciothermic reduction of their pentoxides," *Transactions of the Indian Institute of Metals*, vol. 21, no. 3, p. 450, 1968.

[6] P. V. Gel'D and G. P. Shveikin, "The kinetics of vacuum reduction of niobium Pentoxide by carbon in the uses of vacuum in metallurgy," *Zeitschrift für Metallkunde*, vol. 55, p. 432, 1964.

[7] A. Awasthi, Y. J. Bhatt, N. Krishnamurthy, Y. Ueda, and S. P. Garg, "Reduction of niobium and tantalum pentoxides by silicon in vacuum," *Journal of Alloys and Compounds*, vol. 315, no. 1-2, pp. 187–192, 2001.

[8] K. C. G. Candioto and C. A. Nunes, "Nb-20%Ta alloy powder by the hydriding—dehydriding technique," *International Journal of Refractory Metals and Hard Materials*, vol. 24, no. 6, pp. 413–417, 2006.

[9] T. H. Okabe, I. Park, K. T. Jacob, and Y. Waseda, "Production of niobium powder by electronically mediated reaction (EMR) using calcium as a reductant," *Journal of Alloys and Compounds*, vol. 288, no. 1-2, pp. 200–210, 1999.

[10] Y. Purushotham, T. Balaji, A. Kumar, and T. L. Prakash, "A metallothermic route for producing capacitor grade tantalum powder," *Transactions of the Indian Institute of Metals*, vol. 55, no. 6, pp. 525–529, 2002.

[11] J.-S. Yoon, "The fabrication of niobium powder by sodiothermic reduction process," *International Journal of Refractory Metals and Hard Materials*, vol. 28, no. 2, pp. 265–269, 2010.

[12] S. Luidold, H. Antrekowitsch, and R. Ressel, "Production of niobium powder by magnesiothermic reduction of niobium oxides in a cyclone reactor," *International Journal of Refractory Metals and Hard Materials*, vol. 25, no. 5-6, pp. 423–432, 2007.

[13] T. H. Okabe, S. Iwata, M. Imagunbai, and M. Maeda, "Production of niobium powder by magnesiothermic reduction of feed preform," *ISIJ International*, vol. 43, no. 12, pp. 1882–1889, 2003.

[14] I. Park, T. H. Okabe, Y. Waseda, H. S. Yu, and O. Y. Lee, "Semi-continuous production of niobium powder by magnesiothermic reduction of Nb_2O_5," *Materials Transactions*, vol. 42, no. 5, pp. 850–855, 2001.

[15] A. K. Suri and P. K. Jana, *Extractive Metallurgy of Niobium*, CRC, 1994.

[16] Y. Lee, H. D. Jang, and K. Y. Park, "solvent extraction of tantalum and niobium using mixer settler," *Journal of Korean Institute of Chemical Engineers*, vol. 29, no. 6, pp. 736–774, 1991.

[17] Y. Waseda and M. Isshiki, *Purification Process and Characterization of Ultra High Purity Metals*, Springer, New York, NY, USA, 2002.

16

The Effect of Microstructure on Stress-Strain Behaviour and Susceptibility to Cracking of Pipeline Steels

A. Mustapha, E. A. Charles, and D. Hardie

School of Chemical Engineering and Advanced Materials, Newcastle University, Newcastle upon Tyne NE1 7RU, UK

Correspondence should be addressed to E. A. Charles, e.a.charles@ncl.ac.uk

Academic Editor: Gerhard Sauthoff

The effect of microstructure on the stress-strain behaviour of pipeline steels was studied. Slow strain rate ($2 \times 10^{-6}\,s^{-1}$) tests were conducted on grade X65 and X100 steels in silicone oil and hydrogen carbonate/carbonate solution. The as-received grade X100 steel at 75°C showed serrated stress-strain curves. The magnitude of the serrations depended upon the strain rate and test temperature. Annealing at 600°C or above removes the serrations, but this increased the susceptibility to transgranular cracking in hydrogen carbonate/carbonate solution at potentials below −800 mV (sce). The removal and reformation of banding in pipeline steels were also studied. Ferrite/pearlite becomes aligned during the hot rolling stage of pipe manufacture and causes directionality in crack propagation and mechanical properties. Heat treatments were carried out which show that banding in grade X65 and X100 can be removed above 900°C. This depends on homogenisation of carbon which also depends on temperature, time, and cooling rate.

1. Introduction

The strength and toughness requirements of pipeline steels may be achieved by appropriate steel production techniques. Different strengthening methods are used by manufacturers to satisfy codes. Mechanical performance and susceptibility to environment-assisted cracking depend on microstructure, and so heat-treated pipeline steels with uniform fine-grained microstructure such as bainite are generally more resistant to stress corrosion cracking, SCC, than those consisting of pearlite and ferrite [1]. Some cracks that are hydrogen-induced have been shown to be associated with the boundary between ferrite and pearlite [2]. Such work suggests that homogenisation of composition and phases in steels improve their resistance to cracking.

Pearlite-ferrite bands in carbon steels are associated with the segregation of substitutional alloying elements which raise or lower the temperature (Ar_3) for the formation of proeutectoid ferrite upon cooling [3]. The segregation occurs during solidification of steel and gives rise to longitudinal bands during hot rolling [3–5]. If the Ar_3 is raised by the solute, then proeutectoid ferrite nucleates first in the solute-rich regions. On the other hand, if the Ar_3 temperature is lowered by the solute, then the nucleation of the proeutectoid ferrite begins in the solute lean regions. In either case, carbon atoms, which diffuse rapidly, are rejected from the proeutectoid ferrite, thus producing carbon-rich regions of austenite, which transform finally to pearlite [3, 4]. In pipeline steels, manganese lowers the Ar_3 temperature and pearlite forms in the regions of higher manganese. Hence band removal in pipelines steels depends on homogenisation of either carbon or manganese. Ferrite and pearlite bands cause directionality in crack propagation [2] and mechanical properties [6].

High-pressure transmission pipelines are often buried in the ground, and a cracking environment may be generated as a result of conversion of ground water to high pH solution during cathodic protection [7]. Hydrogen carbonate/carbonate solutions with a pH range between 9.6–12.3 have been reported [8, 9] to cause cracks that have led to the catastrophic failure of a number of buried pipelines when associated with poorly applied cathodic-protection systems. The high pH solution causes IG cracking whereas a low pH ~6.5 ground water promotes TG cracking. Cracking of buried pipes is therefore a safety consideration for evaluating new pipeline steels.

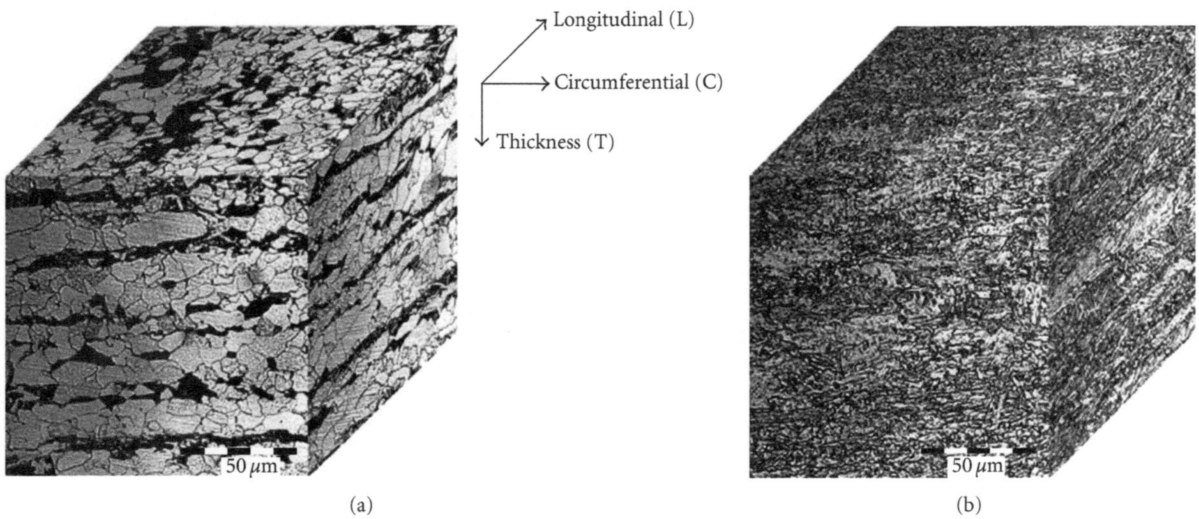

FIGURE 1: Microstructure (magnification 458 times) of (a) grade X65 and (b) grade X100.

The aim of the present work was to study the effect of removal of load-serrations on the resistance of the grade X100 steel to cracking in hydrogen carbonate/carbonate solution using the slow strain rate technique and also to examine the formation and removal of microstructural banding in grade X65 and X100 pipeline steels.

TABLE 1: Chemical composition of the steels (balance iron) in wt%.

Steel Grade	C	Mn	Si	Al	P	S
X65	0.07	1.36	0.19	<0.07	0.013	0.002
X100	0.06	1.82	0.16	0.03	0.011	0.002

2. Experimental

The two pipeline steels tested were supplied in accordance with API-5L specifications and have the compositions given in Table 1. The microalloying and controlled rolling of the fully killed steel achieves high strength by grain refinement. The microstructures of the steels viewed in three different planes: short transverse (thickness), longitudinal, and circumferential are shown in Figure 1. The top view in the figure is the circumferential face of the pipe. The grains in the short transverse and the longitudinal directions are elongated due to rolling. Microstructures of the steels were revealed by mounting 10 mm cube blocks in bakelite, wet-grinding with silicon carbide papers, polishing on diamond cloths down to a 1 μm finish, and finally etching with 2% nitric acid in alcohol (Nital). The average band centre-to-centre spacing and grain size were measured by counting the number of bands or grains across each sample using a travelling microscope.

Slow strain rate test (SSRT) specimens, 90 mm long, were machined to produce parallel gauge lengths of 12.7 mm and of 2.5 mm in diameter, polished to P1200 grit silicon carbide and then cleaned with ethanol. The specimens were cut from the wall of fabricated pipes with their major axes transverse to the longitudinal axis of the pipes. The major banding axis was perpendicular to the specimen surface.

Environment-assisted cracking tests were conducted in 1 M sodium hydrogen carbonate - 0.5 M sodium carbonate solution (pH ~9.5 at room temperature) under potentiostatic control. The specimens were contained in glass cells with external electric heating and were sealed at each end with rubber bungs that permitted the insertion of the specimen, a platinum counterelectrode, a condenser, and a thermocouple for temperature control. A conducting salt bridge led from the solution to an external saturated calomel reference electrode (sce). The solution volume was 85 cm^3. Tests were conducted at a strain rate of $2 \times 10^{-6}\,s^{-1}$, a temperature of 75°C, and over an applied potential range from −800 to −1100 mV (sce). Before SSRT, some specimens were heat-treated at various temperatures in a preheated, nitrogen-purged, tube furnace and then different modes of cooling were applied. The samples were cleaned with ethanol and dried before heat treatment.

The average crack velocity was determined from a metallographic cross-section, by measuring the deepest secondary crack in any one specimen and dividing it by the time for plastic deformation for tests anodic to the free corrosion potential (−890 mV (sce)) where SCC occurs. For cracks formed at potentials cathodic to the free corrosion potential, which may be the result of hydrogen embrittlement, the time from UTS to fracture was used in the calculation of the average crack velocity as here the onset of cracking occurs after necking.

3. Results and Discussion

3.1. Serrations. Tests conducted at a strain rate of $2 \times 10^{-6}\,s^{-1}$ and a temperature of 75°C gave serrated stress-strain curves for the as-received grade X100 steel (Figure 2). The serrations are characterised by a series of sudden load drops followed by an increase with a gradient similar to that of the elastic portion of the curve. The load drop in

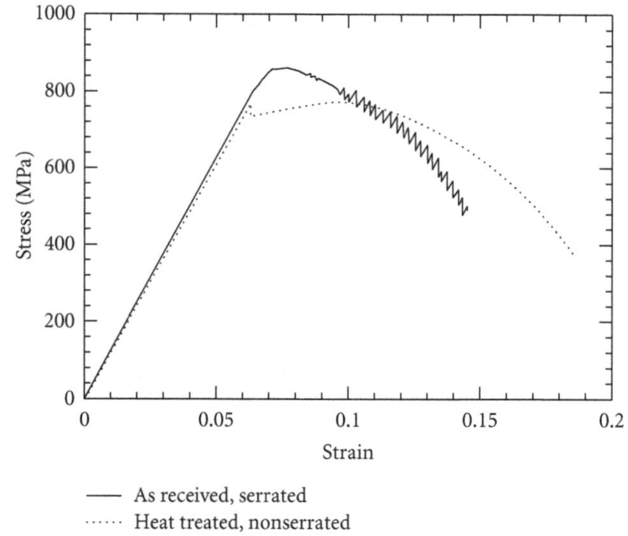

FIGURE 2: Typical SSRT stress-strain curves of grade X100 pipeline steel strained to fracture in silicone oil at 75°C.

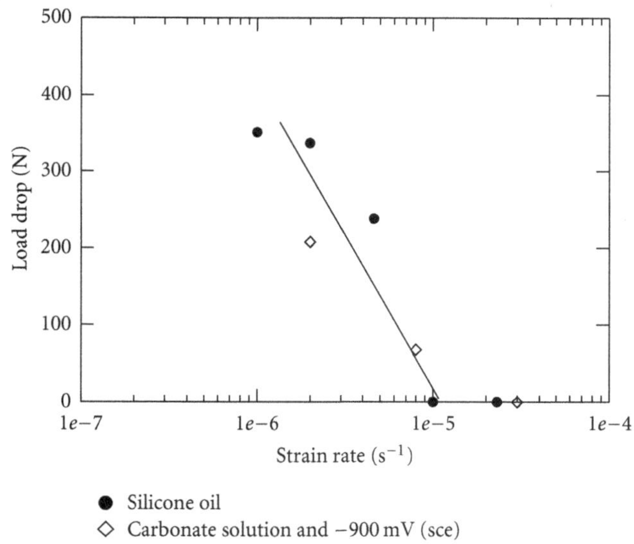

FIGURE 3: Effect of strain rate on maximum load drop due to serrations for tests in both silicone oil and carbonate solution at 75°C.

most cases occurs soon after necking and its magnitude gradually increases to fracture. The observed serrations are similar to the reported [10–12] "Type-B serrations" which rise and fall about the general level of the stress-strain curve. This dynamic strain ageing phenomenon most likely involves the locking of dislocations followed by either unlocking of dislocations to allow them to move and/or the creation of new dislocations to allow the further deformation [13, 14]. The load increases until unlocking and/or new dislocation formation occurs, and the dislocations become separated from obstacles such as solute atoms resulting in a load drop. This process occurs many times and causes serrations in the stress-strain curve during plastic deformation [15, 16].

It has been suggested [15] that interstitial atoms such as carbon and nitrogen interact with the strain field of dislocations in steel and dislocations can be locked in position by the formation of an interstitial "atmosphere" of either carbon or nitrogen in the vicinity of dislocations but may be freed by increasing stress or temperature causing further deformation. In an attempt to further understand the nature of the serration mechanism, a number of SSRTs were conducted and the results show that the serrations are independent of whether tests are conducted in silicone oil or carbonate solution at 75°C since both give serrated curves. The serrations are affected by strain rate (Figure 3) and the size of the maximum load drop due to serrations decreases as the strain rate increases over the range tested. A test at a strain rate of $2.3 \times 10^{-5}\,s^{-1}$ and a temperature of 75°C produced no serrations which agrees with previous work [16–18] and indicates that serrations occur at a matching combination of strain rate and temperature for a low carbon steel. The strain rate for the onset of serrations at 75°C is approximately $1 \times 10^{-5}\,s^{-1}$. SSRT for the grade X65 steel at 75°C gives nonserrated stress-strain curves at $2 \times 10^{-6}\,s^{-1}$ and this is probably due to difference in microstructure and composition.

The serrations in the as-received material were completely removed by air cooling after 30 minutes at 600°C (Figure 2) and the resulting microstructure is shown in Figure 4(a). This heat treatment removes carbon and possibly nitrogen from solution and thus reduces the obstacles to dislocation movement within grains [13].

The process of removal of carbon from solution is quite noticeable at a slightly higher temperature of 700°C (Figure 4(b)). The serrations are also removed by annealing at temperatures higher than 600°C; however, a significant change in mechanical properties (Figure 5) then occurs. The change in microstructure due to the heat treatment at 600°C is a result of the removal of carbon from solution through the formation of globular carbides (sorbite). The change increases the susceptibility of the grade X100 steel to transgranular cracking (Figure 6) in the carbonate solution within a potential range of −800 to −1100 mV (sce). At potentials anodic to the free corrosion potential, −890 mV (sce), the cracking may be designated as SCC whilst at more cathodic potentials hydrogen embrittlement is most likely. The change in SCC cracking mode from intergranular at larger anodic potentials to transgranular as a pipeline steel approaches the free corrosion potential has been reported earlier by Li et al. [19].

3.2. Banding. Removal of the pearlite/ferrite bands in grade X65 steel depends upon the heat treatment temperature and the rate of cooling (Figure 7). A range of cooling rates was tried: furnace cooling is the slowest and water quenching the fastest. Oil and water quenching from 900°C gives a bainitic microstructure whereas air cooling from 1175°C gives a Widmanstätten structure. Air cooling from 1000°C and fan cooling from 950°C both give slightly banded pearlite/ferrite microstructures. A heavily banded pearlite/ferrite structure is obtained by furnace cooling from 1000°C. The nonbanded

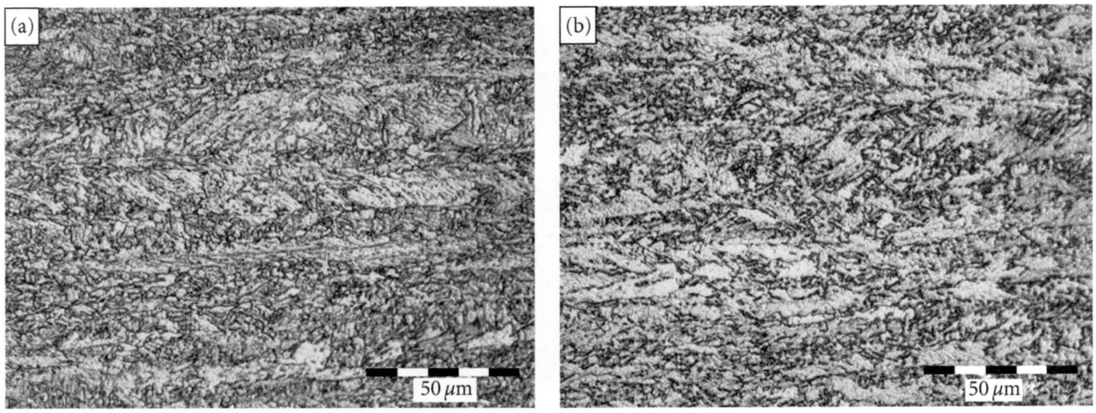

(a) (b)

FIGURE 4: Microstructures of the longitudinal section of the grade X100 air cooled after 30 min. at (a) 600°C and (b) 700°C.

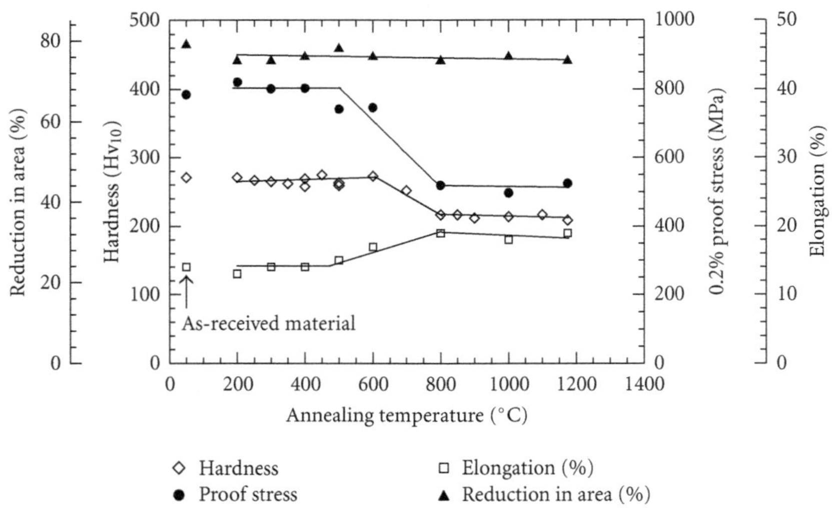

FIGURE 5: Mechanical properties of heat-treated grade X100 line pipe steel. The proof stress and the ductility were measured in silicone oil at 75°C. Average hardness was measured on the longitudinal section at room temperature.

microstructures are obtained by transformation of austenite to a nonequilibrium microstructure by using a faster cooling rate. The faster cooling rate suppresses the enrichment of austenite by carbon diffusion at the austenite/ferrite interface during cooling and this may lead to the precipitation of carbides within the ferrite. This is a process of band removal by hindering the segregation of carbon to manganese-rich regions.

Another approach is to remove banding by homogenisation of manganese through diffusion, as the concentration of Mn in the pearlite band is higher than that of the ferrite band. The result of an attempt to eliminate Mn segregation through diffusion is shown in Figure 8. This demonstrates that as the soak time increases the pearlite centre-to-centre spacing increases. This is probably due to elimination of intermediate bands of lower concentration through diffusion. Other heat treatments considered in an attempt to eliminate banding are summarised in Table 2.

These results indicate that band formation and removal from the pipeline steel depends on heat treatment temperature, soak time, and cooling rate. Slow cooling from austenitic temperatures gives intense banding (provided the average grain size is less than approximately twice the average band spacing) and the intensity decreases with increasing cooling rate. Banding is completely removed by quenching or air cooling from a higher temperature (Figure 7). The fast cooling suppresses the diffusion of carbon from the region of low manganese concentration to the region of higher manganese concentration during the proeutectoid ferrite formation. Band removal by homogenisation of manganese through diffusion requires a longer time and a higher temperature than for carbon (Figure 8 and Table 2). An almost complete homogenisation of manganese was achieved by furnace cooling after 5 hours at 1175°C, and this removed banding completely through grain growth. The average grain size was refined by air cooling after 20 minutes at 950°C without the return of banding (Figure 9(b)).

These results are consistent with the trends in the literature concerning the removal of banding through quenching or air cooling from higher temperature and agree with

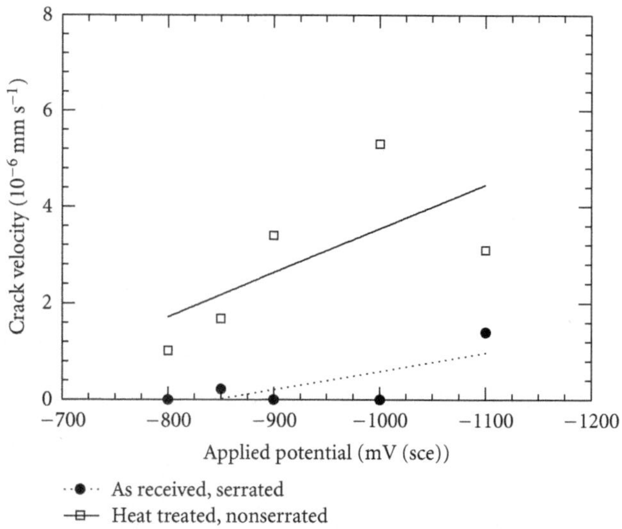

Figure 6: Crack velocities of the as-received grade X100 compared with that of specimens air cooled after 30 min at 600°C.

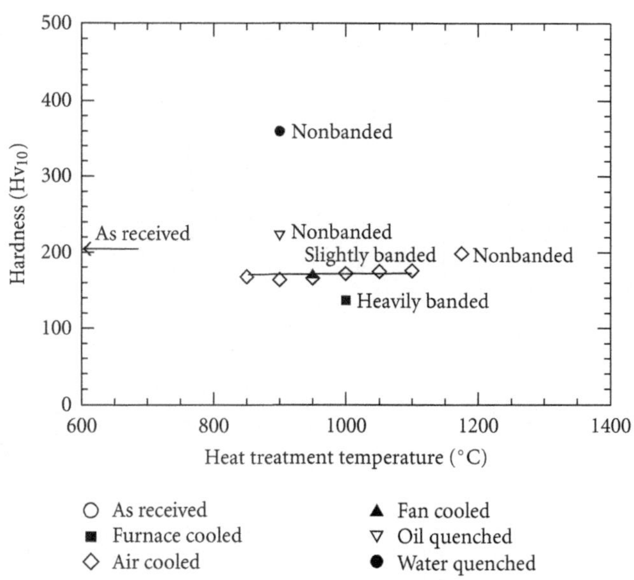

Figure 7: Effect of heat treatment on average hardness and banding in the longitudinal section for the grade X65 pipeline steel. Heating time for each heat treatment was 20 minutes.

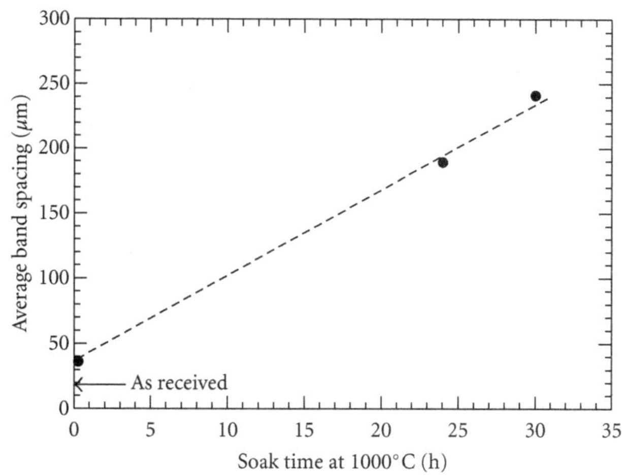

Figure 8: Effect of soak time on the average centre-to-centre spacing of pearlite bands on the longitudinal section of grade X65 steel; specimens were furnace cooled.

the work of Wilms [20] and Grossterlinden et al. [21]. Although the homogenisation of manganese through high-temperature diffusion agrees with earlier work [22–24], in this study homogenisation was achieved at a lower temperature. Homogenisation at temperatures greater than 1200°C may result in overheating the steel, and therefore the removal of banding above this temperature was considered impractical [21]. In this study, band removal through high-temperature homogenisation followed by normalisation was reexamined with a third heat treatment involving furnace cooling after 20 minutes at 1000°C. Widely spaced traces of bands reappeared (Figure 10) indicating that the homogenisation of manganese is not complete. In a slow cooling process, little differences in concentration of manganese favour segregation of carbon but the threshold difference has not been established. When these findings from the grade X65 were applied to the grade X100 steel similar results were produced, although here the average band spacing is smaller than that of the grade X65 steel. The results suggest that a key to band removal is knowledge of average band spacing and the diffusion coefficient of manganese in steels. A time for manganese homogenisation was based on the diffusion coefficient of manganese in a carbon steel [20] but the present work suggests that the diffusion rate of manganese in the grade X65 pipeline steel is slightly lower.

A comparison of SSRT fracture surfaces of banded and nonbanded specimens demonstrates that banding causes fracture surface ovality (ratio of two perpendicular diameters on fracture surface). The ovality is more severe in the as-received material than in the heat-treated specimens because the rolling effect is partially removed by the annealing. The grains are equiaxed and the carbon distribution is fairly uniform because of air cooling from higher temperatures. It is the pearlite bands that contribute to the ovality rather than the segregation of the manganese. The pearlite/ferrite banding creates a layered arrangement of hard and soft regions and, during cross-sectional reduction, it is easier to squeeze the softer ferrite bands than the pearlite bands. So, the reduction in area is greater in a direction perpendicular to banding than parallel to it. Therefore, the elimination of the segregation is important in the removal of directionality to achieve isotropy in reduction area.

4. Conclusions

(1) Slow strain rate tests on as-received grade X100 at 75°C give a serrated load-extension curve. The serrations at 75°C and a strain rate of 2×10^{-6} s^{-1} were removed by air cooling after 30 minutes at 600°C (simulated tempered region of HAZ). This

Table 2: Effect of heat treatment on pearlite bands in the "LT" face of grade X65 steel.

Temperature (°C)	Soak time	Mode of cooling	Microstructure	Grain size (μm)	No. of treatments
950	60 h	FC	Banded with average spacing* (AS) 201 μm	24	1
1175	1 h	FC	Slight traces of banding with large pearlite areas and some Widmanstätten	29	1
1175 then 1000	1 h then 20 min	FC	Banded with AS 60 μm	12	2
1175	5 h	FC	Not banded with ferrite/pearlite and some Widmanstätten	50	1
1175 then 950	5 h then 20 min	FC then AC	Not banded (ferrite/pearlite)	9	2
1175 then 950 then 1000	5 h then 20 min then 20 min	FC then AC then FC	Banded with AS 130 μm	10	3
1200 then 1000	20 min then 20 min	FC	Banded with AS 52 μm	13	2

*As-received average pearlite band spacing and average grain size: 19 and 6 μm, respectively.

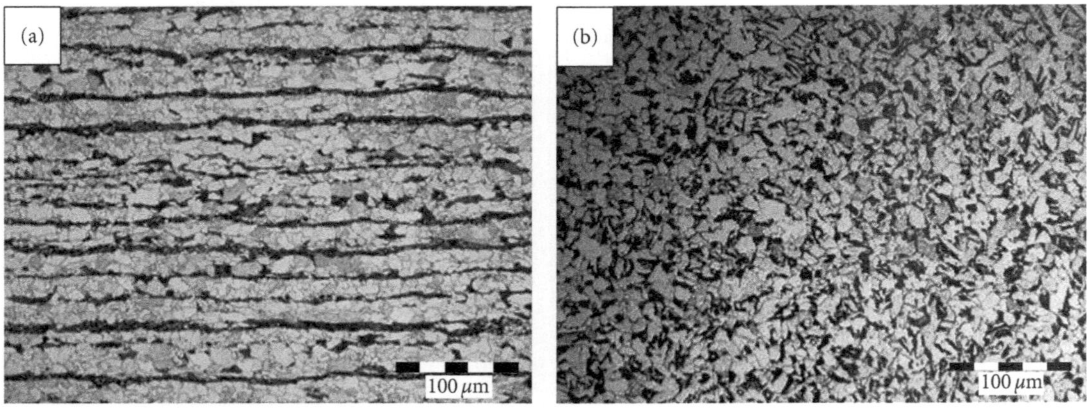

(a) (b)

Figure 9: (a) Banding in the as-received grade X65 steel (b) was removed by furnace cooling after 5 hours at 1175°C and then air cooled after 20 minutes at 950°C.

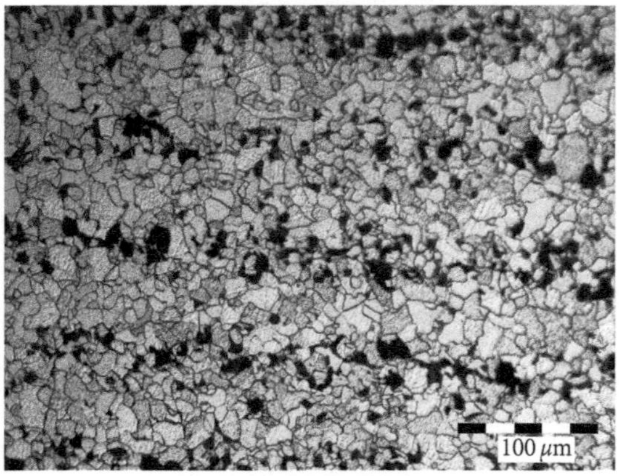

Figure 10: X65 steel furnace cooled after 5 hours at 1175°C, then air cooled after 20 minute at 950°C and again furnace cooled after 20 minutes at 1000°C.

heat treatment increases the susceptibility to transgranular cracking at potentials more negative than −800 mV (sce) when compared with the as-received microstructure. The removal of the serrations and the increase in susceptibility are due to precipitation of carbides from solution as a result of the heat treatment.

(2) Furnace cooling after 5 hours at 1175°C eliminated banding in grade X65 pipeline steel but increased the average grain size. This was refined without the return of banding by air cooling after 20 minutes at 950°C. Thus, banding was removed at a relatively low temperature but with a longer soak time. Quenching in oil or water from 900°C removes banding but the resulting microstructures are harder than the as-received ones.

Acknowledgment

This paper was supported by Petroleum Technology Development Fund (PTDF), Nigeria, to whom the authors express their gratitude.

References

[1] R. N. Parkins, *A Review of Stress Corrosion Cracking of High Pressure Gas Pipelines*, NACE Corrosion, Houston, Tex, USA, 2000.

[2] D. Hardie, E. A. Charles, and A. H. Lopez, "Hydrogen embrittlement of high strength pipeline steels," *Corrosion Science*, vol. 48, no. 12, pp. 4378–4385, 2006.

[3] J. S. Kirkaldy, J. von Destinon-Forstmann, and R. J. Brigham, "Simulation of banding in steels," *Canadian Metallurgical Quarterly*, vol. 1, pp. 59–81, 1962.

[4] S. W. Thompson and P. R. Howell, "Factors influencing ferrite/pearlite banding and origin of large pearlite nodules in a hypoeutectoid plate steel," *Materials Science and Technology*, vol. 8, no. 9, pp. 777–784, 1992.

[5] G. Krauss, "Solidification, segregation, and banding in carbon and alloy steels," *Metallurgical and Materials Transactions B*, vol. 34, no. 6, pp. 781–792, 2003.

[6] F. A. Khalid, M. Farooque, A. Ul Haq, and A. Q. Khan, "Role of ferrite/pearlite banded structure and segregation on mechanical properties of microalloyed hot rolled steel," *Materials Science and Technology*, vol. 15, no. 10, pp. 1209–1215, 1999.

[7] E. A. Charles and R. N. Parkins, "Generation of stress corrosion cracking environments at pipeline surfaces," *Corrosion*, vol. 51, no. 7, pp. 518–527, 1995.

[8] R. N. Parkins and R. R. Fessler, "Stress corrosion cracking of high-pressure gas transmission pipelines," *International Journal of Materials in Engineering Applications*, vol. 1, no. 2, pp. 80–96, 1978.

[9] J. Q. Wang and A. Atrens, "SCC initiation for X65 pipeline steel in the "high" pH carbonate/bicarbonate solution," *Corrosion Science*, vol. 45, no. 10, pp. 2199–2217, 2003.

[10] B. J. Brindley and P. J. Worthington, "Yield-point phenomena in substitutional alloys," *Metallurgical Review*, vol. 15, no. 145, pp. 101–114, 1970.

[11] S. Xu, X. Q. Wu, E. H. Han, and W. Ke, "Effects of strain rate and temperature on tensile properties of a SA508 Cl.3 steel for reactor pressure vessel," *Journal of Materials Science*, vol. 44, no. 11, pp. 2882–2889, 2009.

[12] B. H. Tian, H. X. Li, Y. G. Zhang, and C. Q. Chen, "Types of serrations in Al-Li single crystals," *Journal of Materials Science Letters*, vol. 16, no. 8, pp. 611–612, 1997.

[13] A. Cottrell, *An Introduction to Metallurgy*, chapter 21, Edward Arnold, 1980.

[14] W. C. Leslie, *The Physical Metallurgy of Steels*, chapter 2, McGraw-Hill, 1982.

[15] A. H. Cottrell, "The strength of metals," *Journal of Chemical, Metallurgical, and Mining Society of South Africa*, vol. 53, no. 6, pp. 177–181, 1953.

[16] R. J. Baird, "The effects of strain-ageing due to interstitial solutes on the mechanical properties of metals," *Metallurgical Review*, vol. 16, no. 18, pp. 1–18, 1971.

[17] J. D. Atkinson and J. Yu, "The role of dynamic strain-ageing in the environment assisted cracking observed in pressure vessel steels," *Fatigue and Fracture of Engineering Materials and Structures*, vol. 20, no. 1, pp. 1–12, 1997.

[18] I. S. Kim and S. S. Kang, "Dynamic strain aging in SA508-class 3 pressure vessel steel," *International Journal of Pressure Vessels and Piping*, vol. 62, no. 2, pp. 123–129, 1995.

[19] G. Li, G. Zhang, J. Zhou, C. Huang, and W. Yang, "Characteristics and mechanism of high pH stress corrosion cracking of pipeline steel X70," *Key Engineering Materials*, vol. 353–358, no. 1, pp. 219–222, 2007.

[20] G. R. Wilms, "Note on the hardness changes in tempering mild steel," *Journal of the Australian Institute of Metals*, vol. 20, no. 3, pp. 168–169, 1975.

[21] R. Grossterlinden, R. Kawalla, U. Lotter, and H. Pircher, "Formation of pearlitic banded structures in ferritic-pearlitic steels," *Steel Research*, vol. 63, no. 8, pp. 331–336, 1992.

[22] J. D. Lavender and F. W. Jones, "An investigation on banding," *Journal of the Iron and Steel Institute*, vol. 163, no. 1, pp. 14–17, 1949.

[23] R. A. Grange, "Effect of microstructural banding in steel," *Metallurgical Transactions A*, pp. 417–126, 1971.

[24] W. A. Spitzig, "Effect of sulfides and sulfide morphology on anisotropy of tensile ductility and toughness of Hh hot-rolled C-Mn steels," *Metallurgical Transactions A*, vol. 14, no. 2, pp. 471–484, 1983.

17

Investigations on the Plastic Instability in an HCP Mg-Li Alloy

C. Wang,[1,2] Y. B. Xu,[2] and E. H. Han[1]

[1] *Environmental Corrosion Center, Institute of Metal Research, Chinese Academy of Sciences, Shenyang 110016, China*
[2] *Shenyang National Laboratory for Materials Science, Institute of Metal Research, Chinese Academy of Sciences, Shenyang 110016, China*

Correspondence should be addressed to C. Wang, wangcas@gmail.com

Academic Editor: Michael E. Kassner

Plastic instability is reported in hexagonal close-packed (HCP) LA41 magnesium alloy during tensile tests. Serration amplitude associated with plastic instability is measured to increase with increasing strain and decrease with increasing strain rate. The model of dynamic strain aging (DSA) is proposed to be the controlling mechanism. Moreover, it is reported that annealing could reduce the flow instability, which has potential practical applications.

1. Introduction

Plastic instability associated with the production of Portevin-Le Chatelier (PLC) bands, which is also know as serrated flow or jerky flow, is a topic of current theoretical and engineering interest. Lots of studies in this area are on aluminum alloys; however, due to soaring research on magnesium alloys, similar results have been reported in Mg-Ag [1], Mg-Gd [2], Mg-Y-Nd [3], Mg-Nd [4], and Mg-Al-Zn [5, 6] alloys, respectively.

Microscopically, plastic instability was believed to result from intensive interactions between mobile dislocations and solutes atoms. Cottrell [7] assumed that solute atoms had sufficient mobility to move with mobile dislocations and serrated flow was caused by the drag effect of solute atoms. However, other finding revealed that solute atoms were not expected to have such mobility [8]. Later investigations proposed that the dynamic strain aging (DSA) process during deformation might be the major controlling mechanism [9–14]. Recently, Picu [15] has suggested a new perspective by considering the effect of solute clusters, which are formed on forest dislocations, on the strength of dislocation junctions to shed more light on this long-lasting yet intriguing issue.

Macroscopic features, such as spatiotemporal localization of plastic deformation and propagation of deformation bands, were also investigated systematically. Experimentally, Chmelík et al. [16] monitored *in situ* collective dislocation motion by acoustic emission (AE) and laser extensometry techniques. They concluded that nucleation and propagation of PLC bands were dominated by different dislocation processes. Shabadi et al. [17] and Zhang et al. [18] examined dynamics deformation bands by more sophisticated laser speckle technique. Dablij and Zeghloul [19] characterized the PLC effect by measuring deformation band strains and band propagation velocities. These macroscopic features are also explained theoretically. Particularly, Lebyodkin et al. [20] simulated localized deformation bands patterns by a discrete model based on statistical behavior of stress drops, while Ananthakrishna's extended model [21] explained most of the spatiotemporal features and identified three types of bands. Lebyodkins's mesoscopic model [22], however, pinpointed the dependence of spatiotemporal behavior of strain localizations on deformation conditions.

This paper is intended to report the plastic instability in an HCP magnesium-lithium alloy. The dependence of serration amplitude, which is a major parameter tailoring plastic instability, on strain and strain rate will be discussed in details. Mechanical properties of alloys of the same family will also be compared. It will be shown that plastic instability could be removed through additional annealing treatment, and possible explanations for the disappearance of serrations will also be suggested.

TABLE 1: Chemical compositions of the three alloys under investigation (wt.%).

	Li	Al	Mg
LA41	4.32	0.97	Balance
LA21	2.43	0.89	Balance
LA40	4.06	—	Balance

2. Experimental Metals

Three types of Mg-Li alloys, LA41, LA40, and LA21, were investigated in this study. The chemical compositions of the three alloys are listed in Table 1. The samples were obtained after extrusion with an 8 : 1 ratio at 673 K and were subsequently homogenized at 623 K for 1 hour to remove internal stresses. Following that, the samples were quenched in air at room temperature (the as-received state). A batch of LA41 alloy was heat-treated by an extended annealing at 523 K for 240 hours (the as-annealed state).

Tensile specimens have the dimension of 5 × 25 mm by diameter and length, respectively. Uniaxial tensile experiments were performed on an MTS-880 machine at 293 K. Extensometer was attached to gauge length in order to gather accurate strain values. Care was taken to minimize the machine effects. Stress-strain curves were recorded and plotted at a speed of 10 points per second.

The reagent used for microstructure revelation was composed of 3 g picric acid, 10 mL acetic acid, 10 mL distilled water, and 50 mL ethyl alcohol. Microstructure was characterized by an environmental scanning electron microscope (ESEM, Philips XL30 FEG) under the acceleration voltage of 10 KV. Transmission electron microscope (TEM, JEM 2000FXII) samples having diameters of 3 mm were thinned in a solution of 30 mL nitric + 70 mL methanol in a twin jet electropolisher to the final thickness of 50 microns. TEM examinations were conducted at the voltage of 200 KV.

3. Results and Discussion

3.1. Mechanism for the Onset of Plastic Instability. Figure 1 shows the engineering strain-stress curves of the as-received LA41 at three different strain rates. These curves share similar profiles, with linear-plastic regions, strain-hardening regimes, and stress plateaus. Plastic instability is observed to appear after an incubation strain (ε_c). It can be seen that the as-received LA41 exhibits a negative strain rate sensitivity (SRS), since the flow stress is continuously lowered with increasing strain rate.

The inset of Figure 1 shows the locally magnified stress-strain curve under the strain rate of $6.66 \times 10^{-4}\,s^{-1}$. The serration amplitude ($\Delta\sigma$), denoted inside, is defined as the stress difference between the highest point and the lowest one within a serration. Within a serration cycle, a rapid drop follows the gradual stress increase. It should be noted that those high-frequency stress steps are considered as electronic noises because their corresponding amplitude values are generally less than 0.5 MPa.

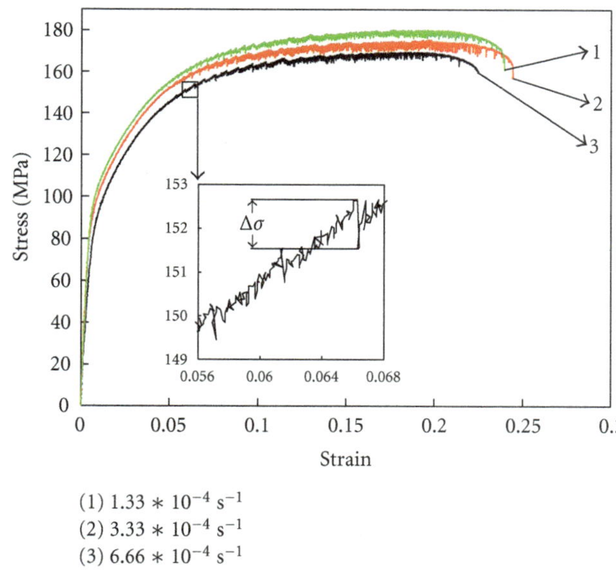

FIGURE 1: Stress-strain curves of the as-received LA41 under different strain rates at 293 K. The inset shows the locally magnified stress-strain curve at the strain rate of $6.66 \times 10^{-4}\,s^{-1}$.

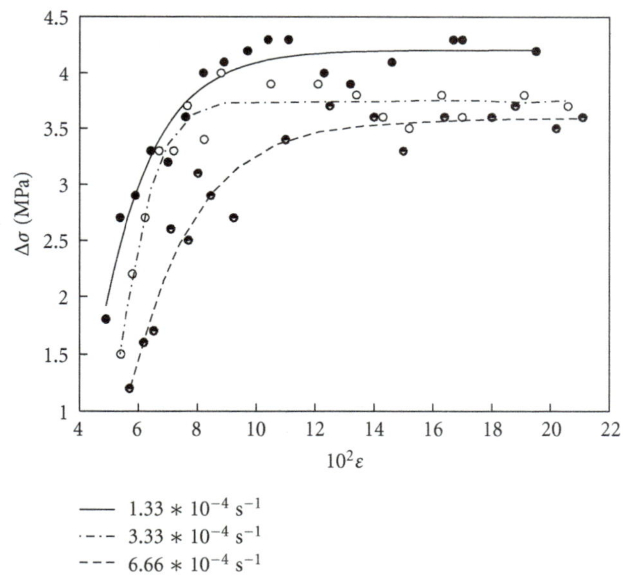

FIGURE 2: Variations of serration amplitude with strain and strain rate of the as-received LA41.

With this definition of serration amplitude, accurate values of $\Delta\sigma$ at various strains under different strain rates were measured by the one-by-one calculation of individual serrations (Figure 2). Through mathematical fitting (corresponding curves), $\Delta\sigma$ is found to increase rapidly at the early stage but get saturated when large strain is achieved. When comparing the influence of the strain rate on serration amplitude, a common trend is that when higher strain rate is high, $\Delta\sigma$ will be decreased to lower values. Summarily, the nature of plastic instability becomes more severe with increasing strain but less severe with increasing strain rate.

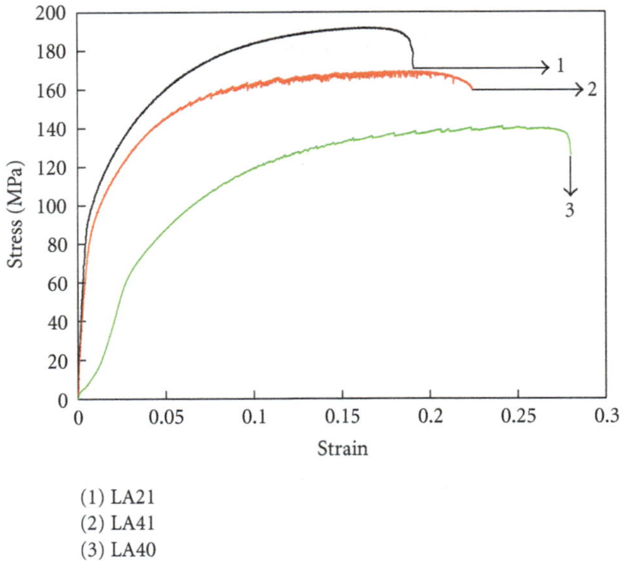

FIGURE 3: Stress-strain curves of three alloys under the strain rate of $6.66 \times 10^{-4}\ s^{-1}$ at 293 K.

Current findings are in good agreement with the characteristics of the DSA model [8, 9, 11, 12, 23–25]. Though DSA is generally a phenomenon observed in dilute alloys, it has been reported to occur in alloys whose solute contents are 20 at.% or even higher [26]. Existing explanations on the PLC effect of magnesium solid solutions are mainly confined to the framework of classical DSA model [1–4, 6]. In fact, LA41 itself is a solid solution [27], and, as mentioned above, this alloy shows negative strain rate sensitivity, which is generated via the operation of the dynamic strain aging effect. Therefore, it is justifiable to apply the DSA model to account for plastic instability in LA41, whose total solute content is about 13 at.%. Diffusion of both Li and Al solute atoms is believed to be the fundamental process of dynamic strain aging of the as-received LA41.

Figure 3 shows the comparison of stress-strain curves of the as-received LA41, LA21, and LA40 (Figure 3). These alloys exhibit different ultimate tensile strength (UTS) and yield strength (YS) due to different content of alloying elements of Al and Li.

It is of great importance to note that serrations are absent in LA21. In a separate paper [28], the activation energy for serrated flow of LA41 is calculated to be 106 $KJ \cdot mol^{-1}$, which is much larger than diffusion activation energy of Al and Li, respectively [29, 30]. That is to say, diffusion of Al and Li atoms could have been provoked before the appearance of serrated flow. Therefore, it can be concluded that plastic instability in the as-annealed LA41 is caused by the dynamic strain aging effect through the diffusion of both Al and Li atoms, while plastic instability in LA40 is caused by the diffusion of Li atoms. It could be further inferred that the absence of serrations in LA21 is probably due to insufficient dynamic strain aging effect of solute atoms, especially of Li atoms.

3.2. Variations of Serration Amplitude. It has been shown that the solute concentration around arrested mobile dislocations is dependent on both strain and strain rate, and the dependence can be described by the following equation [23]:

$$C = C_0\left\{1 + \left[\frac{K_2 \exp(-Q_m/kT)\varepsilon^{m+\beta}}{\dot{\varepsilon}}\right]^{2/3}\right\}, \qquad (1)$$

where C is the solute concentration around arrested mobile dislocations, C_0 is the solute concentration in matrix, K_2 is constant, Q_m is the activation energy for the onset of the serrated flow, m and β are materials constants related to vacancy concentration and mobile dislocation density, respectively, and kT has its usual meaning. Based on this equation, it is easy to know that the higher the solute concentration around the dislocations, the more intense the interaction between solute atoms and arrested dislocations is, and the larger stress required to overcome obstacles will be. Under a fixed temperature, strain and strain rate are the only two variables. Therefore, from (1), it can be demonstrated that, when the strain rate is fixed, the solute concentration will increase dramatically with increasing strain and $\Delta\sigma$ will be enhanced correspondingly. However, when the strain rate is raised, the diffusing process of solute atoms will possibly be slowed down and the concentration of solute atoms in the vicinity of arrested dislocations will be lowered; therefore, the pinning effect of solute atoms will be appreciably reduced and values $\Delta\sigma$ will be decreased as a corollary.

Another point that needs to be noted is that $\Delta\sigma$ is measured to be constant at large plastic strains. At such strains the deformation may have reached the stage of slow work hardening (stress plateau) where dislocation annihilation and rearrangement become substantial and the rate of generation of vacancies may not be as fast as that of in the rapid work-hardening stage. As a result, the diffusion process of Al and Li atoms through vacancies might be reduced, and the concentration of solute atoms around the arrested dislocations will be diluted. Accordingly, the pinning effect of solute atoms could become independent on the strain and the value of $\Delta\sigma$ would remain almost unchanged.

3.3. Disappearance of Serrations. Tensile tests were also performed on the as-annealed LA41 alloy, of which the stress-strain curve is shown in Figure 4. It can be noticed that the as-annealed LA41 alloy does not exhibit plastic instability. In addition, both YS and UTS decrease a lot after annealing, while the plasticity improves to a certain degree.

Figure 5 shows microstructures of as-received LA41 (Figure 5(a)) and as-annealed LA41 (Figure 5(b)). In the as-received state, it can be seen that the grains are homogenized and no twins and precipitates are observed. After annealing, however, a large number of lenticular twins are produced, which often end up with sharp points close to grain boundaries (GBs) and appear in multiple parallel groups intersecting each other. Moreover, bright field TEM image of that alloy presents with almost identical information (Figure 6).

The existence of annealing-induced twins might exert significant influence on the as-annealed LA41. It is believed

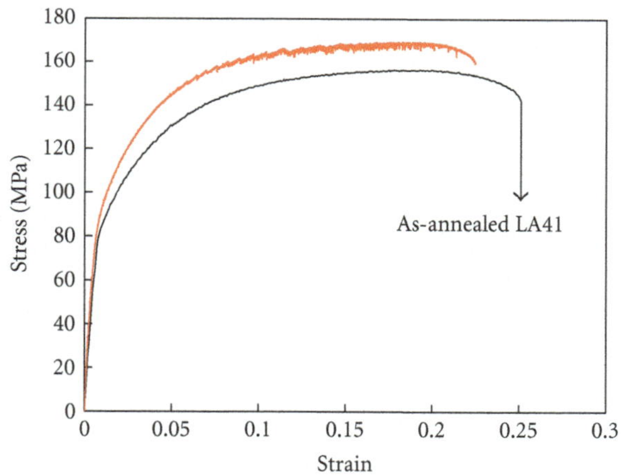

Figure 4: Comparison of stress-strain curves of the as-received and the as-annealed LA41 under the strain rate of $6.66 \times 10^{-4}\,s^{-1}$ at 293 K.

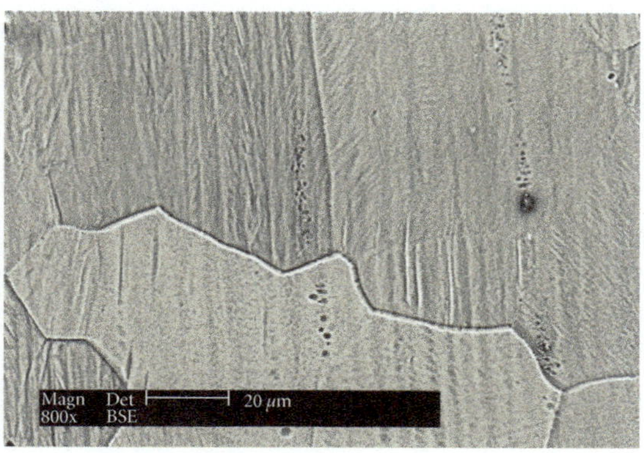

(a)

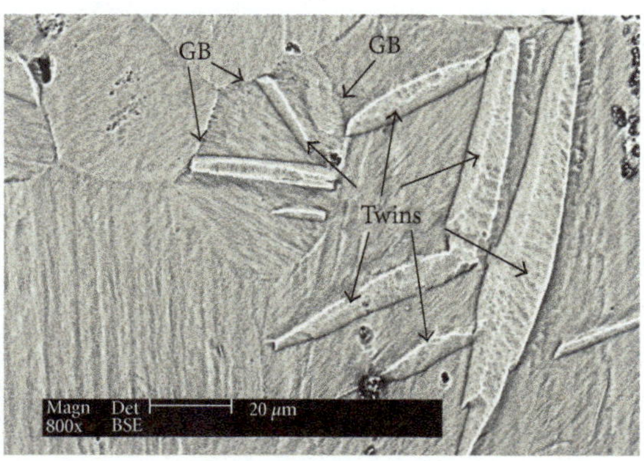

(b)

Figure 5: SEM micrographs of (a) the as-received LA41 and (b) the as-annealed LA41.

that they might act as barriers to mobile dislocations, disabling the DSA process to a large extent. Noting that the annealing temperature is well above 0.4 T_m (T_m is the melting point) and the annealing time is extremely long, it is important to consider the effect of recrystallization, which would lead to appreciable softening of the alloy and bring about lower YS and UTS as compared to the as-received LA41 alloy.

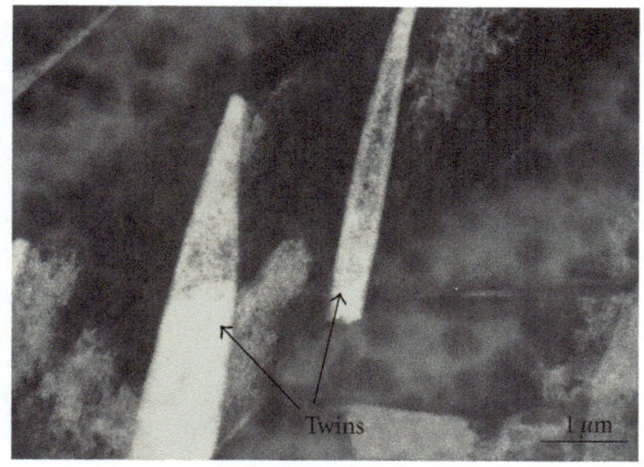

Figure 6: Bright field (BF) morphology of the as-annealed LA41.

4. Conclusion

Plastic instability was observed in HCP LA41 magnesium alloy in the strain rate range from $1.33 \times 10^{-4}\,s^{-1}$ to $6.66 \times 10^{-4}\,s^{-1}$ during tensile tests. Serration amplitude was measured to increase with accumulated strain but decrease with increasing strain rate. The dynamic strain aging model was applied to account for the observed phenomena. It was also demonstrated that additional annealing heat-treatment could remove the plastic instability.

Acknowledgment

Financial support from National Science Foundation of China (Grant no. 50371089) is greatly acknowledged.

References

[1] M. Chaturvedi, D. J. Lloyd, and K. Tangri, "Serrated yielding in magnesium-10 wt.-% Silver alloy," *Metal Science*, vol. 6, no. 1, pp. 16–19, 1972.

[2] X. Y. Fang, D. Q. Yi, and J. F. Nie, "The serrated flow behavior of Mg-Gd(-Mn-Sc) alloys," *Metallurgical and Materials Transactions A*, vol. 40, no. 11, pp. 2761–2771, 2009.

[3] S. M. Zhu and J. F. Nie, "Serrated flow and tensile properties of a Mg-Y-Nd alloy," *Scripta Materialia*, vol. 50, no. 1, pp. 51–55, 2004.

[4] V. Gartnerova, Z. Trojanova, A. Jager, and P. Palcek, "Deformation behaviour of Mg-0.7 wt.% Nd alloy," *Journal of Alloys and Compounds*, vol. 378, no. 1-2, pp. 180–183, 2004.

[5] C. H. Cáceres, C. J. Davidson, J. R. Griffiths, and C. L. Newton, "Effects of solidification rate and ageing on the microstructure and mechanical properties of AZ91 alloy," *Materials Science and Engineering A*, vol. 325, no. 1-2, pp. 344–355, 2002.

[6] C. Corby, C. H. Cáceres, and P. Lukáč, "Serrated flow in magnesium alloy AZ91," *Materials Science and Engineering A*, vol. 387–389, no. 1-2, pp. 22–24, 2004.

[7] A. H. Cottrell, "A note on the Portevin-Le Chatelier effect," *Philosophical Magazine*, vol. 44, no. 355, pp. 829–832, 1953.

[8] P. G. McCormigk, "A model for the Portevin-Le Chatelier effect in substitutional alloys," *Acta Metallurgica*, vol. 20, no. 3, pp. 351–354, 1972.

[9] A. van den Beukel, "Theory of the effect of dynamic strain aging on mechanical properties," *Physica Status Solidi A*, vol. 30, no. 1, pp. 197–206, 1975.

[10] A. W. Sleeswyk, "Slow strain-hardening of ingot iron," *Acta Metallurgica*, vol. 6, no. 9, pp. 598–603, 1958.

[11] R. A. Mulford and U. F. Kocks, "New observations on the mechanisms of dynamic strain aging and of jerky flow," *Acta Metallurgica*, vol. 27, no. 7, pp. 1125–1134, 1979.

[12] A. Van Den Beukel, "On the mechanism of serrated yielding and dynamic strain ageing," *Acta Metallurgica*, vol. 28, no. 7, pp. 965–969, 1980.

[13] P. Penning, "Mathematics of the portevin-le chatelier effect," *Acta Metallurgica*, vol. 20, no. 10, pp. 1169–1175, 1972.

[14] J. M. Robinson and M. P. Shaw, "Microstructural and mechanical influences on dynamic strain aging phenomena," *International Materials Reviews*, vol. 39, pp. 113–122, 1994.

[15] R. C. Picu, "A mechanism for the negative strain-rate sensitivity of dilute solid solutions," *Acta Materialia*, vol. 52, no. 12, pp. 3447–3458, 2004.

[16] F. Chmelík, A. Ziegenbein, H. Neuhäuser, and P. Lukáč, "Investigating the Portevin-Le Chaåtelier effect by the acoustic emission and laser extensometry techniques," *Materials Science and Engineering A*, vol. 324, no. 1-2, pp. 200–207, 2002.

[17] R. Shabadi, S. Kumar, H. J. Roven, and E. S. Dwarakadasa, "Characterisation of PLC band parameters using laser speckle technique," *Materials Science and Engineering A*, vol. 364, no. 1-2, pp. 140–150, 2004.

[18] Q. Zhang, Z. Jiang, H. Jiang, Z. Chen, and X. Wu, "On the propagation and pulsation of Portevin-Le Chatelier deformation bands: an experimental study with digital speckle pattern metrology," *International Journal of Plasticity*, vol. 21, no. 11, pp. 2150–2173, 2005.

[19] M. Dablij and A. Zeghloul, "Portevin-Le Chatelier plastic instabilities: characteristics of deformation bands," *Materials Science and Engineering A*, vol. 237, no. 1, pp. 1–5, 1997.

[20] M. Lebyodkin, Y. Brechet, Y. Estrin, and L. Kubin, "Statistical behaviour and strain localization patterns in the Portevin-Le Chatelier effect," *Acta Materialia*, vol. 44, no. 11, pp. 4531–4541, 1996.

[21] G. Ananthakrishna, "Spatio-temporal features of the Portevin-Le Chatelier effect," *Materials Science and Engineering A*, vol. 400-401, no. 1-2, pp. 210–213, 2005.

[22] M. Lebyodkin, L. Dunin-Barkovskii, Y. Bréchet, L. Kubin, and Y. Estrin, "Kinetics and statistics of jerky flow: experiments and computer simulations," *Materials Science and Engineering A*, vol. 234–236, pp. 115–118, 1997.

[23] C. P. Ling and P. G. McCormick, "The effect of temperature on strain rate sensitivity in an AlMgSi alloy," *Acta Metallurgica Et Materialia*, vol. 41, no. 11, pp. 3127–3131, 1993.

[24] S. Kumar, "Inverse behaviour of the onset strain of serrated flow," *Scripta Metallurgica et Materiala*, vol. 33, no. 1, pp. 81–86, 1995.

[25] M. Abbadi, P. Hähner, and A. Zeghloul, "On the characteristics of Portevin-Le Chatelier bands in aluminum alloy 5182 under stress-controlled and strain-controlled tensile testing," *Materials Science and Engineering A*, vol. 337, no. 1-2, pp. 194–201, 2002.

[26] I. S. Kim and M. C. Chaturvedi, "Serrated flow in Al-5 wt.% Mg alloy," *Materials Science and Engineering*, vol. 37, no. 2, pp. 165–172, 1979.

[27] D. W. Levinson and D. J. McPherson, "Phase relations in magnesium-lithium-aluminum alloys," *Transactions of the American Society for Metals*, vol. 48, pp. 689–705, 1956.

[28] C. Wang, Y. Xu, and E. Han, "PLC effect and its explanation of LA41 magnesium alloys," *Acta Metallurgica Sinica*, vol. 42, no. 2, pp. 191–194, 2006.

[29] W. Gasior and Z. Moser, "Chemical diffusion coefficients in solid al-rich aluminium-magnesium alloys," *Archives of Metallurgy*, vol. 46, no. 4, pp. 455–463, 2001.

[30] W. Gasior and Z. Moser, *Archives of Metallurgy*, vol. 46, p. 269, 2001.

18

Effect of Diacetyl Monoxime Thiosemicarbazone on the Corrosion of Aged 18 Ni 250 Grade Maraging Steel in Sulphuric Acid Solution

T. Poornima,[1] Jagannath Nayak,[2] and A. Nityananda Shetty[3]

[1] *Department of Science and Humanities, PESIT, Bangalore 560085, India*
[2] *Department of Metallurgical and Materials Engineering, National Institute of Technology Karnataka, Surathkal, Srinivasnagar, Karnataka 75025, India*
[3] *Department of Chemistry, National Institute of Technology Karnataka, Surathkal, Srinivasnagar, Karnataka 575025, India*

Correspondence should be addressed to A. Nityananda Shetty, nityashreya@gmail.com

Academic Editor: Sunghak Lee

The corrosion inhibition of the aged 18 Ni 250 grade maraging steel in 0.5 M sulphuric acid by diacetyl monoxime thiosemicarbazone (DAMTSC) at 303–323 K has been investigated by potentiodynamic polarization, EIS, and SEM techniques. Good inhibition efficiency of DAMTSC was revealed even at low concentrations, which increased with the increase in DAMTSC concentration and decreased with the increase in temperature. The activation energies, E_a, as well as other thermodynamic parameters (ΔG^0_{ads}; ΔH^0_{ads}; ΔG^0_{ads}), were evaluated and discussed. The adsorption of DAMTSC on the aged maraging steel surface was found to obey the Langmuir adsorption isotherm model and shows mixed type inhibition behavior.

1. Introduction

Corrosion of structural elements is a major issue for any industry because of the chemical environment of chemical processing. Maraging steels are special class of ultrahigh strength steels that differ from conventional steels in that they are hardened by a metallurgical reaction that does not involve carbon [1]. They derive high strength from age hardening of low carbon, Fe-Ni martensitic matrix [2]. The need for highly reliable substances of high strength and high ductility is gradually increasing with the development of aerospace industry. According to available literature, atmospheric exposure of 18 Ni maraging steel leads to corrosion in a uniform manner and the steel becomes completely rust covered [3]. Pit depths tend to be shallower than high strength steels [4]. Critical and passive current densities increase as the structure is varied from fully annealed to fully aged [5]. Maraging steels are found to be less susceptible to hydrogen embrittlement than common high strength steels owing to significantly low diffusion of hydrogen in them [6]. Several technical papers covering alloy design, material processing, thermo-mechanical treatments, welding, strengthening mechanisms, and so forth, have been published [7]. Search in the literature reveals very little work being reported on the corrosion behavior and on the use of inhibitors in controlling corrosion of maraging steel except some of our publications [8, 9]. Metals and alloys frequently come in contact with acid solutions during cleaning, pickling, descaling, acidising, and so forth. Materials used in acid environment should have good corrosion resistance. The present work is intended to study the corrosion inhibition of aged 18 Ni 250 grade maraging steel in 0.5 M sulphuric acid medium using DAMTSC as inhibitor.

The majority of the well known inhibitors are organic compounds containing heteroatom, such as O, N, or S and multiple bonds or aromatic rings, which allow an adsorption on the metal surface through lone pair of electrons and/or pi electrons present in these molecules [10, 11]. Sulphur containing compounds are preferred for H_2SO_4 while those containing nitrogen act more effectively in HCl [12]. The inhibition efficiency should increase in the order O < N < S < P [13]. The molecules that contain both nitrogen and

Table 1: Composition of the aged maraging steel specimen.

Element	Composition
C	0.015%
Ni	17–19%
Mo	4.6–5.2%
Co	7–8.5%
Si	0.1%
O	30 ppm
H	2.0 ppm
Ti	0.3–0.6%
Al	0.005–0.15%
Mn	0.1%
P	0.01%
S	0.01%
N	30 ppm
Fe	Balance

sulphur in their structures are of particular importance, since these provide an excellent inhibition compared with the compounds that contain only sulphur or nitrogen [14]. Thiosemicarbazones and their derivatives have continued to be subject of extensive investigation in chemistry and biology owing to their broad spectrum of antitumor [15], antimalarial [16], antiviral [17], antibacterial [18], antifungal [19], and many other applications including corrosion inhibition of metals [20–23].

2. Experimental Methods

2.1. Materials. The maraging steel samples (M 250 grade) in aged condition were taken from plates. The solution annealed and air cooled plates were subjected to aging treatment at 480 ± 5°C for 3 hours and air cooled. Percentage composition of 18 Ni 250 grade aged maraging steel samples is given in Table 1. Cylindrical test coupons were cut from the plate and sealed with epoxy resin in such a way that, the area exposed to the medium is 0.503 cm^2. These coupons were polished as per standard metallographic practice, belt grinding followed by polishing on emery papers, finally on polishing wheel using legated alumina to obtain mirror finish, degreased with acetone, washed with double distilled water, and dried before immersing in the corrosion medium. The inhibitor diacetyl monoxime thiosemicarbazone (DAMTSC) was synthesized as per the reported procedure [24] by reaction between diacetyl monoxime and thiosemicarbazide and its structure is given in Scheme 1.

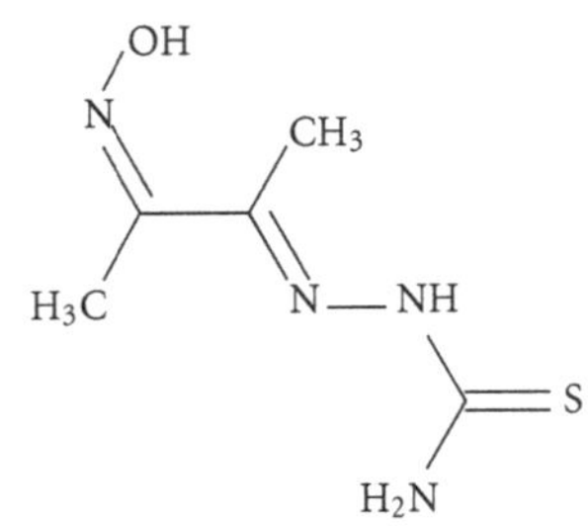

Scheme 1

2.2. Medium. Standard 0.5 M sulphuric acid solution was prepared by diluting analar grade 98% sulphuric acid by using double distilled water. The solutions of inhibitor with 1.14×10^{-4} M, 2.85×10^{-4} M, 5.70×10^{-4} M and 8.55×10^{-4} M concentration were prepared in standard 0.5 M sulphuric acid. Experiments were carried out using calibrated thermostat at temperatures 303 K, 308 K, 313 K, 318 K, 323 K (±0.5 K).

2.3. Electrochemical Measurement

2.3.1. Tafel Polarisation Studies. Electrochemical measurements were carried out by using an electrochemical work station, Auto Lab 30 and GPES software. Tafel plot measurements were carried out using conventional three electrode Pyrex glass cell with platinum counter electrode and saturated calomel electrode (SCE) as reference electrode. All the values of potential are referred to the SCE. Finely polished maraging steel specimens were exposed to corrosion medium of sulphuric acid in the presence and absence of inhibitor at different temperatures and allowed to establish a steady state open circuit potential. The potentiodynamic current-potential curves were recorded by polarizing the specimen to −250 mV cathodically and +250 mV anodically with respect to open circuit potential (OCP) at scan rate of 1 mV s^{-1}.

2.3.2. Electrochemical Impedance Spectroscopy Studies (EIS). The corrosion behavior of the maraging steel specimen was also obtained from EIS technique using electrochemical work station, Auto Lab 30, and FRA software. In EIS technique a small amplitude ac signal of 10 mV and frequency spectrum from 100 kHz to 0.01 Hz was impressed at the OCP and impedance data were analyzed using Nyquist plots. The charge transfer resistance, R_{ct} was extracted from the diameter of the semicircle in Nyquist plot.

In all the above measurements, at least three similar results were considered and their average values are reported.

2.3.3. Scanning Electron Microscopy (SEM) Analysis. The SEM images were recorded using JEOL JSM-6380 LA analytical scanning electron microscope.

3. Results and Discussion

3.1. Tafel Polarization Measurement. The corrosion inhibition of aged maraging steel investigated in 0.5 M sulphuric acid containing various concentration of DAMTSC inhibitor at different temperatures using Tafel polarization technique and Tafel plots at 303 K are shown in Figure 1. The valuable potentiodynamic polarization parameters including corrosion potential (E_{corr}), corrosion current density (I_{corr}), anodic and cathodic slopes (b_a and b_c), and corrosion rate

(C.R) were calculated from Tafel plots and are summarized in Table 2. The C.R is calculated using

$$\text{corrosion rate (mm/y)} = \frac{3270(\text{E.W})I_{corr}}{D}, \quad (1)$$

where 3270 is a constant that defines the unit of corrosion rate, I_{corr} = corrosion current density in A/cm^2, D = density of the corroding material, 8.23 g/cm^3, E.W = equivalent weight of corroding material (atomic weight/oxidation number), 29.2 [25]. The percentage inhibition efficiency (IE%) was calculated from

$$\text{IE\%} = \frac{I_{corr} - I_{corr(inh)}}{I_{corr}} \times 100, \quad (2)$$

where I_{corr} and $I_{corr(inh)}$ are corrosion current densities in the absence and in the presence of inhibitor, respectively. The value of corrosion current density (I_{corr}) decreased effectively even with low concentration of DAMTSC and inhibition efficiency increases with increase inhibitor concentration. The presence of inhibitor does not shift the E_{corr} remarkably, implying that DAMTSC acts as mixed type inhibitor affecting both anodic and cathodic reactions [23]. According to Li and others [26], if the displacement in corrosion potential is more than ±85 mV with respect to corrosion potential of the blank, the inhibitor can be considered as a cathodic or anodic type. But the maximum displacement in this study is less than 10 mV, which indicates that DAMTSC is a mixed type inhibitor. According to Cao [27] if the shift in E_{corr} is negligible the inhibition is most probably caused by a geometric blocking effect of the adsorbed inhibitive species on the surface of corroding metal. The anodic and cathodic Tafel slopes remain almost unchanged upon addition of inhibitor. Thus the adsorbed inhibitor acts by simple blocking of active sites for anodic and cathodic processes, thereby reducing the surface area for corrosion without affecting the corrosion mechanism [28, 29]. The parallel cathodic Tafel curves suggest that the hydrogen evolution is activation controlled and the reduction mechanism is not affected by the presence of the inhibitors [30].

For anodic polarization curves of specimen with DAMTSC (in Figure 2), it seems that at the working electrode potential, higher than −200 mV/SCE, the presence of inhibitor does not change the current versus potential characteristics. At this potential, the anodic current increases steeply and a flat region is observed. This potential can be defined as desorption potential [31]. The phenomenon may be due to the obvious metal dissolution, which leads to desorption of the inhibitor molecule from the electrode surface. In this case, the desorption rate of the inhibitor is higher than its adsorption rate, so the corrosion current increases more obviously with rising potential [32]. At still higher polarization potential the anodic current density change drastically, resulting in sharp increase in Tafel slope. The Tafel slope of the third portion is nearly equal to that in the absence of DAMTSC. Above desorption potential coverage of inhibitor decreases rapidly. Similar conclusion was reported by Qu et al. [33].

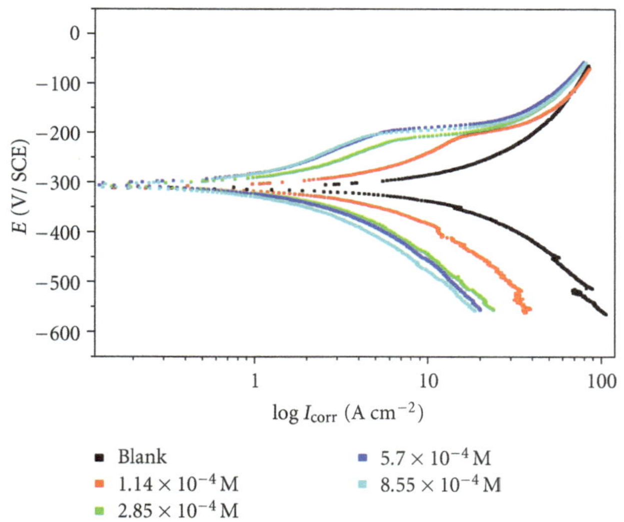

FIGURE 1: Tafel polarization curves for the aged maraging steel in 0.5 M sulphuric acid containing different concentrations inhibitor.

3.2. Electrochemical Impedance Spectroscopy. The results of potentiodynamic polarization experiments were compared with the results of impedance measurements, since EIS has emerged as probably the most powerful technique currently available for identifying corrosion reaction mechanisms. In order to get more information about the corrosion inhibition of aged maraging steel specimens in 0.5 M sulphuric acid containing different concentrations of DAMTSC, EIS measurements were carried out at different temperatures and they are displayed as Nyquists plots. The Nyquist plots obtained for aged samples of maraging steel specimens in 0.5 M sulphuric acid with and without various concentrations of DAMTSC are as shown in Figure 2.

It is clear from the figures that the shapes of the impedance plots for inhibited specimen are different from those of uninhibited one. The Nyquists plots of aged maraging steel in 0.5 M H_2SO_4 without inhibitor are characterized by two time constants with a depressed capacitive semicircle at high frequency region (HF) followed by an inductive loop at low frequency (LF) region. These are not perfect semicircles, because the Nyquist plots obtained in the real system represent a general behavior where the double layer at the metal solution interface does not behave as an ideal capacitor [34]. The depressed capacitive loop with its centre below the real axis often refers to the frequency dispersion of interfacial impedance which has been attributed to the roughness, inhomogeneity of the solid surfaces, and adsorption of inhibitors [10]. The HF capacitive loop was attributed to charge transfer of the corrosion process and time constant of the electric double layer [35]. The LF inductive loop can be attributed to relaxation process obtained by adsorbed sulphate ions and protons [36]. It may also be attributed to the redissolution of the passivated surface at low frequencies [37]. In the present case, the low frequency inductive loop in inhibitor free acid medium can be attributed to surface dissolution process at low frequencies. Based on the shape

TABLE 2: Results of Tafel polarization studies on aged maraging steel in 0.5 M sulphuric acid with and without different concentrations of DAMTSC at various temperatures.

Temp (K)	Conc. of inhibitor ($\times 10^{-4}$) (M)	E_{corr} (V)	$I_{corr} \times (10^{-4})$ (A cm^{-2})	b_c (V dec^{-1})	b_a (V dec^{-1})	C.R in mm Y^{-1}	IE (%)
303	0.00	−0.310	75.26	0.107	0.103	87.30	
	1.14	−0.310	27.18	0.103	0.102	31.53	63.9
	2.85	−0.306	11.72	0.095	0.106	13.60	84.4
	5.70	−0.306	9.57	0.106	0.113	11.10	87.3
	8.55	−0.303	7.73	0.093	0.111	8.97	89.7
308	0.00	−0.312	85.43	0.122	0.127	99.10	
	1.14	−0.310	32.07	0.127	0.098	37.20	62.5
	2.85	−0.309	21.64	0.098	0.112	25.10	74.7
	5.70	−0.304	13.19	0.112	0.115	15.30	84.6
	8.55	−0.307	10.99	0.115	0.118	12.75	87.1
313	0.00	−0.309	96.55	0.118	0.112	112.00	
	1.14	−0.309	37.67	0.112	0.086	43.70	61.0
	2.85	−0.307	26.21	0.086	0.084	30.40	72.9
	5.70	−0.308	17.76	0.084	0.106	20.60	81.6
	8.55	−0.304	14.40	0.106	0.086	16.70	85.1
318	0.00	−0.312	106.90	0.086	0.117	124.00	
	1.14	−0.305	50.95	0.117	0.098	59.10	56.5
	2.85	−0.305	36.81	0.098	0.095	42.70	68.6
	5.70	−0.308	24.22	0.095	0.089	28.10	79.3
	8.55	−0.301	19.05	0.089	0.098	22.10	83.8
323	0.00	−0.311	117.24	0.098	0.123	136.00	
	1.14	−0.303	58.97	0.123	0.095	68.40	49.7
	2.85	−0.305	42.84	0.095	0.091	49.70	63.5
	5.70	−0.303	30.09	0.091	0.088	34.90	74.3
	8.55	−0.301	25.95	0.088	0.088	30.10	77.9

of Nyquist plots we have decided to use equivalent circuit as given in Figure 3(a) which has been previously used to model iron/acid interface [38]. It consists of solution resistance (R_s), charge transfer resistance (R_{ct}), and constant phase element (CPE). The surface roughness, degree of polycrystallinity, and also anion adsorption lead to capacitance dispersion at solid electrodes and hence the real iron/acid interface system deviates from ideal capacitive behavior, which can be empirically represented by CPE [39]. The point of intersection between inductive loop and the real axis represents ($R_s + R_{ct}$). R_s represents solution resistance due to the ohmic resistances of corrosion product films and the solution enclosed between the working electrode and the reference electrode. R_{ct} represents the charge transfer resistance whose value is a measure of electron transfer across the surface and is inversely proportional to corrosion rate [40].

The Nyquists plots of aged maraging steel in 0.5 M H_2SO_4 containing various concentrations of DAMTSC consists of two semicircles at high and middle frequency range and an inductive loop at low frequency range. These plots are analysed by an equivalent circuit, which is represented in Figure 3(b). The smaller HF capacitive loop suggested that some sort of film exists on this alloy in the studied solution which can be attributed to film formation by adsorbed inhibitor while larger middle frequency capacitive loop can be attributed to charge transfer reaction [41–44]. The resistance values corresponding to high frequency and middle frequency semicircles are represented as R_f and R_{ct}, respectively, in equivalent circuit Figure 3(b). Both R_f and R_{ct} increase with increase in inhibitor concentration. The sum of R_f and R_{ct} are taken for calculation of inhibition efficiency. The capacitance of film (C_f) was represented by CPE_1 and double layer capacitance (C_{dl}) was represented as CPE_2. All these values are shown in Table 3. The low frequency inductive loop in inhibited acid solution might be attributed to the surface dissolution, indicating that the alloy still dissolved by the direct charge transfer at the inhibitor adsorbed steel surface [45]. The fact that this inductive loop cannot be observed at maximum concentration of DAMTSC supports this view. The study at higher temperature reveals single semicircle at low concentrations of inhibitor and appearance of two semicircles only at high concentration of

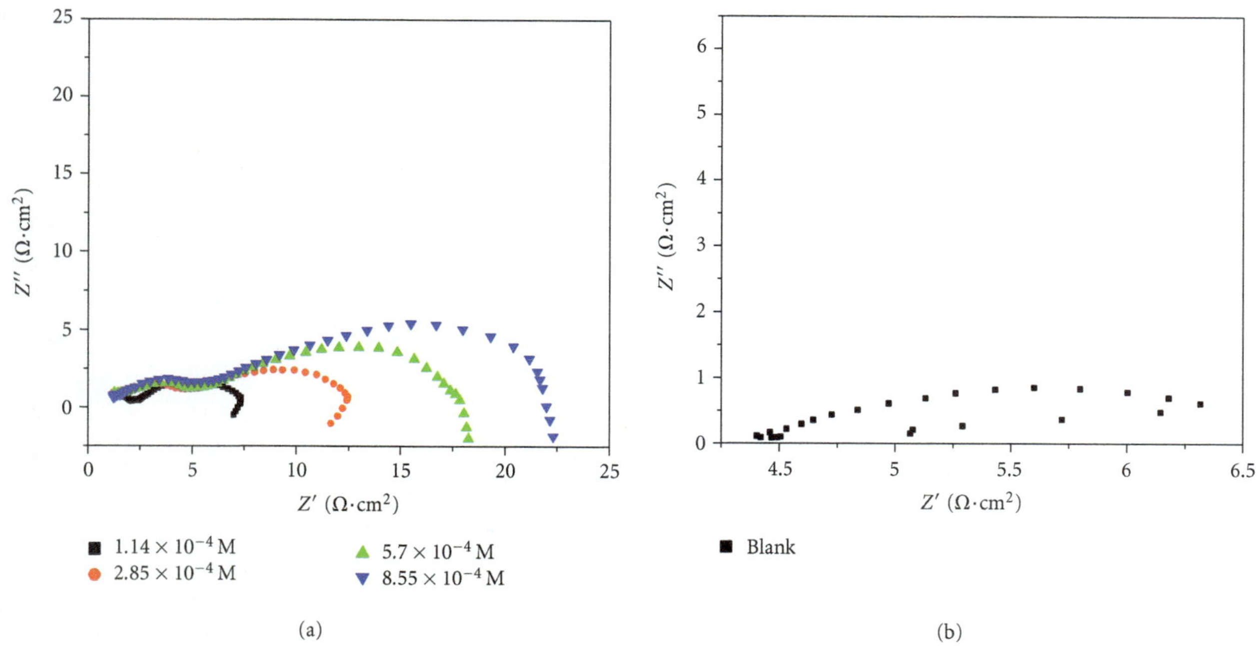

FIGURE 2: Nyquist plots for the aged maraging steel specimen in 0.5 M sulphuric acid (a) containing different concentrations of inhibitor, (b) without inhibitor.

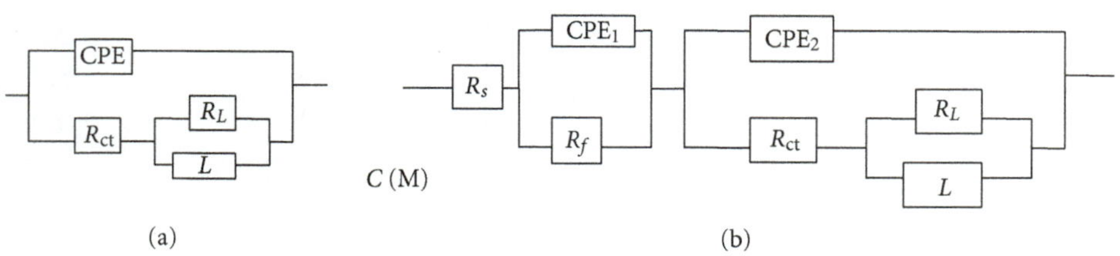

FIGURE 3: Equivalent circuit used to fit experimental EIS data for the corrosion of aged maraging steel specimen in sulphuric acid (a) without inhibitor, (b) with inhibitor.

inhibitor, this can be attributed to the fact that formation of inhibitor film is favorable only at higher inhibitor concentration above 303 K.

The impedance of CPE is expressed as

$$Z_{\mathrm{CPE}} = \frac{1}{Y_0 (jw)^n}, \quad (3)$$

where the amplitude Y_0 and n are frequency independent, and ω is the angular frequency for which $-Z''$ reaches its maximum values, n is dependent on the surface morphology: $-1 \le n \le 1$. The HF loops have depressed semicircular appearance, with $0.5 \le n \le 1$, which is often referred to as frequency dispersion as a result of the non homogeneity or roughness [46].

The double layer capacitances C_{dl}, for a circuit including CPE were calculated from the following equation [47]:

$$C_{\mathrm{dl}} = Y_0 (w_{\max})^{n-1}, \quad (4)$$

where $w_{\max} = 2\pi f_{\max}$, and $f_{\max}$ is the frequency at which the imaginary component of the impedance is maximal. According to the expression of the double layer capacitance presented in the Helmholtz model [48],

$$C_{\mathrm{dl}} = \frac{\varepsilon \varepsilon_0}{d} S, \quad (5)$$

where d is the thickness of the film, S is the surface area of the electrode, ε_0 is the permittivity of air, and ε is the local dielectric constant. The value of C_{dl} decreases due to adsorption of inhibitor molecules, which displaces water molecules originally adsorbed on the mild steel surface and decreases the active surface area. The values of double layer capacitance decreases with increase in inhibitor concentration indicating that inhibitor molecules function by adsorption at the metal/solution interface, leading to protective film on the alloy surface, and decreasing the extent of dissolution reaction [49].

The corrosion current density I_{corr} can be calculated using charge transfer resistance R_{ct}, together with Stern-Geary equation [40]:

$$I_{\mathrm{corr}} = \frac{b_a b_c}{2.303 R_{\mathrm{ct}} A (b_a + b_c)}. \quad (6)$$

TABLE 3: EIS data of aged maraging steel in 0.5 M sulphuric acid with and without different concentrations of DAMTSC at various temperatures.

Temp. (K)	Conc. of inhibitor ($\times 10^{-4}$) (M)	R_f Ohm·cm^2	R_{ct} Ohm·cm^2	R_{total} Ohm·cm^2	C_f (10^{-6}) (F)	C_{dl} (10^{-6}) (F)	IE (%)
303	0.00		2.8	2.80		14490	
	1.14	1.17	5.66	6.83	874	9920	59.0
	2.85	5.37	9.38	14.75	799	9762	81.0
	5.70	6.12	13.5	19.62	653	1590	85.7
	8.55	7.89	17.8	25.70	509	1163	89.1
308	0.00		2.30	2.30		87620	
	1.14		5.10	5.10		10112	54.9
	2.85		8.87	8.87		7566	74.2
	5.70	5.6	11.2	16.8	954	6347	86.3
	8.55	7.19	11.6	18.7	765	4850	87.7
313	0.00		1.6	1.60		113500	
	1.14		4.32	3.32		23724	51.8
	2.85		5.75	5.75		20880	72.2
	5.70	3.43	7.34	10.8	2417	11340	85.2
	8.55	4.2	7.9	12.1	1450	10800	86.7
318	0.00		1.4	1.4		226570	
	1.14		2.7	2.7		31326	48.1
	2.85		4.8	4.8		27667	70.8
	5.70	3.25	5.05	8.3	2625	23471	83.1
	8.55	3.40	5.3	8.7	1562	20125	83.9
323	0.00		1.3	1.3		541000	
	1.14		2.3	2.3		33540	43.5
	2.85		3.5	3.5		24500	62.9
	5.70	1.44	4.16	5.6	2966	20760	76.8
	8.55	2.20	4.90	7.1	1695	15600	81.7

The charge transfer resistance R_{ct} and double layer capacitance C_{dl} were determined by the analysis of Nyquist plot and their values are given in Table 3. The inhibition efficiency in different concentrations of DAMTSC was calculated from the charge transfer resistance according to

$$\mathrm{IE}\% = \left(\frac{R_{ct} - R_{ct}^{0}}{R_{ct}}\right) \times 100, \quad (7)$$

where R_{ct} and R_{ct}^{0} represent the charge transfer resistance in the presence and absence of inhibitor. The values of R_{ct} increases with increase in inhibitor concentration and the results indicate that charge transfer process mainly controls the corrosion process. The inhibition efficiency values obtained by the electron impedance method are in good agreement with the ones obtained by Tafel's method.

3.3. Effect of Temperature. The effect of temperature on the inhibited acid-metal reaction is highly complex because many changes occur on the metal surface, such as rapid etching and desorption of the inhibitor and the inhibitor itself, and in some cases may undergo decomposition and/or rearrangement [10]. However it facilitates the calculation of many thermodynamic functions for the inhibition and/or the adsorption processes which contribute in determining the type of adsorption of the studied inhibitors. In the present study, with increase in solution temperature, corrosion potential (E_{corr}), anodic Tafel slope (b_a), and cathodic Tafel slope (b_c) values are not affected much. This indicates that increase in temperature does not change the mechanism of corrosion reaction. But I_{corr} and hence the corrosion rate of the specimen increases with increase in temperature for both blank and inhibited solutions. This may be attributed to the fact that the hydrogen evolution overpotential decreases with increase in temperature that leads to increase in cathodic reaction rate [50]. The inhibition efficiency decreases with increase in temperature which indicates desorption of inhibitor molecules [51]. This is suggestive of physisorption [52]. But this decrease is small at higher concentration of inhibitor.

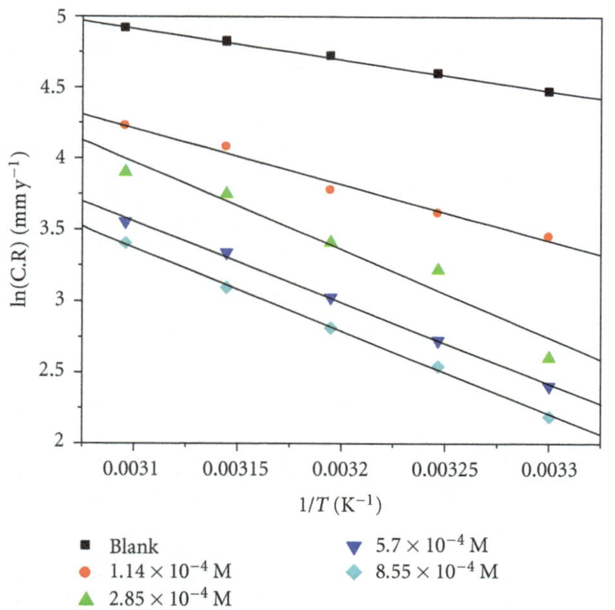

Figure 4: Arrhenious plots for the dissolution of aged maraging steel in 0.5 M sulphuric acid containing different concentrations of inhibitor.

The apparent activation energy (E_a) for the corrosion process in the presence and absence of inhibitor can be calculated using Arrhenius law equation [10],

$$\ln(\text{C.R}) = B - \left(\frac{E_a}{RT}\right), \quad (8)$$

where B is a constant which depends on the metal type and R is the universal gas constant. The plot of ln(corrosion rate) versus reciprocal of absolute temperature $1/T$ gives straight line whose slope $= -E_a/R$ gives activation energy for the corrosion process. The Arrhenius plots for the specimen are shown in Figure 4.

The entropy and enthalpy of activation for the dissolution of alloy, ΔH_a and ΔS_a, were calculated from transition state theory equation [10]

$$\text{C.R} = \left(\frac{RT}{Nh}\right)\exp\left(\frac{\Delta S_a}{R}\right)\exp\left(\frac{-\Delta H_a}{RT}\right), \quad (9)$$

where h is Plank's constant, N is Avagadro's number. A plot of ln(corrosion rate/T) versus $1/T$ gives straight line with slope $= -\Delta H_a/R$ and intercept $= \ln(R/Nh) + \Delta S_a/R$. The calculated values of E_a, ΔH_a, and ΔS_a are given in Table 4. The plot of ln(corrosion rate/T) versus $1/T$ for aged samples of maraging steel is shown in Figure 5.

The effect of chemically stable surface active inhibitors is to increase the energy of activation and to decrease the surface area available for corrosion [53]. The value of activation energy (E_a) in 0.5 M sulphuric acid solution containing inhibitor is greater than that without inhibitor. The extent of increase is proportional to the inhibitor concentration, indicating that the energy barrier for the corrosion reaction increases with increase in DAMTSC concentration.

Table 4: Activation parameters for the corrosion of aged maraging steel in 0.5 M sulphuric acid with and without different concentrations of DAMTSC.

Conc. of inhibitor ($\times10^{-4}$) (M)	E_a kJ mol^{-1}	ΔH_a kJ mol^{-1}	ΔS_a J mol^{-1} K^{-1}
0	18.1	15.5	−156.7
1.14	32.7	30.1	−117.3
2.85	47.3	44.6	−77.7
5.70	48.4	45.8	−75.7
8.55	51.1	48.4	−62.5

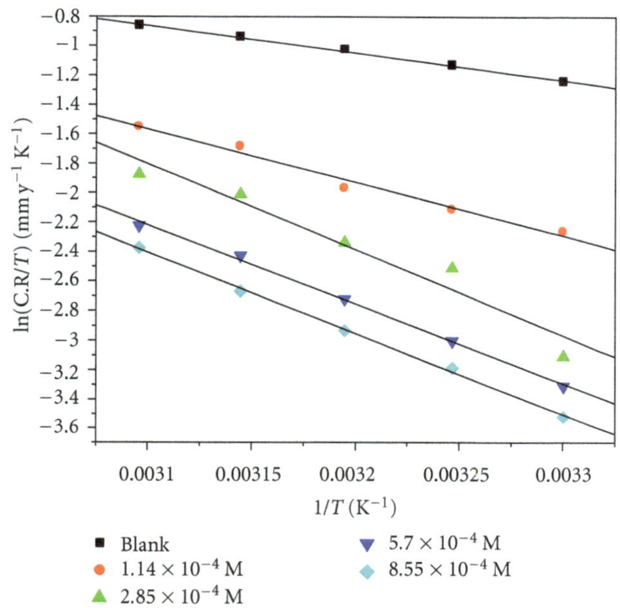

Figure 5: ln(corrosion rate/T) versus $1/T$ for aged maraging steel in 0.5 M sulphuric acid containing different concentrations of inhibitor.

The increase in apparent activation energy E_a may be interpreted as due to physical adsorption [54] of the inhibitor, which results in increase in surface coverage with the increase in the concentration of the inhibitor. The decrease in the inhibition efficiency of DAMTSC with increase in temperature can be attributed to an appreciable decrease in the adsorption of the inhibitor on the metal surface with increase in temperature and a corresponding increase in corrosion rate due to the fact that greater area of metal is exposed to the acid [55].

The values of entropy of activation in the absence and presence of inhibitor are large and negative. This implies that the activated complex in the rate determining step represents an association rather than dissociation step, resulting in a decrease in randomness on going from reactants to activated complex [56, 57]. The entropy of activation values are more for inhibited solutions than that for the uninhibited solutions. This suggested that an increase in randomness occurred on going from reactants to the activated complex. This might be the results of the adsorption of organic inhibitor molecules from the acidic solution which could be regarded as a quasi-substitution process between the organic compound in the aqueous phase and water molecules at electrode surface [58, 59]. In this situation, the adsorption of organic inhibitor is accompanied by desorption of water molecules from the surface. Thus the increasing in entropy of activation was attributed to the increase in solvent entropy [60].

3.4. Adsorption Isotherm. Organic corrosion inhibitors are known to decrease metal dissolution via adsorption on the metal/corrodent interface to form a protective film which separates the metal surface from the corrosive medium. The adsorption route is usually regarded as a substitution process between the organic inhibitor in the aqueous solution [$Inh_{(sol)}$] and water molecules adsorbed at the metal surface [$H_2O_{(ads)}$] as follows [61]:

$$\mathrm{Inh}_{(\mathrm{sol})} + \chi \mathrm{H_2O}_{(\mathrm{ads})} \longleftrightarrow \mathrm{Inh}_{(\mathrm{ads})} + \chi \mathrm{H_2O}_{(\mathrm{sol})}, \qquad (10)$$

where χ represents the number of water molecules replaced by one molecule of adsorbed inhibitor. The adsorption bond strength is dependent on the composition of the metal, corrodent, inhibitor structure, concentration, and orientation as well as temperature. Basic information on the interaction between the inhibitor and alloy surface can be provided by adsorption isotherm. In order to obtain isotherm, the linear relation between surface coverage (θ) value and C_{inh} must be found. The surface coverage θ is given by

$$\theta = \frac{\mathrm{IE}\%}{100}, \qquad (11)$$

where IE% is percentage inhibition efficiency as calculated using (2).

Attempts were made to fit the degree of surface coverage (θ) values to various isotherms including Langmuir, Temkin, Frumkin, and Flory-Huggins isotherms. By far the best fit is obtained with the modified Langmuir adsorption isotherm. Langmuir isotherm for monolayer chemisorption is given by

$$\frac{C_{\mathrm{inh}}}{\theta} = \frac{1}{K_{\mathrm{ads}}} + C_{\mathrm{inh}}, \qquad (12)$$

where C_{inh} is the concentration of inhibitor, K_{ads} is the equilibrium constant for adsorption process, and θ is the degree of the surface coverage which is calculated using (11).

This model has also been used for other inhibitor systems [61]. The plot of C_{inh}/θ versus C_{inh} gives a straight line with intercept $1/K_{\mathrm{ads}}$ as shown in Figure 6.

The values of standard free energy of adsorption are related to K_{ads} by the relation (13).

$$K_{\mathrm{ads}} = \frac{1}{55.5} e^{(-\Delta G^0_{\mathrm{ads}}/RT)}, \qquad (13)$$

where the value 55.5 is the concentration of water in solution in M (mol/L), R is the universal gas constant and T is absolute temperature [61]. The thermodynamic data obtained from adsorption isotherm are tabulated in Table 5. The correlation coefficient (R^2) was used to choose the isotherm

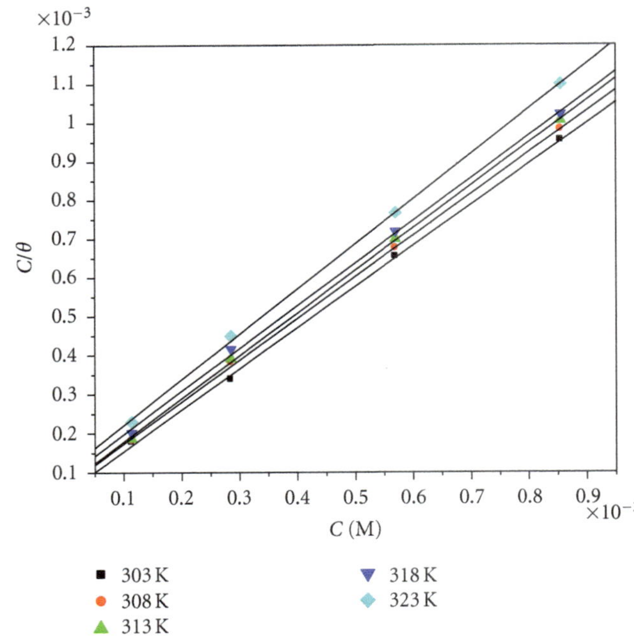

FIGURE 6: Langmuir adsorption isotherm of DAMTSC on aged maraging steel in 0.5 M sulphuric acid at different temperatures.

that best fit experimental data [10]. The linear regression coefficients are close to unity and the slopes of straight lines are nearly unity, suggesting that the adsorption of DAMTSC obeys Langmuir's adsorption isotherm and there is negligible interaction between the adsorbed molecules [26]. The high values of K_{ads} for the studied inhibitor indicate strong adsorption of inhibitor on the alloy surface. Negative values of $\Delta G^0_{\mathrm{ads}}$ are characteristic feature of strong spontaneous adsorption for the studied compounds, which also reflect the high values of inhibition. The negative $\Delta G^0_{\mathrm{ads}}$ values calculated from (13) are consistent with the spontaneity of the adsorption process and the stability of the adsorbed layer on the maraging steel surface. Generally, the standard free energy values of $-20\,\mathrm{kJ\,mol^{-1}}$ or less negative are associated with an electrostatic interaction between charged molecules and charged metal surface, resulting in physisorption and those of $-40\,\mathrm{kJ\,mol^{-1}}$ or more negative involve charge sharing or transfer from the inhibitor molecules to the metal surface to form a coordinate covalent bond, resulting in chemisorption [62]. The $\Delta G^0_{\mathrm{ads}}$ values obtained for the studied inhibitor on the alloy surface in 0.5 M sulphuric acid are −33 to −35 kJ/mol. These values predict both physical and chemical adsorption of inhibitor [63]. A plot of $\Delta G^0_{\mathrm{ads}}$ versus T was used to calculate heat of adsorption $\Delta H^0_{\mathrm{ads}}$ and the standard adsorption entropy $\Delta S^0_{\mathrm{ads}}$ according to the following thermodynamic basic equation [10]:

$$\Delta G^0_{\mathrm{ads}} = \Delta H^0_{\mathrm{ads}} - T\Delta S^0_{\mathrm{ads}}. \qquad (14)$$

The thermodynamic data obtained for DAMTSC using Langmuir adsorption isotherm are depicted in Table 5.

The values of thermodynamic parameters for the adsorption of inhibitors can provide valuable information about the mechanism of corrosion inhibition. While an

Table 5: Thermodynamic parameters for the adsorption of DAMTSC on aged maraging steel surface in 0.5 M sulphuric acid.

Temperature (°C)	K (mol^{-1} L)	$-\Delta G^0_{ads}$ kJ mol^{-1}	R^2	Slope	ΔH^0_{ads} kJ mol^{-1}	ΔS^0_{ads} J mol^{-1} K^{-1}
303	20227	35.1	0.9997	1.05		
308	14931	34.3	0.9997	1.07	−62.25	90
313	14391	34.2	0.9997	1.09		
318	11246	33.6	0.9995	1.09		
323	9476.9	33.2	0.9997	1.14		

endothermic adsorption process ($\Delta H^0_{ads} > 0$) is attributed unequivocally to chemisorption [64], an exothermic adsorption process ($\Delta H^0_{ads} < 0$) may involve either physisorption or chemisorption or a mixture of both the processes. In an exothermic process, physisorption is distinguished from chemisorption by considering the absolute value of adsorption enthalpy. Typically, the enthalpy of a physisorption process is below 41.86 kJ mol^{-1}, while that of a chemisorption process approaches 100 kJ mol^{-1} [65]. In the present case, the calculated value of ΔH^0_{ads} is −69.014 kJ mol^{-1}, which is an intermediate case [63], probably involving both physisorption and chemisorption. The ΔS^0_{ads} value is large and negative, indicating that decrease in disordering takes place on going from reactant to alloy adsorbed species. This can be attributed to the fact that adsorption is an exothermic process and is always accompanied by decrease in entropy.

3.5. Mechanism of Corrosion Inhibition. In order to predict the type of adsorption, the corrosion mechanism of iron must be known. It was suggested that the anodic dissolution of iron in sulphuric acid solutions follow the consecutive or BDD mechanism [66] as reported by [41]. Iron electro dissolution in acidic sulphate solution depends primarily on the adsorbed intermediate $FeOH_{ads}$ according to the mechanism reproduced below:

$$FeH_2O_{ads} \Longleftrightarrow FeOH_{ads} + H^+ + e \quad (15)$$

$$FeOH_{ads} \longrightarrow FeOH^+ + e \quad (16)$$

$$FeOH^+ + H^+ \Longleftrightarrow Fe^{2+} + 2e \quad (17)$$

The following stages are possible for the hydrogen evolution reaction on iron [67]:

$$Fe + H^+ + e^- \longrightarrow FeH \quad (18)$$

$$FeH + FeH \longrightarrow 2Fe + H_2 \quad (19)$$

$$FeH + H^+ + e^- \longrightarrow Fe + H_2 \quad (20)$$

The corrosion rate of iron in H_2SO_4 solutions is controlled by both hydrogen evolution reaction and iron dissolution.

An anodic oxidation mechanism in the presence of inhibitor may be similar to the mechanism discussed by McCafferty et al. [68–70] as reported by [71]. Consider

$$Fe + H_2O \Longleftrightarrow Fe \cdot H_2O_{ads} \quad (21)$$

$$Fe \cdot H_2O_{ads} + Y \Longleftrightarrow FeOH^-_{ads} + H^+ + Y \quad (22)$$

$$Fe \cdot H_2O_{ads} + Y \Longleftrightarrow FeY_{ads} + H_2O \quad (23)$$

$$FeOH^-_{ads} \longrightarrow FeOH_{ads} + e \text{ (rate determining step)} \quad (24)$$

$$FeY_{ads} \Longleftrightarrow FeY^+_{ads} + e \quad (25)$$

$$FeOH_{ads} + FeY^+_{ads} \Longleftrightarrow FeOH^+ + FeY_{ads} \quad (26)$$

$$FeOH^+ + H^+ \Longleftrightarrow Fe^{2+} + H_2O \quad (27)$$

where Y is inhibitor molecule.

According to the above mechanism, displacement of some adsorbed water molecules on the metal surface by inhibitor species to yield the adsorbed intermediate FeY_{ads} in step (23) reduces the amount of the species $FeOH^-_{ads}$ available for the rate determining step. The inhibitor used may form dominating adsorbed intermediate which results in inhibitive effects. This conclusion is in line with those of Obot and Obi-Egbedi [72], and Shoesmith [73]. Aged maraging steel results from precipitation of intermetallics. Since these intermetallics have different composition, their electrochemical behavior is expected to be different compared to matrix [74]. Also there will be strain fields around these coherent precipitates as a result of lattice mismatch between the precipitate and the matrix due to the difference in the crystal structure and lattice parameters. These strain fields in combination with the galvanic effect due to the composition difference leads to the enhanced corrosion of aged maraging steel in acid medium. Considering the inhomogeneous nature, the surface of alloy is generally characterized by multiple adsorption sites having activation energies and heats of adsorption. Inhibitor molecules may thus be adsorbed more readily at surface active sites having suitable adsorption enthalpies.

The adsorption mechanism for a given inhibitor depends on factors such as, the nature of metal, the corrosive medium, the pH, and the concentration of the inhibitor as well as the functional groups present in its molecule, since different groups are adsorbed to different extents. For instance S-containing substances have been shown to preferentially

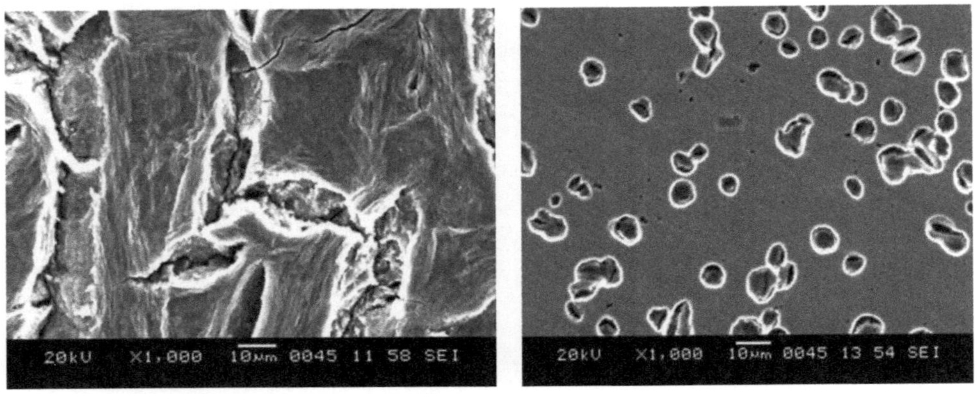

(a) (b)

Figure 7: SEM image of the aged maraging steel after immersion in (a) 0.5 M sulphuric acid without inhibitor, (b) 0.5 M sulphuric acid containing DAMTSC.

chemisorb on the surface of iron in acidic medium, whereas N-containing substances tend to favor physisorption [75]. It is reported that protective value of sulphur containing compounds is superior to that of nitrogen containing compounds. This may be due to greater polarizability of sulphur atom and the presence of two electron pairs available for coordination [22].

The inhibition effect of DAMTSC in sulphuric acid solution can be explained as follows: DAMTSC might be protonated in the acid solution as follows:

$$\text{DAMTSC} + x\text{H}^{+} \longrightarrow [\text{DAMTSCH}_x]^{x+} \quad (28)$$

Thus in aqueous acidic solution, DAMTSC exists partly in the form of protonated species and partly as neutral molecules. Generally two modes of adsorption could be considered. Taking into consideration of many reported results Hackerman et al. [76] concluded that, the surface charge of iron in sulphuric acid solution at the free corrosion potential is positive. Therefore, the adsorption of protonated species would be through electrostatic attraction between the positively charged molecule and the negatively charged sulphate ions adsorbed on the metal surface via physisorption. In the study of the effect of inorganic anions and organic compounds on corrosion inhibition of mild steel in various acids, Li et al. [77] reported the degree of adsorption of anions was in the order $I^- > Br^- > Cl^- > SO_4^{2-} > ClO^{4-}$. The sulphate ions with low degree of adsorption could not cover the metal surface fully and therefore there is always possibility of the neutral inhibitor molecules occupying the vacant adsorption sites on the metal surface via the chemisorption mode of protection of corroding surface [78]. The possible bonding sites of the neutral inhibitor are the free electron pairs on sulphur and nitrogen atoms, as well as oxygen, which could interact with vacant d orbital of iron [78].

3.6. Scanning Electron Microscope Studies (SEM). The scanning electron microscope images were recorded to establish the interaction of acid medium with the metal surface. The surface morphology of the aged samples was examined by SEM immediately after the sample is subjected to corrosion tests in H_2SO_4 medium in the absence and in the presence of inhibitor. The SEM image of corroded aged maraging steel sample in Figure 7(a) shows degradation of alloy in the absence of inhibitor. The attack by H_2SO_4 is seen to be more at grain boundary since these regions are highly susceptible to corrosion. In aged samples the intermetallic precipitation at grain boundary may be responsible for the higher rate of corrosion. Figure 7(b) represents SEM image of aged maraging steel after the corrosion tests in a medium of sulphuric acid containing DMBTSC. The image clearly shows the adsorbed layer of inhibitor molecules on the alloy surface thus protecting the metal from corrosion.

4. Conclusions

Potentiodynamic polarization and electrochemical impedance methods were used to evaluate the ability of DAMTSC to inhibit corrosion of aged maraging steel in 0.5 M sulphuric acid. The principle conclusions are as follows.

(1) The corrosion of aged maraging steel in 0.5 M sulphuric acid is significantly reduced by the addition of DAMTSC even at low concentration.

(2) The inhibition efficiency increases with increase in inhibitor concentration.

(3) The inhibition efficiency decreases with increase in temperature of medium.

(4) DAMTSC acts as mixed type inhibitor, affecting both anodic and cathodic reactions.

(5) The adsorption of DAMTSC on aged maraging steel surface obeys Langmuir's adsorption isotherm model.

(6) The negative value of ΔG^0_{ads} obtained indicates that DAMTSC adsorbed spontaneously on the aged maraging steel. The adsorption process is exothermic and accompanied by decrease in entropy.

(7) The ΔG^0_{ads} and ΔH^0_{ads} values predict both physisorption and chemisorption of inhibitor on alloy surface.

(8) SEM images revealed protection of alloy surface in sulphuric acid medium by the adsorption of DAMTSC.

(9) The inhibition efficiencies obtained from potentiodynamic polarization and EIS techniques are in reasonably good agreement.

References

[1] K. Y. Sastry, R. Narayanan, C. R. Shamantha et al., "Stress corrosion cracking of maraging steel weldments," *Materials Science and Technology*, vol. 19, no. 3, pp. 375–381, 2003.

[2] K. Rohrbach and M. Schmidt, *Properties and Selection: Irons, Steels, and High-Performance Alloys*, vol. 1 of *ASM Handbook*, 10th edition, 1990.

[3] W. W. Kirk, R. A. Covert, and T. P. May, "Corrosion behaviour of high-strength steels in marine environment," *Metals Engineering Quarterly*, vol. 8, pp. 31–38, 1968.

[4] S. W. Dean and H. R. Copson, "Stress corrosion behaviour of maraging steel in natural environments," *Corrosion*, vol. 21, pp. 95–103, 1965.

[5] Data bulletin on 18%Ni maraging steel. The International Nickel Company, INC, 1964.

[6] J. Rezek, I. E. Klein, and J. Yhalom, "Electrochemical properties of protective coatings on maraging steel," *Corrosion Science*, vol. 39, pp. 385–397, 1997.

[7] P. P. Sinha, "Design and development of new variety of 250 grade stainless steel maraging steel," *Transactions of The IIM*, vol. 35, article 2, 1982.

[8] T. Poornima, N. Jagannatha, and A. Nityananda Shetty, "Studies on corrosion of annealed and aged 18 Ni 250 grade maraging steel in sulphuric acid medium," *Portugaliae Electrochimica Acta*, vol. 28, no. 3, pp. 173–188, 2010.

[9] T. Poornima, J. Nayak, and A. Nityananda Shetty, "3,4-Dimethoxybenzaldehydethiosemicarbazone as corrosion inhibitor for aged 18 Ni 250 grade maraging steel in 0.5 M sulfuric acid," *Journal of Applied Electrochemistry*, vol. 41, no. 2, pp. 223–233, 2011.

[10] F. Bentiss, M. Lebrini, and M. Lagrenee, "Thermodynamic characterizationof metal dissolution and inhibitor adsorption processes in mild steel/2. 5-bis(n-thienyl)-1. 3. 4-thiadiazoles/hydrochloric acidsystem," *Corrosion Science*, vol. 47, pp. 2915–2931, 2005.

[11] M. Lebrini, F. Bentiss, H. Vezin, and M. Lagrenée, "The inhibition of mild steel corrosion in acidic solutions by 2,5-bis(4-pyridyl)-1,3,4-thiadiazole: structure-activity correlation," *Corrosion Science*, vol. 48, no. 5, pp. 1279–1291, 2006.

[12] P. Lowmunkhong, D. Ungthararak, and P. Sutthivaiyakit, "Tryptamine as a corrosion inhibitor of mild steel in hydrochloric acid solution," *Corrosion Science*, vol. 52, no. 1, pp. 30–36, 2010.

[13] S. Sankarapapavinasam, F. Pushpanaden, and M. F. Ahmed, "Piperidine, piperidones and tetrahydrothiopyrones as inhibitors for the corrosion of copper in H_2SO_4," *Corrosion Science*, vol. 32, no. 2, pp. 193–203, 1991.

[14] G. Schmitt, "Application of inhibitors for acid media. Report Prepared for the European Federation of Corrosion Working Party on Inhibitors," *British Corrosion Journal*, vol. 19, pp. 165–176, 1984.

[15] S. Singh, F. Athar, and A. Azam, "Synthesis, spectral studies and in vitro assessment for antiamoebic activity of new cyclooctadiene ruthenium(ii) complexes with 5-nitrothiophene-2-carboxaldehyde thiosemicarbazones," *Bioorganic and Medicinal Chemistry Letters*, vol. 15, pp. 5424–5428, 2005.

[16] B. O. Renata, M. S. F. Elaine, P. P. S. Rodrigo, A. A. Anderson, U. K. Antoniana, and L. Z. Carlos, "Synthesis and antimalarial activity of semicarbazone and thiosemicarbazone derivatives," *European Journal of Medicinal Chemistry*, vol. 43, p. 1984, 2008.

[17] J. Easmon, G. Heinisch, W. Holzer, and B. Rosenwirth, "Novel thiosemicarbazones derived from formyl- and acyldiazines: synthesis, effects on cell proliferation, and synergism with antiviral agents," *Journal of Medicinal Chemistry*, vol. 35, no. 17, pp. 3288–3296, 1992.

[18] O. Efanga Offiong and S. Martelli, "Antifungal and antibacterial activity of 2-acetylpyridine-(4-phenylthiosemicarbazone) and its metal (II) complexes," *Farmaco*, vol. 47, no. 12, pp. 1543–1554, 1992.

[19] O. E. Offiong and S. Martelli, "Synthesis, antibacterial and antifungal activity of metal (II) complexes of 2-acetylpyridine thiosemicarbazones," *Farmaco*, vol. 48, no. 6, pp. 777–793, 1993.

[20] S. T. Arab, "Inhibition action of thiosemicabazone and some of it is ρ-substituted compounds on the corrosion of iron-base metallic glass alloy in 0.5 M H_2SO_4 at 30°C," *Materials Research Bulletin*, vol. 43, pp. 510–521, 2008.

[21] B. A. Abd El-Nabey, E. Khamis, G. E. Thompson, and J. L. Dawson, "Effect of temperature on the inhibition of the acid corrosion of steel by benzaldehyde thiosemicarbazone: impedance measurements," *Surface and Coatings Technology*, vol. 28, no. 1, pp. 83–91, 1986.

[22] E. E. Ebenso, U. J. Ekpe, B. I. Ita, O. E. Offiong, and U. J. Ibok, "Effect of molecular structure on the efficiency of amides and thiosemicarbazones used for corrosion inhibition of mild steel in hydrochloric acid," *Materials Chemistry and Physics*, vol. 60, no. 1, pp. 79–90, 1999.

[23] K. Stanly Jacob and G. Parameswaran, "Corrosion inhibition of mild steel in hydrochloric acid solution by Schiff base furoin thiosemicarbazone," *Corrosion Science*, vol. 52, no. 1, pp. 224–228, 2010.

[24] A. Riose and M. Valcarcel, "Homogeneous precipitation of palladium dimethylglyoximate by interchange reactions of C=N groups," *Analyst*, vol. 107, pp. 737–743, 1982.

[25] M. G. Fontana, *Corrosion Engineering*, McGraw-Hill, Singapore, 3rd edition, 1987.

[26] W. H. Li, Q. He, S. T. Zhang, C. L. Pei, and B. R. Hou, "Some new triazole derivatives as inhibitors for mild steel corrosion in acidic medium," *Journal of Applied Electrochemistry*, vol. 38, no. 3, pp. 289–295, 2008.

[27] C. Cao, "On electrochemical techniques for interface inhibitor research," vol. 38, no. 12, pp. 2073–2082, 1996.

[28] A. R. El-Sayed, A. M. Shaker, and H. M. Abd El-Lateef, "Corrosion inhibition of tin, indium and tin-indium alloys by adenine or adenosine in hydrochloric acid solution," *Corrosion Science*, vol. 52, no. 1, pp. 72–81, 2010.

[29] S. S. Abd El-Rehim, H. H. Hassan, and M. A. Amin, "Corrosion inhibition study of pure Al and some of its alloys in 1.0 M HCl solution by impedance technique," *Corrosion Science*, vol. 46, pp. 5–25, 2004.

[30] W. Li, Q. He, C. Pei, and B. Hou, "Experimental and theoretical investigation of the adsorption behaviour of new triazole derivatives as inhibitors for mild steel corrosion in acid media," *Electrochimica Acta*, vol. 52, pp. 6386–6394, 2007.

[31] W. J. Lorenz and F. Mansfeld, "Determination of corrosion rates by electrochemical DC and AC methods," *Corrosion Science*, vol. 21, pp. 647–672, 1982.

[32] A. A. Aksut, W. J. Lorenz, and F. Mansfeld, "The determination of corrosion rates by electrochemical d.c. and a.c. methods—2. Systems with discontinuous steady state polarization behavior," *Corrosion Science*, vol. 22, pp. 611–619, 1982.

[33] Q. Qu, Z. Hao, L. Li, W. Bai, Y. Liu, and Z. Ding, "Synthesis and evaluation of Tris-hydroxymethyl-(2-hydroxybenzylidenamino)-methane as a corrosion inhibitor for cold rolled steel in hydrochloric acid," *Corrosion Science*, vol. 51, no. 3, pp. 569–574, 2009.

[34] G. Quartarone, L. Ronchin, A. Vavasori, C. Tortato, and L. Bonaldo, "Inhibitive action of gramine towards corrosion of mild steel in deaerated 1.0 M hydrochloric acid solutions," *Corrosion Science*, vol. 64, pp. 82–89, 2012.

[35] I. Ahamad, R. Prasad, and M. A. Quraishi, "Adsorption and inhibitive properties of some new Mannich bases of Isatin derivatives on corrosion of mild steel in acidic media," *Corrosion Science*, vol. 52, no. 4, pp. 1472–1481, 2010.

[36] M. A. Veloz and I. Gonzalez, "Electrochemical study of carbon steel corrosion in buffered acetic acid solutions with chlorides and H_2S," *Electrochimica Acta*, vol. 48, pp. 135–144, 2002.

[37] E. M. Sherif and S. M. Park, "Effects of 1, 4-naphthoquinone on aluminum corrosion in 0.50 M sodium chloride solutions," *Electrochimica Acta*, vol. 51, pp. 1313–1321, 2006.

[38] A. Popova, E. Sokolova, S. Raicheva, and M. Christov, "AC and DC study of the temperature effect on mild steel corrosion in acid media in the presence of benzimidazole derivatives," *Corrosion Science*, vol. 45, no. 1, pp. 33–58, 2003.

[39] E. Barsoukov and J. R. Macdonald, *Impedance Spectroscopy Theory, Experiment, and Applications*, John Wiley & Sons, 2nd edition, 2005.

[40] A. El-Sayed, "Phenothiazine as inhibitor of the corrosion of cadmium in acidic solutions," *Journal of Applied Electrochemistry*, vol. 27, pp. 193–200, 1997.

[41] A. Popova, M. Christov, and A. Vasilev, "Mono- and dicationic benzothiazolic quaternary ammonium bromides as mild steel corrosion inhibitors—part 2: electrochemical impedance and polarisation resistance results," *Corrosion Science*, vol. 53, no. 5, pp. 1770–1777, 2011.

[42] B. Seshu, A. K. Bhatnagar, A. Venugopal, and V. S. Raja, "Electrochemical corrosion behaviours of $Fe_{68}Ni_{14-x}Mo_xSi_2B_{16}$ metallic glasses in 1N HCl and 1N H_2SO_4," *Journal of Materials Science*, vol. 32, pp. 2071–2075, 1997.

[43] S. T. Arab and K. M. Emran, "Structure effect of some thiosemicarbazone derivatives on the corrosion inhibition of $Fe_{78}B_{13}Si_9$ glassy alloy in Na_2SO_4 solution," *Materials Letters*, vol. 62, no. 6-7, pp. 1022–1032, 2008.

[44] H. Ashassi-Sorkabi, T. A. Aliyev, S. Nasiri, and R. Zarepoor, *Electrochimica Acta*, vol. 52, p. 5240, 2007.

[45] M. A. Amin, K. F. Khaled, and S. A. Fadl-Allah, "Testing validity of the Tafel extrapolation method for monitoring corrosion of cold rolled steel in HCl solutions—experimental and theoretical studies," *Corrosion Science*, vol. 52, no. 1, pp. 140–151, 2010.

[46] H. H. Hassan, E. Abdelghani, and M. A. Amin, "Inhibition of mild steel corrosion in hydrochloric acid solution by triazole derivatives—part I. Polarization and EIS studies," *Electrochimica Acta*, vol. 52, pp. 6359–6364, 2007.

[47] C. H. Hsu and F. Mansfeld, "Concernng the conversion of the constant phase element parameter Y0 into a capacitance," *Corrosion*, vol. 57, no. 9, pp. 747–748, 2001.

[48] E. McCafferty and N. Hackerman, "Double layer capacitance of iron and corrosion inhibition with polymethylene diamines," *Journal of The Electrochemical Society*, vol. 119, pp. 146–154, 1972.

[49] F. Bentiss, M. Traisnel, and M. Lagrenee, "The substituted 1,3,4-oxadiazoles: a new class of corrosion inhibitors of mild steel in acidic media," *Corrosion Science*, vol. 42, no. 1, pp. 127–146, 2000.

[50] L. Larabi, Y. Harek, O. Benali, and S. Ghalem, "Hydrazide derivatives as corrosion inhibitors for mild steel in 1 M HCl," *Progress in Organic Coatings*, vol. 54, p. 261, 2005.

[51] M. Schorr and J. Yahalom, "The significance of the energy of activation for the dissolution reaction of metal in acids," *Corrosion Science*, vol. 12, no. 11, pp. 867–868, 1972.

[52] M. M. Solomon, S. A. Umoren, I. I. Udosoro, and A. P. Udoh, "Inhibitive and adsorption behaviour of carboxymethyl cellulose on mild steel corrosion in sulphuric acid solution," *Corrosion Science*, vol. 52, no. 4, pp. 1317–1325, 2010.

[53] L. Antropov, "A correlation between kinetics of corrosion and the mechanism of inhibition by organic compounds," *Corrosion Science*, vol. 7, pp. 607–620, 1967.

[54] E. F. E. Sherbini, "Effect of some ethoxylated fatty acids on the corrosion behaviour of mild steel in sulphuric acid solution," *Materials Chemistry and Physics*, vol. 60, pp. 286–290, 1999.

[55] T. Szauer and A. Brand, "On the role of fatty acid in adsorption and corrosion inhibition of iron by amine-fatty acid salts in acidic solution," *Electrochimica Acta*, vol. 26, pp. 1257–1260, 1981.

[56] M. K. Gomma and M. H. Wahdan, "Schiff bases as corrosion inhibitors for aluminium in hydrochloric acid solution," *Materials Chemistry and Physics*, vol. 39, pp. 209–213, 1995.

[57] J. Marsh, *Advanced Organic Chemistry*, Wiley Eastern, New Delhi, India, 3rd edition, 1988.

[58] M. Sahin, S. Bilgic, and H. Yilmaz, "The inhibition effects of some cyclic nitrogen compounds on the corrosion of the steel in NaCl mediums," *Applied Surface Science*, vol. 195, pp. 1–7, 2002.

[59] B. Ateya, B. E. El-Anadouli, and F. M. El-Nizamy, "The adsorption of thiourea on mild steel," *Corrosion Science*, vol. 24, pp. 509–515, 1984.

[60] E. E. Oguzie, V. O. Njoku, C. K. Enenebeaku, C. O. Akalezi, and C. Obi, "Effect of hexamethylpararosaniline chloride (crystal violet) on mild steel corrosion in acidic media," *Corrosion Science*, vol. 50, no. 12, pp. 3480–3486, 2008.

[61] M. Hosseini, S. F. L. Mertens, and M. R. Arshadi, "Synergism and antagonism in mild steel corrosion inhibition by sodium dodecylbenzenesulphonate and hexamethylenetetramine," *Corrosion Science*, vol. 45, no. 7, pp. 1473–1489.

[62] A. K. Singh and M. A. Quraishi, "Effect of Cefazolin on the corrosion of mild steel in HCl solution," *Corrosion Science*, vol. 52, pp. 152–160, 2010.

[63] W. Durnie, R. De Marco, A. Jefferson, and B. Kinsella, "Development of a structure-activity relationship for oil field corrosion inhibitors," *Journal of the Electrochemical Society*, vol. 146, no. 5, pp. 1751–1756, 1999.

[64] S. Martinez and I. Stern, "Thermodynamic characterization of metal dissolution and inhibitor adsorption processes in the low carbon steel/mimosa tannin/sulfuric acid system," *Applied Surface Science*, vol. 199, pp. 83–89, 2002.

[65] J. O. M. Bockris, M. A. Devanathan, and K. Muller, "Thermodynamic characterization of metal dissolution and inhibitor adsorption processes in the low carbon steel/mimosa tannin/sulfuric acid system," *Proceedings of the Royal Society A*, vol. 274, pp. 55–79, 1963.

[66] M. A. V. Devanathan and Z. Stachurski, "The mechanism of hydrogen evolution on iron in acid solutions by determination of permeation rates," *Journal of The Electrochemical Society*, vol. 111, p. 619, 1964.
[67] E. J. Kelly, "Iron dissolution and hydrogen evolution reactions in acidic sulfate solutions," *Journal of The Electrochemical Society*, vol. 112, pp. 124–131, 1965.
[68] A. Frignani, M. Tassinari, and G. Trabanelli, "Impedance measurements on Armco iron in acid solution inhibited by S-containing additives," *Electrochimica Acta*, vol. 34, no. 8, pp. 1259–1263, 1989.
[69] E. McCafferty and N. Hackerman, "Kinetics of iron corrosion in concentrated acidic chloride solutions," *Journal of The Electrochemical Society*, vol. 119, pp. 999–1009, 1972.
[70] H. Ashassi-Sorkhabi and S. A. Nabavi-Amri, "Corrosion inhibition of carbon steel in petroleum/water mixtures by N-containing compounds," *Acta Chimica Slovenica*, vol. 47, no. 4, pp. 507–517, 2000.
[71] P. C. Okafor, M. E. Ikpi, I. E. Uwah, E. E. Ebenso, U. J. Ekpe, and S. A. Umoren, "Inhibitory action of Phyllanthus amarus extracts on the corrosion of mild steel in acidic media," *Corrosion Science*, vol. 50, no. 8, pp. 2310–2317, 2008.
[72] I. B. Obot and N. O. Obi-Egbedi, "Adsorption properties and inhibition of mild steel corrosion in sulphuric acid solution by ketoconazole: experimental and theoretical investigation," *Corrosion Science*, vol. 52, no. 1, pp. 198–204, 2010.
[73] D. W. Shoesmith, *Metals Handbook*, vol. 13, 9th edition, 1987.
[74] A. Fragnani and G. Trabanelli, "Influence of organic additives on the corrosion of iron-based amorphous alloys in dilute sulfuric acid solution," *Corrosion*, vol. 55, pp. 653–660, 1999.
[75] A. Popova, E. Sokolova, S. Raicheva, and M. Christov, "AC and DC study of the temperature effect on mild steel corrosion in acid media in the presence of benzimidazole derivatives," *Corrosion Science*, vol. 45, no. 1, pp. 33–58, 2003.
[76] N. Hackerman, E. S. Snavely Jr., and J. S. Payne Jr., "Effects of anions on corrosion inhibition by organic compounds," *Journal of The Electrochemical Society*, vol. 113, pp. 677–981, 1966.
[77] X. Li, S. Deng, H. Fu, and G. Mu, "Inhibition effect of 6-benzylaminopurine on the corrosion of cold rolled steel in H_2SO_4 solution," *Corrosion Science*, vol. 51, no. 3, pp. 620–634, 2009.
[78] A. K. Satpati and P. V. Ravindran, "Electrochemical study of the inhibition of corrosion of stainless steel by 1,2,3-benzotriazole in acidic media," *Materials Chemistry and Physics*, vol. 109, pp. 352–359, 2008.

19

Internal Demagnetizing Factor in Ferrous Metals

Jenő Takács,[1] György Kovács,[2] and Lajos K. Varga[3]

[1] *Department of Engineering Science, University of Oxford, Parks Road, Oxford OX1 3PJ, UK*
[2] *Department of Materials, Physics of Eotvos Lorand University, P.O. Box 32, Budapest 1518, Hungary*
[3] *Research Institute for Solid State Physics and Optics, P.O. Box 49, Budapest 1525, Hungary*

Correspondence should be addressed to Jenő Takács, jenotakacs@aol.com

Academic Editor: Chih-Ming Chen

Modelling the saturation major loop of a ferrous metal produces the intrinsic magnetization parameters; fitting the measured commutation curve, however, can yield different results. The relation of the intrinsic loci of the vertices of the minor loops ($N_D = 0$) to the experimental curve ($N_D \neq 0$) is investigated. The two-way transformation between the two curves is formulated in closed mathematical form with the help of the internal demagnetization factor, N_D. The method is applied to four ferrous metals, with widely different intrinsic properties (soft nonoriented Fe-Si steel, normalized low carbon steel, and Finemet in nanocrystalline and amorphous state) supporting the predictions of the proposal. The developed relationship is model independent and it is shown that the N_D factor depends linearly on coercivity based on experimental evidence.

1. Introduction

A large number of the measurements of ferrous substances are aimed at finding the intrinsic material properties [1] of the tested ferrous sample (as defined by Fiorillo). Due to the ever-presence of demagnetization field, various measuring methods have been developed to minimize its effect. The most commonly accepted way is to make the sample turn into a closed magnetic circuit, such as a toroid or an Epstein square [1–3]. Although these two methods are not completely free from the ever-present internal demagnetization, they suffer the least from it [1]. Researchers went into great length to include the internal demagnetization force into current models like Preisach, Jiles, Stoner-Wohlfarth [2, 4, 5], and so forth, leading to complicated, so called, dynamic versions.

The saturated major hysteresis loop of the sample carries all the intrinsic magnetic parameters directly recoverable from the measured data. Within this loop lie the unhysteretic loci of the vertices of the symmetrical minor loops, the only curve, which belongs to both the ascending and descending branches of the hysteresis loops [5, 6].

A proposal is put forward in this paper to show the relationship between the intrinsic curve and the loci of vertices of the measured minor loops. This relationship between the two curves, independent of models, is formulated in closed mathematical form and its prediction is verified by the experimental data obtained from four different ferrous samples.

Once the intrinsic locus (M_{01} for $N_d = 0$) is modelled from saturation or minor loop data, by using any of the static models, the measured curve ($N_d \neq 0$) can be calculated from the proposed formulation below with optimization of the N_d parameter value as specified by Jiles [2] and used by Fiorillo [1].

2. The Intrinsic Loci of Vertices

The intrinsic commutation curve ($N_D = 0$) is the locus of the return points or the maxima of the set of symmetrical minor loops. It is a single-valued function and in spite of having no hysteresis, it carries all the hysteretic properties of the ferrous material obtained from the saturation data [7].

M_0 is the arithmetic mean of the ascending and descending magnetization functions, M_u and M_d, with equal application to $N_D = 0$ case as well.

$$M_0 = \frac{(M_u + M_d)}{2}. \tag{1}$$

Due to the difference between M_0 and the measured commutation curve, hysteresis loop modelling based solely on the measured commutation curve can produce different parameters, when no molecular interaction is assumed [6, 8, 9].

3. The Effective Field and Its Implications

For the description of the effect of the internal demagnetizing field we will use the concept of the effective field, which is analogous to the Weiss mean field as defined and used by other authors [2, 10, 11]. Although in simple cases, proportionality is assumed between $\mathbf{H_D}$ interaction field and $\mathbf{M}$ magnetization vector, this can only be regarded as a linear approximation [12, 13] expressed as $H_D = N_D M$ in scalar form, leading to

$$H_{\text{eff}} = H + N_D M, \tag{2}$$

where N_D is the internal demagnetization factor.

For simplicity we will use normalised quantities in further calculations, where the lower case letters will represent the normalised quantities of the physical equivalents, denoted by the same capital letters.

m_0 denote the normalised M_0 function in (1). With the normalized effective field, h_{eff}, m_0 can be expressed as:

$$m_0 = f(h_{\text{eff}}) = f(h + n_D m_0), \tag{3}$$

where n_D is also normalized. (1 Tesla = 8.10^5 A/m. The internal demagnetization factor has unity dimension only when the magnetization is measured in A/m.).

The first derivative of m_0 by h in (3) leads to an expression, which shows a character similar to the feedback in an electrical circuit [14].

This expression describes a well-known relationship between the inherent (μ_i) and the effective (μ_{eff}) permeabilities [1, 2].

For most magnetic substances the value of N_D is small in the order of $\sim -10^{-5}$ with unity dimension [1]. Consider

$$\frac{dm_0}{dh} = \frac{dm_0/dh_{\text{eff}}}{1 - n_D(dm_0/dh_{\text{eff}})}. \tag{4}$$

By using the hysteretic model [6] (see also the appendix), the integral of (4) by h leads us to the following expression for m_0, when

$$\frac{dm_0}{dh} = \sum_{k=1}^{n} \frac{a_k}{2}\left[\frac{\sec h^2(h_{k\text{ueff}})}{1 - n_D a_k \sec h^2(h_{k\text{ueff}})} + \frac{\sec h^2(h_{k\text{deff}})}{1 - n_D a_k \sec h^2(h_{k\text{deff}})}\right]. \tag{5}$$

After integration:

$$m_0 = \sum_{k=1}^{n} \frac{a_k}{2}\left\{-\sqrt{a_k}\frac{\text{Arc}\tanh[(\sqrt{a_k n_D}\tanh[\alpha_k(h - h_{ck})]/\sqrt{-1 + a_k n_D})]}{\sqrt{n_D(-1 + a_k n_D)}} - \sqrt{a_k}\frac{\text{Arc}\tanh[(\sqrt{a_k n_D}\tanh[\alpha_k(h + h_{ck})]/\sqrt{-1 + a_k n_D})]}{\sqrt{n_D(-1 + a_k n_D)}}\right\}. \tag{6}$$

When the integration constants are appropriately chosen, this form of m_0 is equivalent to the one given in (A.5).

The intrinsic locus is entirely a theoretical concept. It was introduced for the free (Gibbs) energy calculations [2]. It assumes zero internal demagnetization in a system, where a m moment can freely move around under the influence of H external excitation field without any hindrance from the interaction between the magnetic moments.

We must remind the reader that, for various ferromagnetic substances, this internal demagnetization constant N_D is given traditionally in a numerical value with unity dimension (i.e., when both the H and B measured in A/m). When different unitary system is used, N_D has a different physical dimension and must be normalised (see n_D as normalised N_D).

4. Experimental Verification

To verify the predictions of the proposed method, it was applied initially to two ferrous materials with very different characters. The first was a soft steel NO Fe-Si with 67.5 A/m coercivity [15], shown in Figure 1. The second material was a normalized low carbon steel (AISI 1040) with coercivity of 450 A/m (see Figure 5). Following the excellent results, the same experiment was also repeated later on two other samples; Finemet in nanocrystalline and amorphous (as cast) state. The detailed data of those samples are not included in this paper due to its limited size. All measurements were carried out under identical conditions by using triangular excitation of $f = 0.001$ Hz on toroid samples with geometrical details as follows: $D_e = 25$ mm, $D_{\text{int}} = 15$ mm and thickness $d = 0.5$ mm [1]. In making the identical toroid samples, extreme care was taken to avoid any changes in magnetic properties due to mechanical handling of the materials.

For numerical calculations we used the hyperbolic model for its simplicity and speed and the Mathematica program interactively. A brief summary of the model is given in the appendix. For further details we refer the reader to the literature [6, 16, 17].

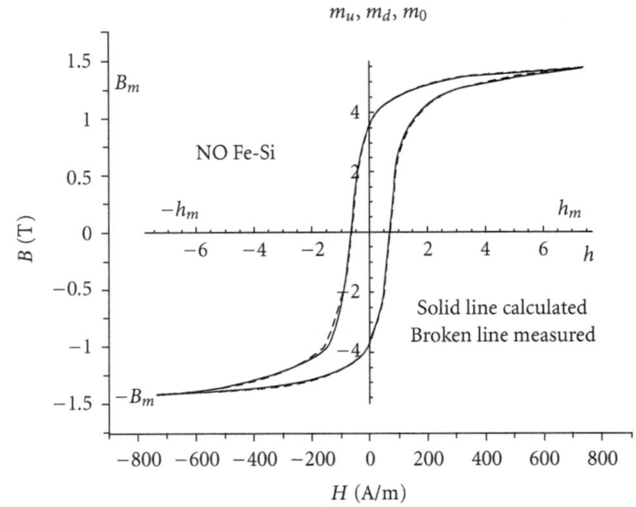

FIGURE 1: The measured and the modelled major loop of NO Fe-Si soft steel.

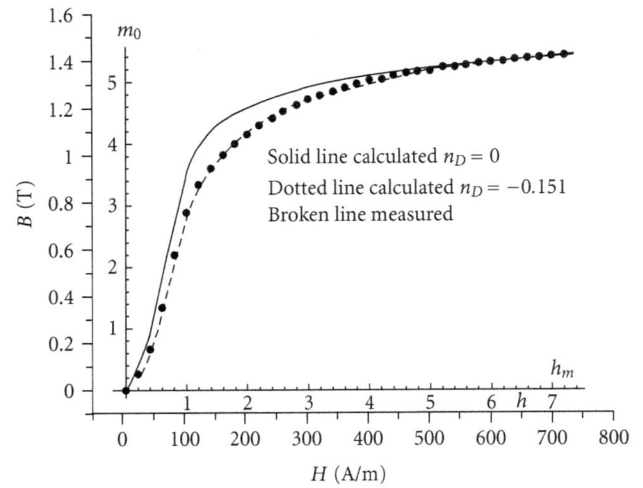

FIGURE 2: Measured (broken line), modelled (dotted line), and intrinsic (solid line) commutation curves.

4.1. NO Fe-Si. Starting with conveniently chosen parameter values at the beginning with subsequent changes of the parameters, new curves are calculated and compared with the measured one.

When the iteration produced the best fit to the measured curve, the normalized and the equivalent physical values can be easily read from the two coordinate systems (normalization) as shown in Figures 1, 2, and 3.

The first sample (see Figure 1) was modelled with the following normalized and physical parameters:

$a_1 = 3.18$, $a_2 = 1.45$, $a_3 = 1.09$,

$\alpha_1 = 2.75$, $\alpha_2 = 0.56$, $\alpha_3 = 0.134$,

$h_{c1} = 0.675$, $h_{2c} = 0.57$, $h_{c3} = 2.2$,

$h_m = 7.35$ equivalent to:

$A_1 = 0.842$ T, $A_2 = 0.384$ T, $A_3 = 0.288$ T.

$H_{c1} = 67.5$ A/m, $H_{c2} = 57$ A/m, $H_{c3} = 220$ A/m, $H_m = 735$ A/m.

with normalization of $1\,h = H\,100$ A/m and $1\,m = M\,0.2647$ T.

The measured and the modelled m_0 curves for $n_D = -0.151$ ($N_D = -57$ A/Tm, or -7.10^{-5}). $n_D = 0$ are depicted in Figure 2. For the symbols see the appendix.

In order to check the accuracy of the transformation, between the two curves, 25 of the minor loops were measured with maximum field excitation values between $h_m = 5.01$, $Hm = 501$ A/m and $h_m = 0.152$, $Hm = 15.2$ A/m. For all the minor loops measured, the corresponding loops were calculated for the reduced maximum magnetization at $n_D = 0$ and $n_D = -0.151$.

For clarity, only one of the minor loops is shown in Figure 3 for the peek excitation field value of 111.3 A/m.

All the calculated loops had an excellent fit to the equivalent measured loops for the same N_D value. The proposed method is applicable to all measured hysteretic data, where the external demagnetization field is eliminated or reduced to a negligible level.

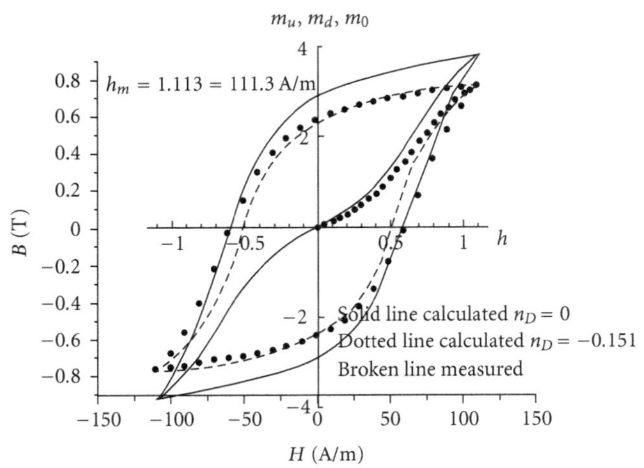

FIGURE 3: Calculated NO Fe-Si minor hysteresis loops for $n_D = 0$ (solid line) and $n_D = -0.151$ (dotted line), fitted to the measured loop (broken line) with internal demagnetization.

4.2. Low Carbon Steel Toroid (AISI 1040). For the second sample, we selected a toroid from low carbon steel with coercivity of 450 A/m. The use of the same experimental setup has yielded the following parameters in normalized and physical units:

$a_1 = 2.68$, $a_2 = 2.08$, $a_3 = 1.6$,

$\alpha_1 = 3.52$, $\alpha_2 = 0.47$, $\alpha_3 = 0.09$,

$h_{c1} = 0.45$, $h_{c2} = 0.44$, $h_{c3} = 0.9$ and $h_m = 7.35$,

equivalent to:

$A_1 = 0.64$ T, $A_2 = 0.5$ T, $A_3 = 0.38$ T,

$H_{c1} = 450$ A/m, $H_{c2} = 440$ A/m, $H_{c3} = 900$ A/m, and $H_m = 7350$ A/m,

with normalization of $1\,h = H\,1000$ A/m and $1\,m = M\,0.24$ T.

The measured and the modelled hysteresis loops are depicted in Figure 4 with the intrinsic and the measured

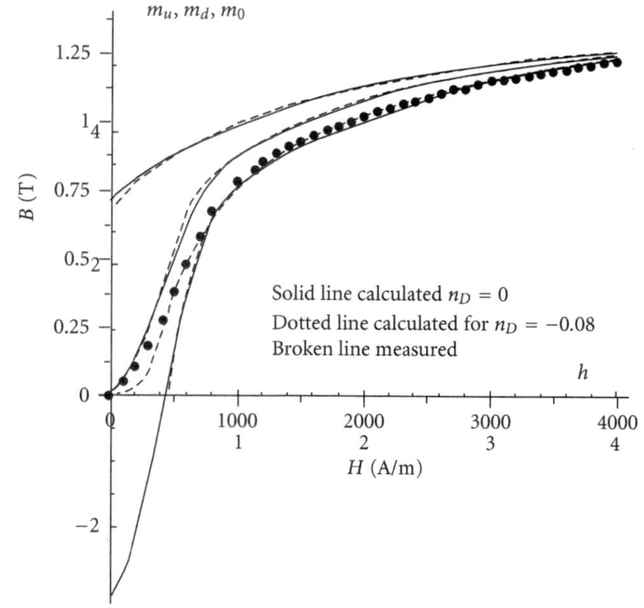

FIGURE 4: Measured (broken line) and modelled (solid line) hysteresis and the commutation curves for $n_D = 0$ and $n_D = -0.08$ (dotted line) in the first quadrant for the low carbon steel (AISI 1040).

commutation curves. The curves are shown in the first quadrant only for better visual resolution.

Following the calculation of the intrinsic parameters, the measured commutation curve was modelled with $n_D = -0.08$ ($N_D = -333.3$ A/Tm or $-4.1.10^{-4}$), which yielded the best result, giving an excellent fit to the measured curve.

5. N_d as a Function of Coercivity

The experimental results have indicated that N_D is greatly dependent on the coercivity of the sample. To verify this dependency, four samples, listed under Section 4, were tested for this purpose, with coercivity H_c ranging between 1.6 A/m and 450 A/m. Figure 5 depicts the relationship between N_D and H_c.

The graph, in Figure 5, was constructed by using the parameters of the samples listed below, whose coercivity values change in steps approximately by an order of magnitude:

Finemet: $H_c = 1.6$ A/m, $N_D = -0.729$ A/Tm or $-0.9 \cdot 10^{-6}$,

Finemet in amorphous state: $H_c = 12.6$ A/m, $N_D = -4.96$ A/Tm or $-6.2 \cdot 10^{-6}$,

NO Fe-Si: $H_c = 67.5$ A/m, $N_D = -57$ A/Tm or $-7 \cdot 10^{-5}$,

Low carbon steel (AISI 1040): $H_c = 450$ A/m, $N_D = -333$ A/Tm or $-4 \cdot 16.\ 10^{-4}$.

The graph shows N_D linear dependency on H_c coercivity and based on experimental evidence, it can be approximated as

$$N_D \approx \mu_0 \frac{H_c}{B_m}, \tag{7}$$

where B_m represents the saturation magnetization.

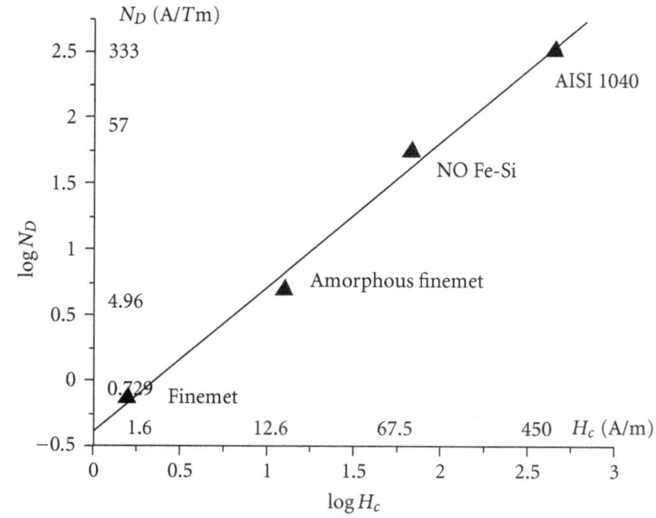

FIGURE 5: N_D as the function of coercivity H_c.

6. Conclusions

The relation between the intrinsic ($N_D = 0$) and the measured ($N_D \neq 0$) loci of the vertices of the minor loops was investigated. By using the effective field, this relationship was formulated. The method was subjected to tests on four magnetic materials with widely ranging magnetic properties. The test results show that the mathematical approach, presented here, describes the relation well. The N_D factor linear dependence on coercivity, based on empirical evidence, was also demonstrated. The paper shows that magnetic parameters (H_c and B_m) N_D can be estimated very close to the real value.

The proposal's aim is to recover the intrinsic magnetization properties from the measured commutation curves.

Appendix

The characteristic equations of the hyperbolic model in canonic form:

$$m_u = \sum_{k=1}^{n} [a_k f_{uk} + f_{0k}(h_m)], \tag{A.1}$$

$$m_d = \sum_{k=1}^{n} [a_k f_{dk} - f_{0k}(h_m)], \tag{A.2}$$

$$f_{uk,dk} = \tanh[\alpha_k(h \mp h_{ck})], \tag{A.3}$$

$$f_{0k} = \sum_k \frac{(f_{dk} - f_{uk})}{2}, \tag{A.4}$$

$$m_{0k} = \sum_k \frac{(f_{dk} + f_{uk})}{2}. \tag{A.5}$$

Here m_u and m_d signify the ascending and descending magnetization, respectively, h represents the field excitation, h_{ck} is the coercivity of the kth process. a_k is the amplitude of the magnetization, α_k is a scaling factor, and f_{ok} is the integration constant, while h_m is the maximum field excitation, common to all. The index k refers to the individual processes and n is the number of total processes involved (for most substances $n = 3$).

The parameters are calculated by changing the model parameters until the best fit to the measured curve is achieved. When the iteration gives the best fit, the normalized and the equivalent physical values can be read from the two coordinate systems (normalized and measured), as shown in Figures 1, 2, and 3.

References

[1] F. Fiorillo, *Measurement and Characterization of Magnetic Materials*, Elsevier, Oxford, UK, 2004.

[2] D. Jiles, *Introduction to Magnetism and Magnetic Materials*, Chapman and Hall, New York, NY, USA, 1998.

[3] T. Nakata, N. Takahashi, K. Fujiwara, M. Nakano, Y. Ogura, and K. Matsubara, "An improved method for determining the DC magnetization curve using a ring specimen," *IEEE Transactions on Magnetics*, vol. 28, no. 5, pp. 2456–2458, 1992.

[4] F. Preisach, "Uber die magnetische Nachwirkung," *Zeitschrift für Physik*, vol. 94, pp. 227–302, 1935.

[5] E. C. Stoner and E. P. Wohlfarth, "A mechanism of magnetic hysteresis in heterogeneous alloys," *IEEE Transactions on Magnetics*, vol. 27, no. 4, pp. 3475–3518, 1991.

[6] J. Takacs, "The everett integral and its analytical approximation," in *Advanced Magnetic Materials*, L. Malkinsk, Ed., pp. 203–230, Intech, 2012.

[7] J. Takâcs, "A phenomenological mathematical model of hysteresis," *Compel*, vol. 20, no. 4, pp. 1002–1014, 2001.

[8] I. Meszaros, Private communication on m_0.

[9] H. Hauser, Y. Melikhov, and D. C. Jiles, "Examination of the equivalence of ferromagnetic hysteresis models describing the dependence of magnetization on magnetic field and stress," *IEEE Transactions on Magnetics*, vol. 45, no. 4, pp. 1940–1949, 2009.

[10] G. Bertotti, *Hysteresis in Magnetism*, Academic Press, San Diego, Calif, USA, 1998.

[11] D. C. Jiles and D. L. Atherton, "Theory of ferromagnetic hysteresis," *Journal of Magnetism and Magnetic Materials*, vol. 61, no. 1-2, pp. 48–60, 1986.

[12] J. A. Brug and W. P. Wolf, "Demagnetizing fields in magnetic measurements. I. Thin discs," *Journal of Applied Physics*, vol. 57, no. 10, pp. 4685–4694, 1985.

[13] A. Benabou, J. V. Leite, S. Clénet, C. Simão, and N. Sadowski, "Minor loops modelling with a modified Jiles-Atherton model and comparison with the Preisach model," *JMMM*, vol. 320, no. 20, pp. e1034–e1038, 2008.

[14] R. G. Harrison, "Physical theory of ferromagnetic first-order return curves," *IEEE Transactions on Magnetics*, vol. 45, no. 4, pp. 1922–1939, 2009.

[15] L. K. Varga, Gy. Kovacs, and J. Takacs, "Anhysteretic and biased first magnetization curves for Finemet-type toroidal samples," *JMMM*, vol. 320, no. 3-4, pp. L.26–L.29, 2008.

[16] J. Takacs, "Barkhausen instability and its implication in T(x) modelling of hysteresis," *Compel*, vol. 24, no. 4, pp. 1180–1190, 2005.

[17] J. Takacs, *Mathematics of Hysteretic Phenomena*, Wiley, Berlin, Germany, 2003.

Surface Modification of Light Alloys by Low-Energy High-Current Pulsed Electron Beam

X. D. Zhang,[1] S. Z. Hao,[1] T. Grosdidier,[2] J. X. Zou,[3] B. Gao,[4] B. Bolle,[2] N. Allain-Bonasso,[2] Y. Qin,[1] X. N. Li,[1] and C. Dong[1]

[1] *Key Laboratory of Materials Modification, Department of Materials Engineering, Dalian University of Technology, Dalian 116024, China*
[2] *Laboratoire d'Etude des Microstructures et de Mécanique des Matériaux (LEM3), CNRS UMR 7239, Université Paul Verlaine-Metz, Ile du Saulcy, 57045 Metz, France*
[3] *Shanghai Engineering Research Center of Mg Materials and Applications, School of Materials Science and Engineering, Shanghai Jiao Tong University, Shanghai 200240, China*
[4] *School of Materials Science and Engineering, Shenyang University, Shenyang 200240, China*

Correspondence should be addressed to X. D. Zhang, zhangxiangdong0803@gmail.com

Academic Editor: Stefano Gialanella

This paper reviews results obtained by the research groups developing the low-energy high-current pulsed electron beam (LEHCPEB) in Dalian (China) and Metz (France) on the surface treatment of light alloys. The pulsed electron irradiation induces an ultra-fast thermal cycle at the surface combined with the formation of thermal stress and shock waves. As illustrated for Mg alloys and Ti, this results in deep subsurface hardening (over several 100 μm) which improves the wear resistance. The analysis of the top surface melted surface of light alloys also often witnesses evaporation and condensation of chemical species. This phenomenon can significantly modify the melt chemistry and was also suggested to lead to the development of specific solidification textures in the rapidly solidified layer. The potential use of the LEHCPEB technique for producing thermomechanical treatments under the so-called heating mode and, thus, modify the surface crystallographic texture, and enhance solid-state diffusion is also demonstrated in the case of the FeAl intermetallic compound.

1. Introduction

Light alloys, such as Mg-, Al-, and Ti-based alloys, have attracted increasing attention in the past few decades owing to their low density and, correspondingly, their high strength/ductility ratio. Because of this, light alloys are increasingly used to replace steels in industrial components. However, these light alloys are all facing some serious surface-related disadvantages such as poor wear or corrosion resistance that have strongly limited their potential for some specific industrial applications. Therefore, surface treatment techniques should be applied on these light alloys in order to improve their global performance.

The low-energy high-current pulsed electron beam (LEHCPEB) process is a fairly new surface modification technique [1, 2]. The LEHCPEB sources have been first developed for surface treatment of materials by Proskurovsky and Ozur in Tomsk (Russia) [1–3]. As one kind of high-power charged particle beam, LEHCPEB exhibits essential advantages over pulsed laser and ion beams by its high efficiency, simplicity, and reliability. The pulsed electron irradiation induces (i) a rapid heating and cooling of the surface together with (ii) the formation of thermal stress and stress waves [4, 5]. As a result, improved surface properties of the material, often unattainable with conventional surface treatment techniques, can be obtained fairly easily. This is particularly true for tribological [2, 4, 6, 7] and corrosion properties [7–10]. Proskurovsky et al. [1–4] have carried out pioneer investigations of the LEHCPEB treatment on a series of Al- and Ti-base alloys and the results obtained showed an increase of corrosion and strength properties for the treated alloys. The group in Dalian University of Technology (China) has more recently done research on various light materials such as pure Al [11, 12], pure Ti [13], pure

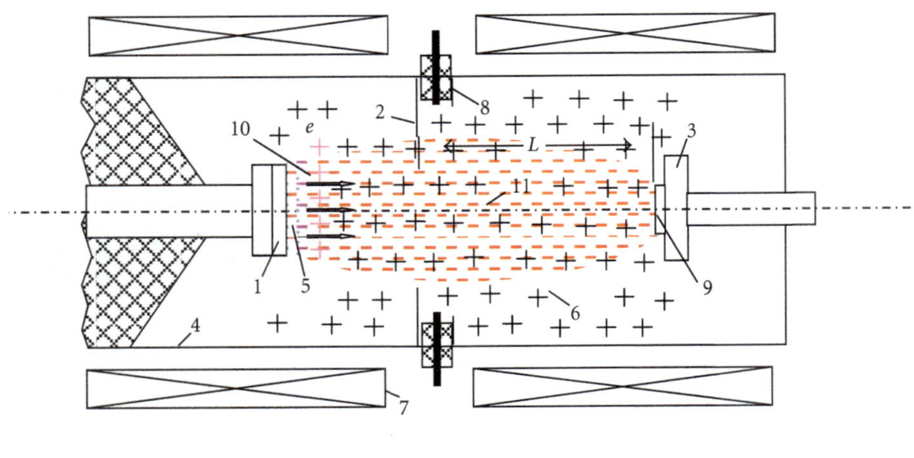

FIGURE 1: Schematic diagram of the LEHCPEB source based on vacuum spark plasma. 1: cathode, 2: anode, 3: collector, 4: vacuum chamber, 5: cathode plasma, 6: anode plasma, 7: solenoid, 8: spark plasma sources, 9: specimen, 10: plasma sheath, 11: electron beam [4].

Mg [14, 15], and Mg alloys [6, 16, 17]. In particular, though collaboration with the University of Metz (France), they have detailed the microstructure evolution of LEHCPEB treated alloys and their corrosion properties [18–23]. Also, the group has investigated the LEHCPEB treatment of a FeAl intermetallic compounds [24, 25], which can be used to replace refractory steels for high-temperature weight saving applications. This paper reviews the main results obtained from their studies on light alloys treated with LEHCPEB in connection to improvements in wear and corrosion resistances as well as specific features related to texture and microstructure modifications, deep hardening, and evaporation phenomenon.

2. The Surface Treatment System

The samples investigated in the present manuscript were treated in Dalian using a "Nadezhda-2" system. A schematic diagram of this LEHCPEB source is given in Figure 1. The "Nadezhda-2" electron-beam source can produce electron beams with the parameters as follows: electron energy 10–40 keV, peak current 10^2–10^3 A/cm^2, pulse duration 0.5–5 μs, energy density 0.5–40 J/cm^2, beam cross-section area 10–50 cm^2, and repeating pulse interval 10 s. The electron beam is generated by an explosive emission graphite cathode. Spark plasma sources are placed evenly in a circle behind the anode, providing anode plasma that conducts the beam to the collector where the specimen to be treated is placed. An external magnetic field is applied to confine the beam to prevent the beam from pinching and dispersing. The accelerating voltage, magnetic fields intensity, and anode-collector distance control the beam energy density. They are simple and reliable in operation.

3. Improvement of Wear and Corrosion Resistances

The pulsed electron irradiation generated by this surface treatment technique induces rapid melting of the surface followed by extremely fast solidification. This process leads to the dissolution of second-phase particles and, after sufficient number of pulses, to the formation of 2-3 μm thick homogeneous melted surfaces [18, 22, 26]. The so-formed homogeneous layers were demonstrated to improve the corrosion properties of dual-phase alloys such as the 316L stainless steel [7, 21] and the D2 steel [18] as well as NiTi shape memory alloys [20]. The LEHCPEB treatment also often induces in steels the formation of nanoscale structures formed by rapid solidification in the melted layer [18, 21, 27, 28]. Some extent of grain refinement was also observed in the subsurface that remained solid of a D2 steel because of the heavy deformation induced by the thermal stress wave accompanying the very fast thermal cycles [29]. In steels, these finer grains together with the strain hardening imparted by the thermal stresses are believed to be the major contributors to surface/subsurface hardening and associated wear resistance improvement [7]. In the following section, we present some results obtained on light alloys in terms of wear and corrosion resistances.

3.1. Mg and Its Alloys. The AZ31 (Mg-3Al-0.5Zn-0.5Mn) is a commercial α (Mg) single-phase Mg alloy while the AZ91 (Mg-8.9Al-0.5Zn-0.2Mn-0.04Si) one consist of a two-phase α (Mg) + β-$Mg_{17}Al_{12}$ mixture. The LEHCPEB surface treatments were carried out for 5, 10, and 15 pulses using electron beam energy densities of 2.5 J/cm^2 and of ~3 J/cm^2 for the number AZ31 and AZ91 alloys, respectively. Wear tests were carried out without lubricant using a ball (WC-Co)-on-flat apparatus. Figure 2 gives the evolution of the wear rate with the number of pulses for the AZ31 alloy while the wear volume and the friction coefficients for the AZ91 alloy are shown in Figure 3. For both alloys, the wear mechanism was abrasive wear and the wear tracks were much wider for the untreated samples than for the treated ones. This indicates that a hardening effect occurred due to the LEHCPEB treatment.

For the AZ31 alloy, the wear rate decreases when increasing the number of LEHCPEB pulses and the best wear

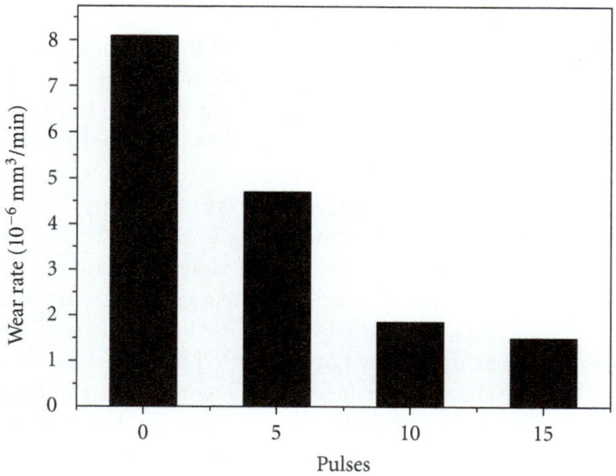

FIGURE 2: Evolution of the wear rate with the number of pulses for the AZ31 Mg alloy [6].

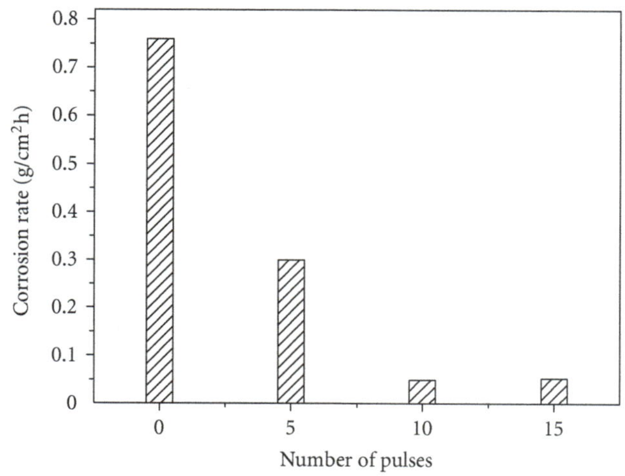

FIGURE 4: Evolution of the corrosion rate with the number of pulses [16].

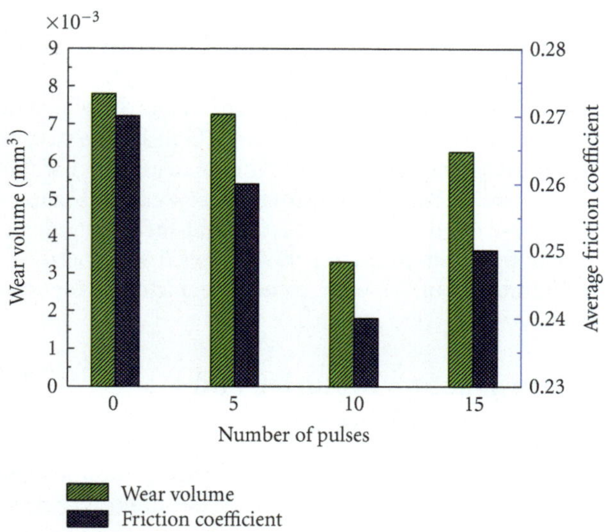

FIGURE 3: Evolution of the wear volume and the friction coefficient with the number of pulse for the AZ91 Mg alloy [17].

resistance was obtained after 15 pulses of treatments. This corresponds to an improvement in wear resistance by a factor of 6.7.

As illustrated in Figure 3, there is also a general trend for both the wear volumes and the friction coefficients of the AZ91-treated samples to decrease after the LEHCPEB treatment. However, the best wear resistance was obtained for the sample treated for 10 pulses only.

In terms of corrosion resistance, it is also the 10 pulsed treated sample that displayed the best behavior. This is illustrated in Figure 4 that gives the evolution of the corrosion rate with the number of pulses for the AZ91 alloy. The immersion corrosion experiments were performed in a 5% NaCl solution for 72 h on both the intial and LEHCPEB-treated AZ91 samples. The corrosion results indicated that the average corrosion rate decreased after the LEHCPEB treatment with the 10-pulsed sample showing the best corrosion resistance (the lowest corrosion rate). It is, therefore, clear that both the best wear and corrosion resistance are obtained after 10 pulses and not after the highest number of pulses (15 pulses). As will be discussed later in the text, this is likely due to some evaporation phenomena taking place more significantly after 15 pulses.

Such improvements of the wear and corrosion resistance can be attributed to the formation of an extended graded structure (a nanograined MgO layer and a melted layer) at the surface layer of the irradiated AZ91 Mg alloy after the LEHCPEB treatment. The thin nanograined MgO layer present on the top surface played an important role of lubricant during wear tests and a protective layer during corrosion test. The melted layer formed beneath the oxide layer consists of Al oversupersaturated α (Mg) solid solution due to the solute trapping effect during the fast solidification process. Such a melted layer is favorable for the formation of a homogenous passive film during the corrosion test and thereby protects the substrate.

3.2. Ti and Its Alloys. LEHCPEB treatments were performed on the commercially pure Ti (CP-Ti) samples [13] with the accelerating voltage 25.2 kV and pulse times 15 and 25. Electrochemical corrosion tests for both the initial and treated samples were carried out using the conventional three-electrode cell in a 5% NaCl water solution.

Figure 5 shows the potentiodynamic polarization curves. There is a significant improvement in corrosion resistance after the LEHCPEB treatments evidenced by a shift of the whole polarization curve towards the region of lower current density and higher potential. The improvement in corrosion performance can be ascribed to two major factors. Firstly, it is well established that ultrafine structures in CP-Ti lead to improved corrosion resistance. This is due to the redistribution of impurities that segregate at grain boundaries and higher reactivity to form a protective oxide layer. Secondly, the LEHCPEB technique is very effective for removing top surface contaminants or self-purification helping to form more homogeneous protective layers.

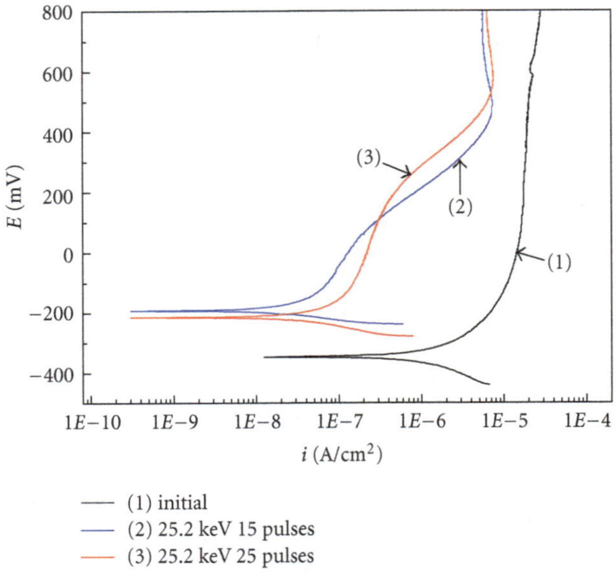

FIGURE 5: Potentiodynamic polarization curves of CP-Ti samples before and after LEHCPEB treatments [9].

4. Deep Hardening Effects

The hardness of a material depends on several factors: the grain size via the Hall-Petch hardening, the alloying elements via the solid solution, and/or precipitation hardenings, the presence of structural defects such as vacancies or dislocations as well as the residual stress state. The understanding of hardening in iron and steels treated by LEHCPEB has already been investigated in detail recently. Two interesting behaviours are generally reported in steels concerning the hardness evolution on the LEHCPEB-treated samples. First, a modification of the hardness is obtained in the melted layer (hardening or softening) [7, 30]. Second, below this melted layer, the hardened subsurface—which extends over slightly more than 100 micron—consists in fact of two successive hardened zones. It is well established that the high dislocation density in the surface layer of the target material usually increases dramatically after the LEHCPEB treatment [31, 32]. For example, Valyaev et al. have shown the dependence of hardness on dislocation densities in pure iron treated by pulsed electron beam treatment [31, 32]. The LEHCPEB irradiation also induced vacancy type of defects that should also favour the increased hardness in steels [33, 34]. The first hardness peak was attributed essentially to the quasi-static thermal stress and the repeated action of the beam induced the formation of microdeformation bands. The second peak—of lower intensity—was related to deeper hardening induced by the thermal stress wave [7]. In some cases, the tensile stress that is created in the rapidly quenched melted layer has been suggested to induce a top surface softening over the top first few microns. For example, Rotshtein et al. [30] and Zou et al. [7] have also observed a decrease in hardness at the top surface layer of a 316LS steel. They considered that this was due to the residual tensile stress generated by the LEHCPEB treatment [7, 30].

Figure 6 gives the evolution of the cross-section microhardness with depth for the LEHCPEB-treated Mg alloys AZ31 and AZ91, pure Mg, and pure Al. Although the hardness was measured from different light alloys treated with different LEHCPEB parameters, the hardness profiles often show some common characteristic features: an overall increase in subsurface hardness with the possibility of some local sharp increase. It is interesting to notice that the increase in hardness is generally obtained over a depth of more than 100 μm for the LEHCPEB-treated samples, sometimes reaching a depth of nearly 500 μm (Figure 6(a)). These increases in hardness must be attributed to the action of the stress waves which can propagate far beyond the heat affected zone (HAZ). Hence, deformation occurs below the HAZ and the hardness can be increased due to the work-hardening effect [6, 14]. In fact, such deformation induced by thermal stress wave propagation was revealed by TEM observations in LEHCPEB treated pure Al. Figure 7 gives a TEM bright field image showing the deformation marks (stacking fault-like contrast) about 1 μm beneath the surface having orientations of about 45° with respect to the surface. Moreover, a direct experimental proof of the existence of the stress far below the surface is illustrated in the TEM image of Figure 8, where two wave fronts are observed about 0.5 mm beneath the surface.

It is usually accepted that the increase in hardness plays the major role for improving the wear resistance of LEHCPEB-treated samples. It is the subsurface hardening that is in fact responsible for the improved wear behaviour and diminution of the wear volumes depicted in Figures 2 and 3.

5. Microstructure Evolution and Evaporation Phenomena

The LEHCPEB surface treatment on light alloys is normally accompanied by significant surface evaporation of chemical species. Under intensive evaporation, a special surface morphology is observed after LEHCPEB treatment. Figure 9 shows a typical SEM surface micrograph of pure Mg irradiated by LEHCPEB with an energy density of about 3 J/cm^2. Two distinct features can be clearly observed. The first one is the wavy aspect of the surface consisting of a succession of hills and valleys at the sample surface, which is different from common feature of many LEHCPEB treated metal surfaces. The second feature is the presence of many isolated particles on the surface, which is due to the condensation of the Mg vapor.

After treatment in evaporation mode, it is expected that the selective evaporation taking place on the surface results in composition changes. Figure 10 gives the electron probe microanalyzer (EPMA) line scan analysis of the melted layer present locally at the surface of the AZ31 Mg alloy samples. The results clearly indicated a depletion of the Mg element and a corresponding increase of Al element in the melted layer. Such a selective evaporation of Mg over Al element is clearly the consequence of the LEHCPEB treatment operating in the evaporation mode at the sample surface.

Results of SNMS (sputter neutral mass spectrometry, VG instrument) analysis are shown in Figure 11 where

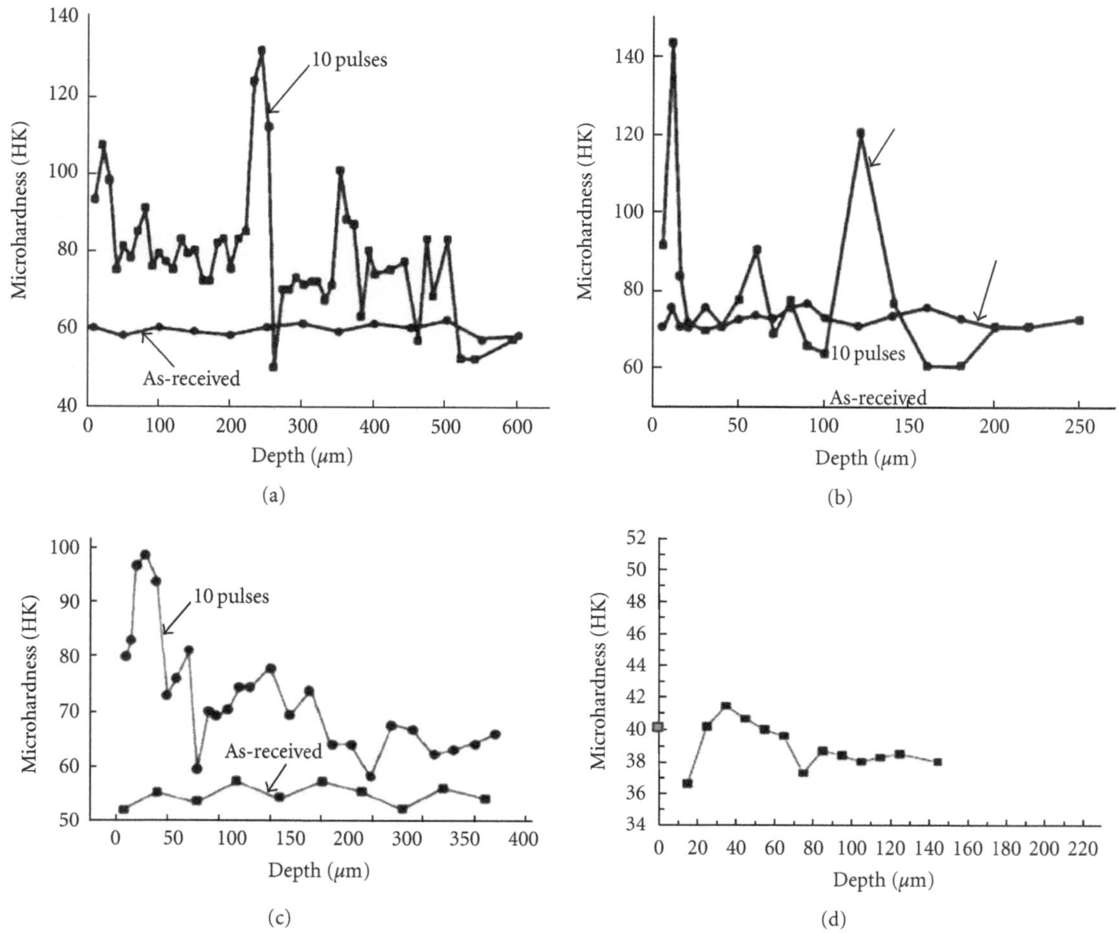

FIGURE 6: Evolution of cross-section microhardness with depth for the LEHCPEB-treated samples: (a) AZ31 Mg alloy (27 kV, 10 pulses); (b) AZ91 Mg alloy (27 kV, 10 pulses); (c) pure Mg (27 kV, 10 pulses); (d) pure Al (28 kV, 1 pulse).

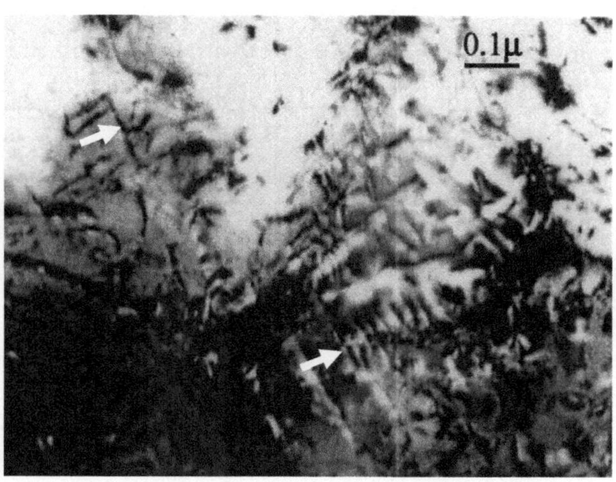

FIGURE 7: The stacking faults-like contrast about 1 μm beneath the surface melted layer [11].

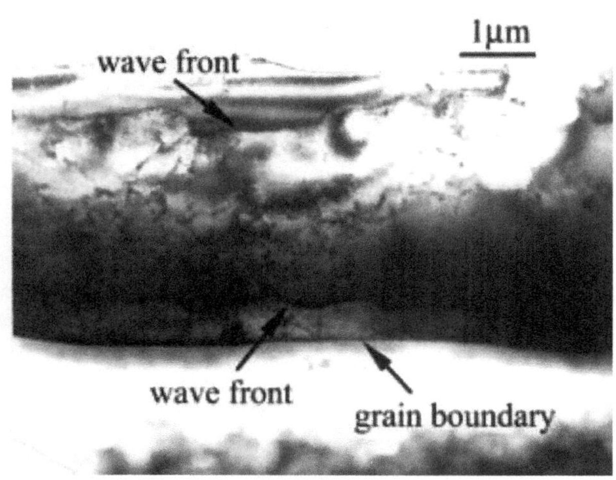

FIGURE 8: Two wavy fronts approximately 0.5 mm beneath and perpendicular to the treated surface. They are parallel to a grain boundary [15].

the evolution of the amount of the major elements—Ti and Al—is plotted as a function of the distance from the surface for the near α titanium TA15 samples treated with 23.4 kV. Chemical modifications are clearly visible at the top surface of the melted layers: the Ti concentrations increase while the Al concentrations decrease when approaching the surface. For the 5 pulsed sample (Figure 11(a)), starting from a depth

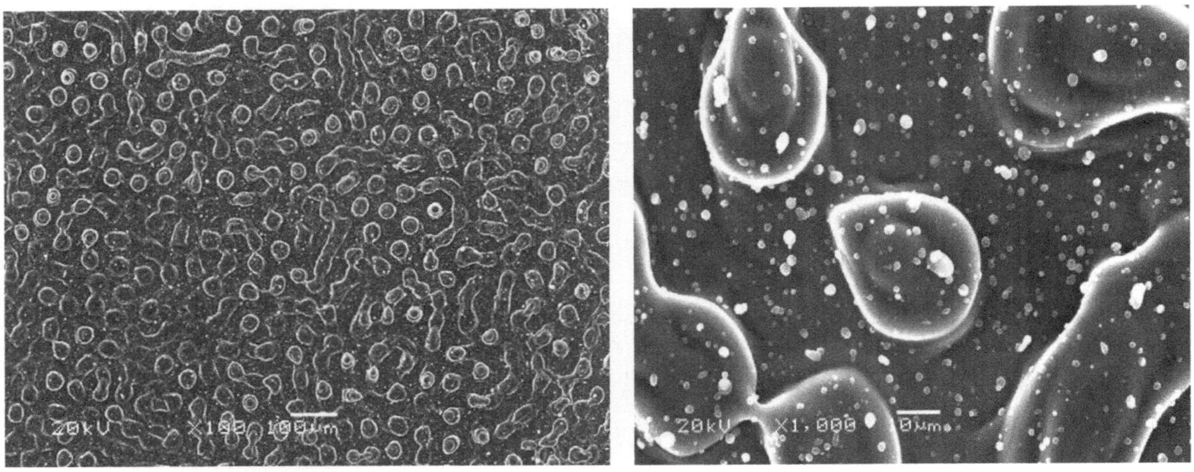

(a) (b)

FIGURE 9: SEM micrographs showing typical surface morphology of pure Mg after evaporation treatment, energy density 3 J/cm^2, 5 pulses [16].

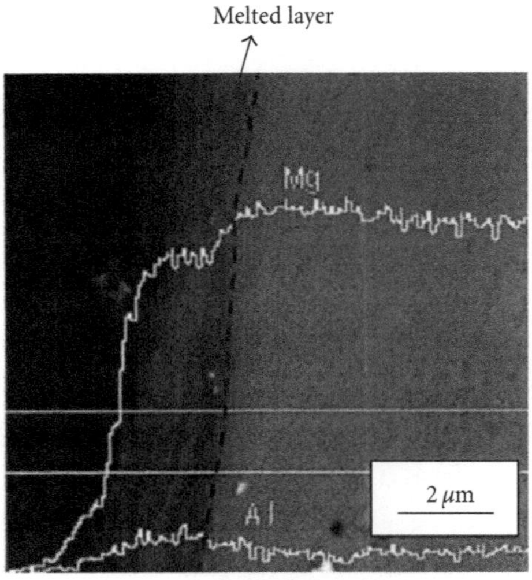

FIGURE 10: EPMA line scan analysis of the melted layer present locally at the surface of the samples after the LEHCPEB treatment 15 pulses [6].

of about 250 nm, the Ti content can reach about 88.5 at%, slightly higher than the average concentration in matrix (87.7 at%). Comparatively, the Al concentration decreases to about 10.5 at% instead of the initial 11.3 at%. For the 10 pulsed sample (Figure 11(b)), the chemical modifications in the surface layer are more pronounced. They start from a depth of about 400 nm and the Ti content reaches the maximum value at about 89 at% at the top surface while the Al concentration decreases to about 9.8 at%. Numerical calculation also confirmed that the Al element is more prone to evaporate than the Ti element. The evaporation of Al is more pronounced at higher number of pulses indicated that more energy input can generate a more significant evaporation phenomenon. This more pronounced change in chemistry modification in terms of magnitude and depth with increasing number of pulses when the evaporation mode is taking place is consistent with previous results on a NiTi alloy [20].

6. Modification of Surface Crystallographic Textures

An interesting effect of the LEHCPEB surface treatment is the modification of the surface crystallographic texture without changing the macroscopic shape of the sample. In a classical manner, this can be done when the top surface that has been melted resolidifies by the nucleation and growth of new grains. The growth of the new grains follows the major thermal gradient (perpendicular to the surface) along which some preferred orientations can grow more rapidly; creating thereby a solidification texture. As previously stated, it is also well established that evaporation can become significant at high energy densities [2, 35]. This phenomenon was suggested to lead to a modification of the atomic attachment along the growing solidification interface leading to the development of specific solidification textures in the rapidly solidified layer [20]. Another interesting phenomenon reported recently is the modification of crystallographic surface texture by treating the material below the onset of melting (the so-called heating mode) [24, 25].

Figure 12 shows the EBSD analysis of the surface of an FeAl alloy before and after the LEHCPEB treatment. The standard triangle is shown inset. The initial extruded bar was obtained from milled powder and the initial sample was cut perpendicular to the extrusion axis. The EBSD map in Figure 12(a) shows that a majority of grains have a green color. This illustrates that the as-extruded bar is characterized by a <110> fiber texture along the extrusion axis [36]. This is typical of extruded body-centered cubic metals. Figure 12(b) shows an EBSD map obtained from the surface of a sample treated for 20 pulses of LEHCPEB under the heating mode [25]. The fraction of large angle grain boundaries (LAGBs) is increased quite substantially due to the creation of new subgrains inside the initial grains, as

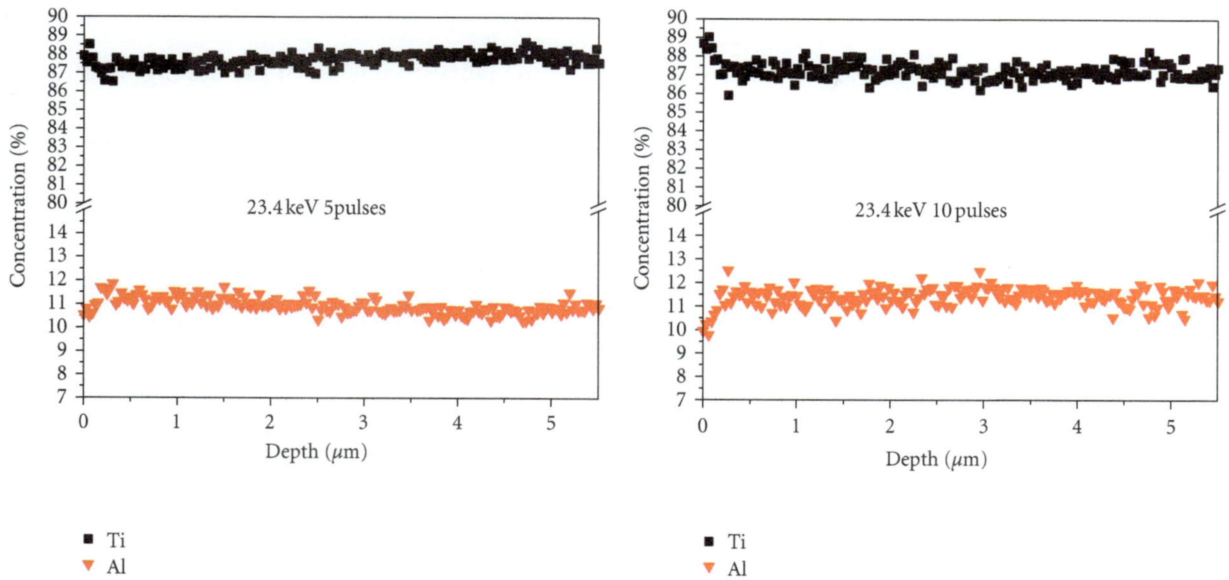

(a) (b)

FIGURE 11: SNMS composition profiles in Ti and Al for the TA15 near α titanium alloy. Experimental conditions are (a) 23.4 KeV 5 pulses; (b) 23.4 KeV 10 pulses.

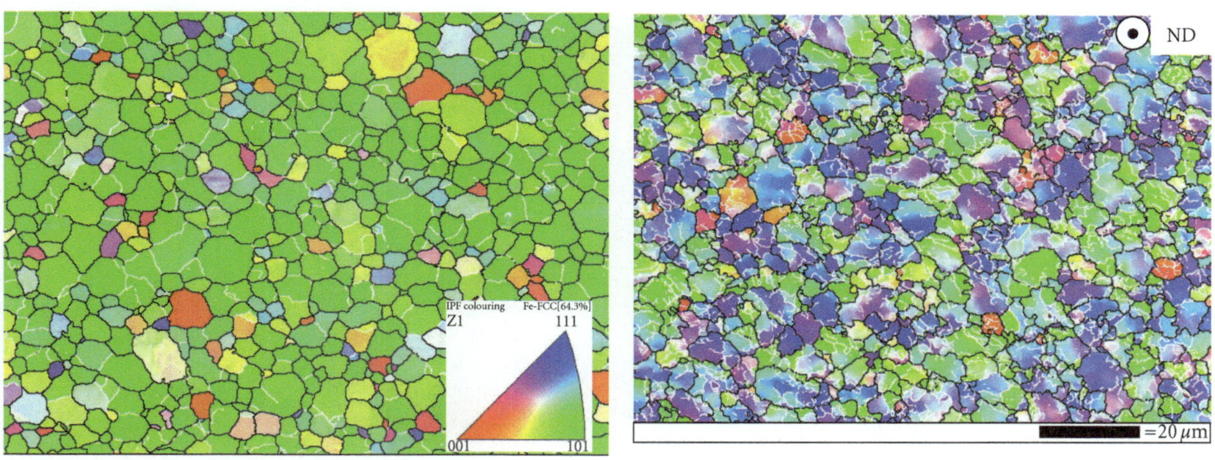

(a) (b)

FIGURE 12: EBSD analysis showing the evolution of microstructure and texture of a FeAl alloy: (a) in the as-received state (extruded) and (b) after 20 pulsed of LEHCPEB treatment.

witnessed by the presence of numerous LAGBs (in white) visible in Figure 12(b). These LAGBs must have been formed by the effect of the thermomechanical treatment associated with the LEHCPEB irradiation. Indeed, the material adjacent to the zone of energy transfer is rapidly heated, leading to the formation of a nonequilibrium temperature field coupled with dynamic stress fields propagating into the material. The maximum amplitude of thermal stress during treatment can reach values much higher than the yield strength of the material [5]. From the comparison of the maps in Figure 12, it is clear that many of the initially green grains (Figure 12(a)) have changed their colour, thereby indicating that the texture at the top surface has also been modified by the LEHCPEB treatment. The blue colour in the map—see the colour code in the inset standard suggests that the orientation of most of the grains has rotated towards <111> // the normal direction (ND) or in its vicinity. This modification is due to the heavy deformation and thermal cycles generated during each pulse that also lead to a complex combination of dynamic recovery and strain-induced grain boundary migration. Clearly, the thermomechanical treatment associated with the LEHCPEB treatment in the heating mode has drastically changed the surface texture and microstructure and, contrary to many processes involving plastic deformation, this is done without changing the macroscopic shape of the sample.

The treatment under the heating mode also modified the mechanical properties of the surface. Indeed, even after only 2 pulses, an increase of about 50 HV in hardness was associated with the LEHCPEB treatment at the surface of the FeAl sample (initially at 300 Hv). This was attributed to the formation of quenched-in vacancies, structural defects such as dislocation and subgrains, as well as the combination of grain refinement and texture modification toward the much stiffer <111> orientation [25]. In addition, LEHCPEB carried out on samples having their surface covered by graphite powder also demonstrates solid-state surface hardening and rapid solid-state surface alloying effects. This led to an increase of more than 100 HV (about 30%) associated with the thermally enhanced solid-state diffusion of C [25]. A final interest of the treatment under the heating mode is that the surface is free of any crater. Thereby, the corrosion resistance remains almost unchanged while other properties such as texture, grain size, and hardness can be modified to a significant extent [24].

As attested by the increasing number of papers published recently from different research groups [37–46], the LEHCPEB surface treatment technique is now spreading worldwide seeking for additional new applications including, for example, surface cleaning, surface alloying [14, 32, 41, 42], surface strengthening, polishing, perforating, homogenization, and pre- or posttreatment.

7. Summary and Conclusion

This article deals with the application of the low-energy high-current pulsed electron beam surface treatment technique on light alloys, such as pure Mg and Ti, Mg and Al alloys as well as FeAl.

The pulsed electron irradiation induces an ultra-fast thermal cycle at the surface combined with the formation of thermal stress and shock waves. The LEHCPEB-modified layer is usually divided into three successive regions having different orders of penetration depth: (i) a melted and rapidly solidified layer on the top surface ($\sim 1\,\mu$m), (ii) a heat-affected zone where solid state phase transformation, deformation and recrystallization may occur ($\sim 10\,\mu$m), and (iii) a stress-wave-affected zone that can be hardened ($\sim 100\,\mu$m).

The detailed analysis of Al and Mg alloys revealed that the plastic deformation associated with the imparted stresses results in deep subsurface hardening that can extend over several 100 μm. To this hardening is associated a significant improvement in the wear resistance.

The analysis of the top surface melted surface of light alloys also often witnesses evaporation and condensation of chemical species. This evaporation phenomenon can significantly modify the melt chemistry. Through second-phase dissolution and microstructure refinement in the melted layer, significant corrosion improvement can also be obtained.

Crystallographic texture modifications are also revealed by detailed analysis of the top surface of LEHCPEB-treated samples. The evaporation process was, for example, suggested to lead to the development of specific solidification textures in the rapidly solidified layer. Comparatively, as demonstrated by the analysis of FeAl samples, the potential use of the LEHCPEB technique for producing thermo-mechanical treatments under the so-called heating mode modifies the surface crystallographic texture without any melting of the surface or any macroscopic change of the sample size.

An additional interest of the heating mode is that the surface is free of any crater. Thereby, the corrosion resistance remains almost unchanged while other properties such as texture, grain size, and hardness can be modified to a significant extent.

References

[1] D. I. Proskurovsky, V. P. Rotshtein, and G. E. Ozur, "Use of low-energy, high-current electron beams for surface treatment of materials," *Surface and Coatings Technology*, vol. 96, no. 1, pp. 117–122, 1997.

[2] D. I. Proskurovsky, V. P. Rotshtein, G. E. Ozur et al., "Pulsed electron-beam technology for surface modification of metallic materials," *Journal of Vacuum Science and Technology A*, vol. 16, no. 4, pp. 2480–2488, 1998.

[3] G. E. Ozur, D. I. Proskurovsky, V. P. Rotshtein, and A. B. Markov, "Production and application of low-energy high-current electron beams," *Laser and Particle Beams*, vol. 21, no. 2, pp. 157–174, 2003.

[4] D. I. Proskurovsky, V. P. Rotshtein, G. E. Ozur, Y. F. Ivanov, and A. B. Markov, "Physical foundations for surface treatment of materials with low energy, high current electron beams," *Surface and Coatings Technology*, vol. 125, no. 1–3, pp. 49–56, 2000.

[5] Y. Qin, C. Dong, Z. Song et al., "Deep Modification of materials by thermal stress wave generated by irradiation of high-current pulsed electron beams," *Journal of Vacuum Science and Technology A*, vol. 27, no. 3, pp. 430–435, 2009.

[6] B. Gao, S. Hao, J. Zou et al., "High current pulsed electron beam treatment of AZ31 Mg alloy," *Journal of Vacuum Science and Technology A*, vol. 23, no. 6, pp. 1548–1553, 2005.

[7] J. X. Zou, K. M. Zhang, S. Z. Hao, C. Dong, and T. Grosdidier, "Mechanisms of hardening, wear and corrosion improvement of 316 L stainless steel by low energy high current pulsed electron beam surface treatment," *Thin Solid Films*, vol. 519, no. 4, pp. 1404–1415, 2010.

[8] J. Zou, K. Zhang, C. Dong, Y. Qin, S. Hao, and T. Grosdidier, "Selective surface purification via crater eruption under pulsed electron beam irradiation," *Applied Physics Letters*, vol. 89, no. 4, Article ID 041913, 2006.

[9] K. M. Zhang, J. X. Zou, T. Grosdidier, C. Dong, and D. Yang, "Improved pitting corrosion resistance of AISI 316L stainless steel treated by high current pulsed electron beam," *Surface and Coatings Technology*, vol. 201, no. 3-4, pp. 1393–1400, 2006.

[10] K. M. Zhang, D. Z. Yang, J. X. Zou, T. Grosdidier, and C. Dong, "Improved in vitro corrosion resistance of a NiTi alloy by high current pulsed electron beam treatment," *Surface and Coatings Technology*, vol. 201, no. 6, pp. 3096–3102, 2006.

[11] S. Hao, S. Yao, J. Guan, A. Wu, P. Zhong, and C. Dong, "Surface treatment of aluminum by high current pulsed electron beam," *Current Applied Physics*, vol. 1, no. 2-3, pp. 203–208, 2001.

[12] J. Zou, Y. Qin, A. Wu, S. Hao, X. Wang, and C. Dong, "Thermal-mechanical simulation of high-current pulsed electron beam surface modification process of pure aluminum," *Nuclear Techniques*, vol. 27, no. 7, p. 519, 2004.

[13] X. D. Zhang, S. Z. Hao, X. N. Li, C. Dong, and T. Grosdidier, "Surface modification of pure titanium by pulsed electron

beam," *Applied Surface Science*, vol. 257, no. 13, pp. 5899–5902, 2011.

[14] T. Zhao, B. Gao, X. Tian, G. Tu, S. Hao, and C. Dong, "Modification and alloying of high purity magnesium surface with high current pulsed electron beam," *Journal of Vacuum Science and Technology*, vol. 28, no. 1, pp. 11–15, 2008.

[15] J. X. Zou, Y. Qin, B. Gao et al., "Surface melting and evaporation of pure Mg bombarded by high current pulsed electron beam," *Transactions of Materials and Heat Treatment*, vol. 26, no. 5, pp. 36–41, 2005.

[16] B. Gao, S. Z. Hao, C. Dong, and G. F. Tu, "Improvement of wear resistance of AZ31 and AZ91HP by high current pulsed electron beam treatment," *Transactions of Nonferrous Metals Society of China (English Edition)*, vol. 15, no. 5, pp. 978–984, 2005.

[17] B. Gao, S. Hao, J. Zou, W. Wu, G. Tu, and C. Dong, "Effect of high current pulsed electron beam treatment on surface microstructure and wear and corrosion resistance of an AZ91HP magnesium alloy," *Surface and Coatings Technology*, vol. 201, no. 14, pp. 6297–6303, 2007.

[18] J. Zou, T. Grosdidier, K. Zhang, and C. Dong, "Mechanisms of nanostructure and metastable phase formations in the surface melted layers of a HCPEB-treated D2 steel," *Acta Materialia*, vol. 54, no. 20, pp. 5409–5419, 2006.

[19] K. Zhang, J. Zou, T. Grosdidier, and C. Dong, "Formation and evolution of craters in carbon steels during low-energy high-current pulsed electron-beam treatment," *Journal of Vacuum Science and Technology A*, vol. 27, no. 5, pp. 1217–1226, 2009.

[20] K. M. Zhang, J. X. Zou, T. Grosdidier et al., "Mechanisms of structural evolutions associated with the high current pulsed electron beam treatment of a NiTi shape memory alloy," *Journal of Vacuum Science and Technology A*, vol. 25, no. 1, pp. 28–36, 2007.

[21] S. Hao, P. Wu, J. Zou, T. Grosdidier, and C. Dong, "Microstructure evolution occurring in the modified surface of 316L stainless steel under high current pulsed electron beam treatment," *Applied Surface Science*, vol. 253, no. 12, pp. 5349–5354, 2007.

[22] J. X. Zou, T. Grosdidier, B. Bolle, K. M. Zhang, and C. Dong, "Texture and microstructure at the surface of an AISI D2 steel treated by high current pulsed electron beam," *Metallurgical and Materials Transactions A*, vol. 38, no. 9, pp. 2061–2071, 2007.

[23] J. X. Zou, K. M. Zhang, T. Grosdidier et al., "Orientation-dependent deformation on 316L stainless steel induced by high-current pulsed electron beam irradiation," *Materials Science and Engineering A*, vol. 483-484, no. 1-2 C, pp. 302–305, 2008.

[24] T. Grosdidier, J. X. Zou, N. Stein, C. Boulanger, S. Z. Hao, and C. Dong, "Texture modification, grain refinement and improved hardness/corrosion balance of a FeAl alloy by pulsed electron beam surface treatment in the "heating mode"," *Scripta Materialia*, vol. 58, no. 12, pp. 1058–1061, 2008.

[25] T. Grosdidier, J. X. Zou, B. Bolle, S. Z. Hao, and C. Dong, "Grain refinement, hardening and metastable phase formation by high current pulsed electron beam (HCPEB) treatment under heating and melting modes," *Journal of Alloys and Compounds*, vol. 504, no. 1, pp. S508–S511, 2010.

[26] J. X. Zou, T. Grosdidier, K. M. Zhang, and C. Dong, "Cross-sectional analysis of the graded microstructure in an AISI D2-steel treated with low energy high-current pulsed electron beam," *Applied Surface Science*, vol. 255, no. 9, pp. 4758–4764, 2009.

[27] Q. F. Guan, P. L. Yang, H. Zou, and G. T. Zou, "Nanocrystalline and amorphous surface structure of 0.45%C steel produced by high current pulsed electron beam," *Journal of Materials Science*, vol. 41, no. 2, pp. 479–483, 2006.

[28] D. Q. Cheng, Q. F. Guan, J. Zhu, D. H. Qiu, X. W. Cheng, and X. T. Wang, "Mechanism of surface nanocrystallization in pure nickel induced by high-current pulsed electron beam," *Acta Physica Sinica*, vol. 58, no. 10, pp. 7300–7306, 2009.

[29] T. Grosdidier, Y. Samih, N. Allain-Bonasso et al., "Grain refinement and hardening induced by heavy deformation using low energy high current pulsed electron beam surface treatment," in *Proceedings of the 5th International Conference on Nanomaterials by Severe Plastic Deformation (NanoSPD '11)*, pp. 499–504, Nanjing, China, Trans Tech Publications Ltd, 2011.

[30] V. P. Rotshtein, D. I. Proskurovsky, G. E. Ozur, Y. F. Ivanov, and A. B. Markov, "Surface modification and alloying of metallic materials with low-energy high-current electron beams," *Surface and Coatings Technology*, vol. 180-181, pp. 377–381, 2004.

[31] A. N. Valyaev, V. S. Ladysev, A. D. Pogrebnjak, A. A. Valyaev, and S. V. Plotnikov, "Comparative analysis of radiation damages, mechanical and tribological properties of α-Fe exposed to intense-pulsed electron and ion beams," *Nuclear Instruments and Methods in Physics Research, Section B*, vol. 161, pp. 1132–1136, 2000.

[32] V. P. Rotshtein, Y. F. Ivanov, A. B. Markov et al., "Surface alloying of stainless steel 316 with copper using pulsed electron-beam melting of film-substrate system," *Surface and Coatings Technology*, vol. 200, no. 22-23, pp. 6378–6383, 2006.

[33] E. H. Lee, J. D. Hunn, N. Hashimoto, and L. K. Mansur, "Hardness and defect structures in EC316LN austenitic alloy irradiated under a simulated spallation neutron source environment using triple ion-beams," *Journal of Nuclear Materials*, vol. 278, no. 2, pp. 266–272, 2000.

[34] D. J. Edwards, E. P. Simonen, and S. M. Bruemmer, "Evolution of fine-scale defects in stainless steels neutron-irradiated at 275°C," *Journal of Nuclear Materials*, vol. 317, no. 1, pp. 13–31, 2003.

[35] X. P. Zhu, H. Suematsu, W. Jiang, K. Yatsui, and M. K. Lei, "Pulsed-ion-beam nitriding and smoothing of titanium surface in a vacuum," *Applied Physics Letters*, vol. 87, no. 9, Article ID 09311, pp. 1–3, 2005.

[36] S. Guessasma, E. Suzon, T. Grosdidier, and F. Wagner, "Static recrystallisation in an ODS FeAl alloy: the effect of particles on texture and anisotropic grain growth developments," *Intermetallics*, vol. 16, no. 8, pp. 1013–1025, 2008.

[37] X.-X. Lu, R.-G. Li, and J. An, "Numerical simulation of thermal-mechanical process of Al-Si-Pb alloy treated by high current pulsed electron beam," *Transactions of the Nonferrous Metals Society of China*, vol. 16, pp. s1433–s1436, 2006.

[38] J. An, X. X. Shen, Y. Lu et al., "Influence of high current pulsed electron beam treatment on the tribological properties of Al-Si-Pb alloy," *Surface and Coatings Technology*, vol. 200, no. 18-19, pp. 5590–5597, 2006.

[39] Y. Daichi, Z. Wang, K. Yamazaki, and S. Sano, "Investigation of the formation and energy density of high-current pulsed electron beams," *Plasma Chemistry and Plasma Processing*, vol. 27, no. 4, pp. 458–472, 2007.

[40] S. Hao, X. Zhang, X. Mei, T. Grosdidier, and C. Dong, "Surface treatment of DZ4 directionally solidified nickel-based superalloy by high current pulsed electron beam," *Materials Letters*, vol. 62, no. 3, pp. 414–417, 2008.

[41] K. M. Zhang, J. X. Zou, T. Grosdidier, C. Dong, and S. Weber, "Ti surface alloying of an AISI 316L stainless steel by low energy high current pulsed electron beam treatment," *Journal of Vacuum Science and Technology A*, vol. 26, no. 6, pp. 1407–1414, 2008.

[42] K. M. Zhang, J. X. Zou, T. Grosdidier, C. Dong, and S. Weber, "Rapid surface alloying by Ti of AISI 316L stainless steel using low energy high current pulsed electron beam," *The European Physical Journal: Applied Physics*, vol. 43, no. 3, pp. 343–347, 2008.

[43] T. Grosdidier, J. X. Zou, B. Bolle, S. Z. Hao, and C. Dong, "Grain refinement, hardening and metastable phase formation by high current pulsed electron beam (HCPEB) treatment under heating and melting modes," *Journal of Alloys and Compounds*, vol. 504, no. 1, pp. S508–S511, 2010.

[44] B. Gao, Y. Hao, G. Tu, W. Shi, F. Yu, and S. Li, "Compounded surface modification of ZK60 Mg alloy by high current pulsed electron beam+micro-plasma oxidation," *Plasma Science and Technology*, vol. 12, no. 1, pp. 67–70, 2010.

[45] F. Xu, G. Tang, G. Guo, X. Ma, and G. E. Ozur, "Influence of irradiation number of high current pulsed electron beam on the structure and properties of M50 steel," *Nuclear Instruments and Methods in Physics Research, Section B*, vol. 268, no. 15, pp. 2395–2399, 2010.

[46] F. Xu, G. Guo, G. Tang et al., "Microstructure modifications and corrosion behaviors of Cr4Mo4V steel treated by high current pulsed electron beam," *Materials Chemistry and Physics*, vol. 126, no. 3, pp. 904–908, 2011.

XRD and DTA Analysis of Developed Agglomerated Fluxes for Submerged Arc Welding

Ajay Kumar,[1] Hari Singh,[2] and Sachin Maheshwari[3]

[1] *Department of Mechanical Engineering, Noida Institute of Engineering and Technology, Greater Noida 201306, India*
[2] *Department of Mechanical Engineering, National Institute of Technology, Kurukshetra, Haryana 136119, India*
[3] *Division of MPAE, Netaji Subhas Institute of Technology, New Delhi 110078, India*

Correspondence should be addressed to Ajay Kumar; ajayagrohi@gmail.com

Academic Editor: Gerhard Sauthoff

A unique study of structural and chemical analysis of crystalline phases in developed agglomerated fluxes was carried out. Thirty-two fluxes were developed by using a mixture of oxides, halides, carbonates, silicates, and ferroalloys for submerged arc welding. The present paper focuses on only ten (out of thirty-two) fluxes which were analyzed by X-ray diffraction (XRD) to know the different types of oxides formed and changed in oxidation number of metallic centers after sintering process at around 850°C. To know the effect of temperature over phase transformation and melting of different compounds, differential thermal analysis (DTA) was carried out from 1000 to 1400°C. This study aims to know the quantity of ions present (percentage) and melting behavior of developed agglomerated fluxes for submerged arc welding process.

1. Introduction

The agglomeration method is used for flux preparation by using a mixture of oxides, halides, carbonates, silicates, and ferroalloys [1]. The chemistry of weld metal is always governed by the electrochemical changes at the weld pool flux interface because the important properties of submerged arc welding fluxes are the function of their physiochemical properties [2–5]. The important chemical reactions take place during heating of ceramic oxide which is the chemical reaction among the different phases [6], phase transformation of minerals [7], formation of crystalline phases [8], and chemical and structural characterization of crystalline phases in agglomerated fluxes for submerged arc welding were observed [9]. Therefore, the developed agglomerated fluxes can be analyzed to know their crystal structure observed during the rise of temperature. The objective of this analysis is to know the crystalline phases as well as to quantify the different types of ions and their distribution in developed agglomerated fluxes (AGF1101–AGF1110) by X-ray diffraction (XRD) and differential thermal analysis (DTA) is used to know the peaks of endothermic temperatures.

2. Experimentation

In the present study, the fluxes have been designed on the basis of binary and ternary phase diagrams for different oxide and fluoride systems [10] and formulated by using response surface methodology technique [11, 12]. In these fluxes, varying amount of oxides and fluorides were added to know the crystalline phases as well as to quantify the different types of ions and their distribution in developed agglomerated fluxes (AGF1101–AGF1110) by X-ray diffraction (XRD) and differential thermal analysis (DTA) was carried out to know the peaks of endothermic temperatures.

CaO-Al_2O_3-SiO_2-based flux systems have been selected for the study as these are the most widely used fluxes at the commercial level. The details of basic constituents, alloying constituents, and their percentage in the developed agglomerated fluxes are reported in Table 1.

Table 1: Constituents and their ranges.

Basic constituents and alloying constituents	CaO	SiO_2	Al_2O_3	MnO	CaF_2	MgO	NiO	Fe-Cr
Amount (wt %)	20	35	10	5–10	10–20	5–15	0–10	0–10

Table 2: Crystalline phases and crystal structure in developed agglomerated fluxes.

Flux number	Crystalline phase	Crystal structure
AFG1101	$CaAl_2SiO_8$ (Dmisteinbergite)	Hexagonal
AFG1101	$Mg_7Fe_7Si_8O_{22}(OH)_2$ (Protoanthrophyllite)	Orthorhombic
AFG1101 AFG1104	$CaAl_2Si_6O_{16} \cdot 5(H_2O)$ (Epistilbite)	Triclinic
AFG1101	$Mg_2Al_2Si_4O_{10}(OH) \cdot 4(H_2O)$ (Palygorskite)	Monoclinic
AFG1101	$Ca_2Fe^{+++}(Fe^{+++}, Al)_2(SiO_4)(Si_2O_7)(O, OH)_2 \cdot H_2O$ (Julgoldite)	Monoclinic
AFG1102	$CaAl_2Si_2O_8$ (Svyatoslavite)	Monoclinic
AFG1102	$K_2CaMg_2(Al, Si)_{36}O_{72} \cdot 28H_2O$ (Mazzite)	Hexagonal
AFG1102 AFG1103 AFG1104	$Mg_2(Al, Fe^{+++})_3Si_3AlO_{10}(OH)_8$ (Sudoite)	Monoclinic
AFG1102	$Mn_7Si_3O_{12}(OH)_2$ (Leucophoenicite)	Monoclinic
AFG1102	$Ca_3Al_2(SiO_4)_{3-x}(OH)_{4x}(x = 0.2–1.5)$ (Hibschite)	Isometric
AFG1104	$KAl_2(Si_3Al)O_{10}(OH, F)_2$ (Muscovite)	Monoclinic
AFG1104 AFG1107 AFG1108	$Ca_2(Mn, Fe^{++})Fe^{+++}Si_5O_{14}(OH)$ (Manganbabingtonite)	Triclinic
AFG1103 AFG1110	$Ca_{10}Mg_4Al_2Si_{11}O_{39} \cdot 4(H_2O)$ (Juanite)	Orthorhombic
AFG1105 AFG1109	$Ca_5Si_2O_7(CO_3)_2$ (Tilleyite)	Monoclinic
AFG1104 AFG1108	$Mg_9(SiO_4)_4(OH, F)_2$ (Hydroxyl clinotumite)	Monoclinic
AFG1108	$Ca_7(SiO_4)_3(OH)_2$ (Chegemite)	Orthorhombic
AFG1110	$Ca_{14}(Mn^{++})_{14}Si_{24}O_{58}(OH)_8 \cdot 2(H_2O)$ (Truscottite)	Triclinic
AFG1110	$CaO \cdot 3(Fe^{++}, Mg, Fe^{+++})_3Si_4Al_4O_{10}(OH)_2 \cdot 4(H_2O)$ (Ferrosaponite)	Monoclinic
AFG1110	$Ca_4Si_2O_6CO_3(OH, F)_2$ (Fukalite)	Orthorhombic
AFG1105	$Mg_5(Fe^{+++})_5(SiO_4)_2(OH, F)_2$ (Chondrodite)	Monoclinic

Table 3: Quantity of oxides in percentage weight for developed agglomerated fluxes.

Serial number	Oxides	Quantity of oxides present (wt.%) in developed agglomerated fluxes									
		AFG1101	AFG1102	AFG1103	AFG1104	AFG1105	AFG1106	AFG1107	AFG1108	AFG1109	AFG1110
1	CaO	8.29	12.48	28.10	5.76	11.47	9.33	4.89	17.21	14.65	24.53
2	SiO_2	58.80	31.16	42.38	5.71	21.19	38.57	21.79	31.27	12.65	33.12
3	Al_2O_3	10.50	22.94	14.30	18.29	24.18	16.66	11.32	23.39	20.00	3.27
4	MgO	7.80	3.37	8.12	14.47	29.78	6.59	5.78	17.34	18.52	3.37
5	MnO	0.02	14.29	—	1.86	—	—	2.32	1.85	—	21.20
6	NiO	0.02	—	—	—	—	—	—	—	—	—
7	FeO	0.91	0.26	0.44	0.89	—	3.91	0.78	0.63	—	4.20
8	Fe_2O_3	6.47	0.58	0.64	3.37	—	—	3.48	2.79	—	1.73
9	Na_2O	0.04	—	—	—	—	—	—	—	—	0.48
10	CO_2	—	—	—	—	5.41	7.82	6.31	—	13.57	2.15
11	MnO_2	—	—	—	—	—	16.66	—	—	—	—

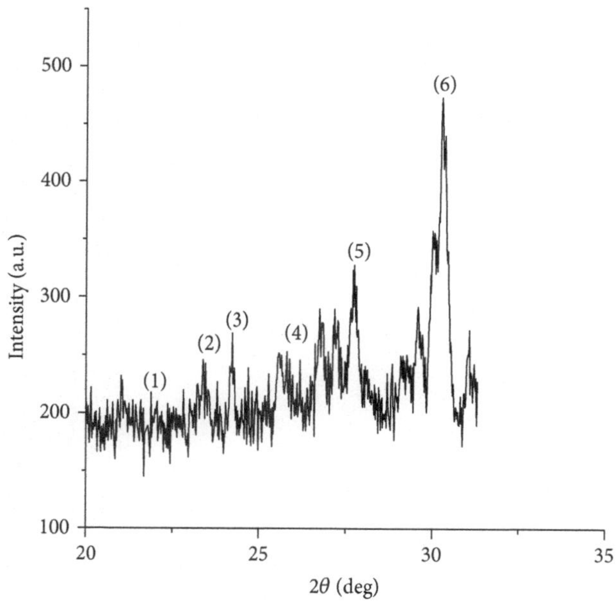

(1) SiO_2
(2) $CaAl_2Si_2O_8$
(3) $Mg_7Fe_7Si_8O_{22}(OH)_2$
(4) $CaAl_2Si_6O_{16}{\cdot}5H_2O$
(5) $Mg_2Al_2Si_4O_{10}(OH){\cdot}4H_2O$
(6) $Ca_2Fe^{+++}(Fe^{+++}, Al)_2SiO_4(Si_2O_7(O, OH)_2{\cdot}H_2O$

FIGURE 1: XRD graph of AGF1101 flux.

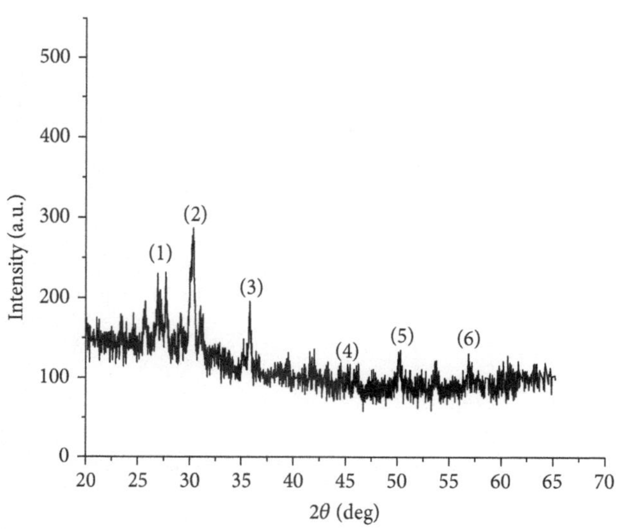

(1) $CaAl_2Si_2O_8$
(2) $K_2CaMg_2Al_{36}Si_{36}O_{72}{\cdot}28(H_2O)$
(3) $Mg_2(Al, Fe^{+++})_3Si_3AlO_{10}(OH)_8$
(4) Fe, Ni
(5) $Mn_7(SiO_4)_3(OH)_2$
(6) $Ca_3Al_2(SiO_4)_3{\cdot}X(OH)_{4X}$ ($X = 0.2$–1.5)

FIGURE 2: XRD graph of AGF1102 flux.

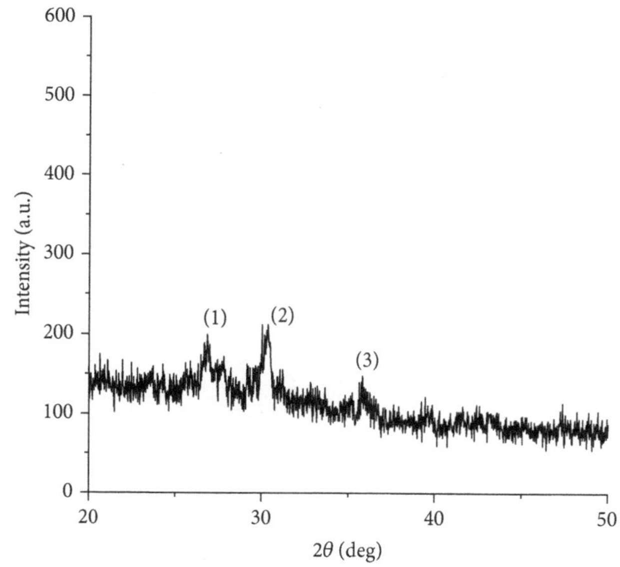

(1) $CaSiO_3$
(2) $Ca_{10}Mg_4Al_2Si_{11}O_{39}{\cdot}4(H_2O)$
(3) $Mg_2(Al, Fe^{+++})_3Si_3AlO_{10}(OH)_8$

FIGURE 3: XRD graph of AGF1103 flux.

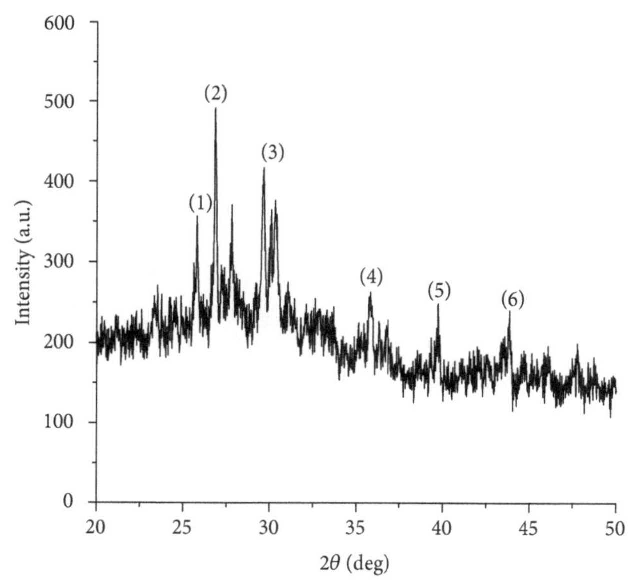

(1) $CaAl_2Si_6O_{16}{\cdot}5(H_2O)$
(2) $KAl_2(Si_3Al)O_{10}(OH, F)_2$
(3) $Ca_2(Mn, Fe^{++})Fe^{+++}Si_5O_{14}(OH)$
(4) $Mg_2(Al, Fe^{+++})_3Si_3Al_{10}(OH)_8$
(5) $Mg_9(SiO_4)_4(OH, F)_2$
(6) Mn_5Si_3

FIGURE 4: XRD graph of AGF1104 flux.

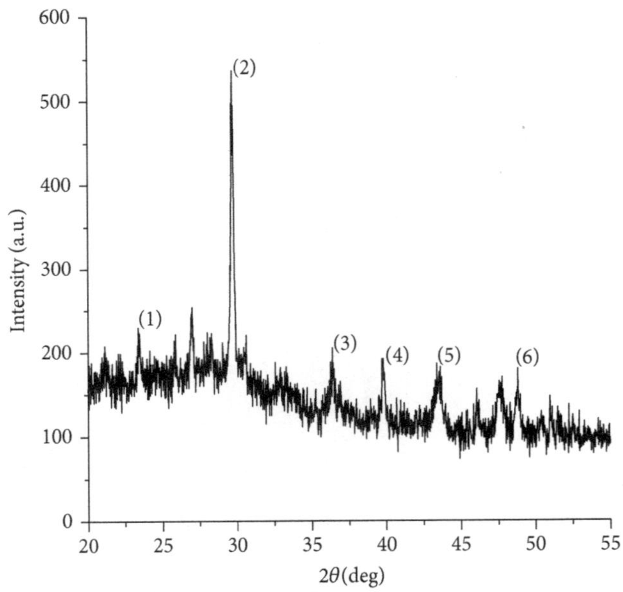

(1) $Mg_4Al_2(OH)_{12}Co_3{\cdot}4(H_2O)$
(2) $Ca_5Si_2O_7(Co_3)_2$
(3) Mg_2SiO_4
(4) $Mg_9(SiO_4)4(OH, F)_2$
(5) Al_2O_3

FIGURE 5: XRD graph of AGF1105 flux.

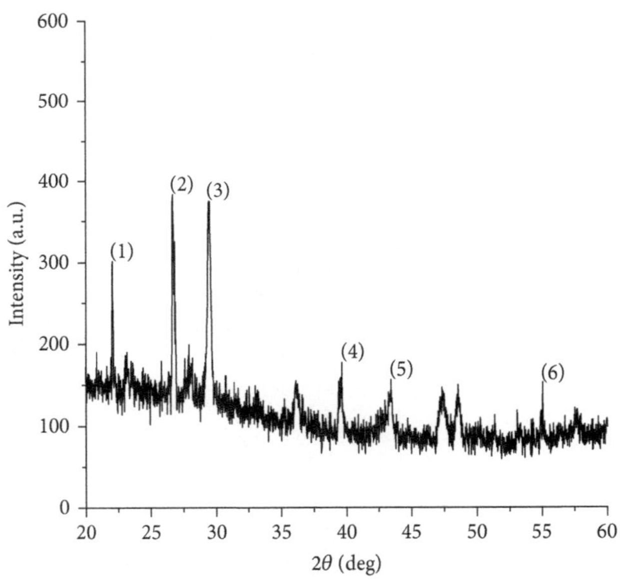

(1) SiO_2
(2) SiO_2
(3) $CaCo_3$
(4) $(Mg, Fe^{++})_5(SiO_4)_2(F, OH)_2$
(5) Al_2O_3
(6) MnO_2

FIGURE 6: XRD graph of AGF1106 flux.

TABLE 4: DTA analysis for endothermic temperatures.

Flux	Peaks of endothermic temperature (°C)
AGF1101	1239.89
AGF1102	1234.25
AGF1103	1233.58
AGF1104	1240.77
AGF1105	1226.99
AGF1106	1334.85
AGF1107	1244.58
AGF1108	1191.77
AGF1109	1245.84
AGF1110	1248.37

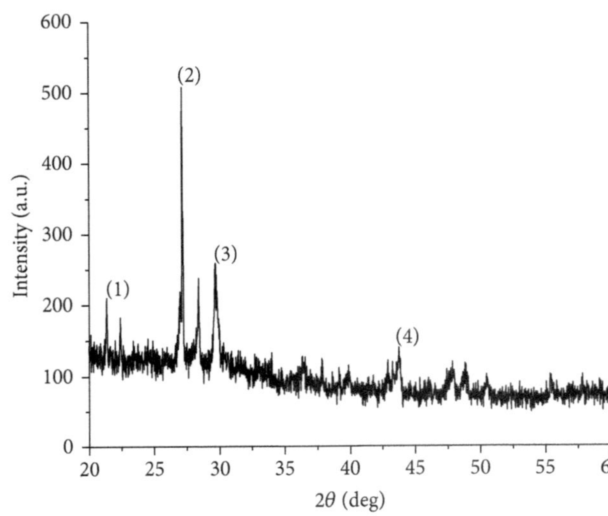

(1) $MgCo_3{\cdot}5(H_2O)$
(2) $Al_2O_3{\cdot}(SiO_2)_{1.3\text{–}2}{\cdot}(H_2O)_{2.5\text{–}3}$
(3) $Ca_2(M, Fe^{++})\,Fe^{+++}\,Si_5O_{14}(OH)$
(4) Ni_2Fe to Ni_3Fe

FIGURE 7: XRD graph of AGF1107 flux.

After ascertaining the compositions and constituents of fluxes, they were prepared by agglomeration method by taking small batches (2 kg) of weighed quantities of powdered chemicals except binder and thoroughly mixed in a ball miller. The prepared granular flux was dried and baked in an electrically heated furnace for 3 hrs at 850°C. The dried flux was allowed to cool down to room temperature before storing it in a moisture free box. The XRD analysis was carried out by diffractometer with the monochromatic light wavelength 1.540600 Å (Cu Kα) for all developed agglomerated flux. The DTA was carried out in an Al_2O_3 crucible from 28 to 1400°C at the heating rate of 10°C/min. Alumina was used as a standard reference for differential thermal analysis.

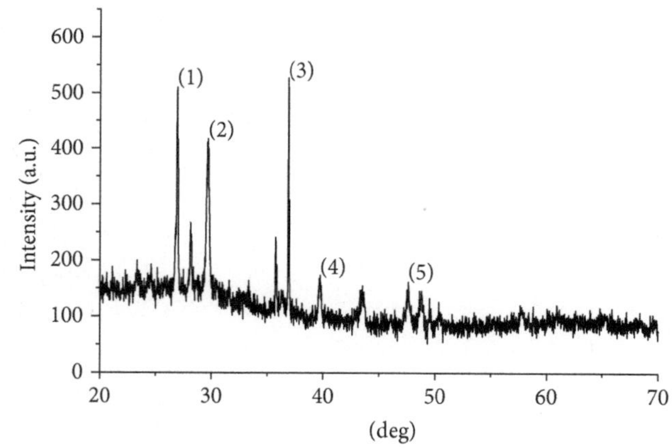

(1) $Al_2O_3{\cdot}(SiO_2)_{1.3\text{-}2}{\cdot}(H_2O)$
(2) $Ca_2(Mn, Fe^{++})Fe^{+++}Si_5O_{14}(OH)$
(3) $MgAl_2O_4$
(4) Mg_5Si_3
(5) $Ca_7(SiO_4)_3(OH)_2$

FIGURE 8: XRD graph of AGF1108 flux.

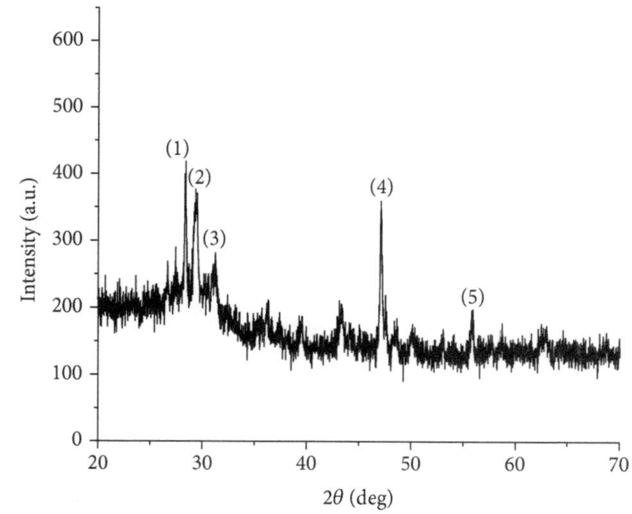

(1) $(Ca, Mn^{++})_{14}Si_{24}O_{58}(OH)_8{\cdot}(H_2O)$
(2) $(Mn^{++})_2SiO_3(OH)_2{\cdot}(H_2O)$
(3) $Ca_4SiO_2O_6(CO_3)(OH, F)_2$
(4) $Ca_{10}Mg_4Al_2Si_{11}O_{39}{\cdot}4(H_2O)$
(5) MnO_2

FIGURE 10: XRD graph of AGF1110 flux.

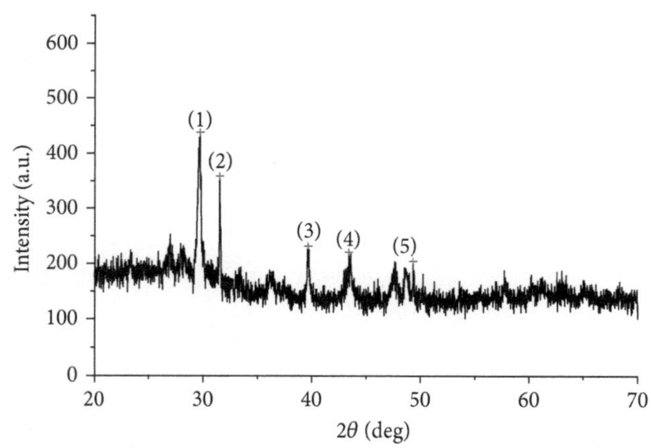

(1) $Ca_5Si_2O_7(Co_3)$
(2) $CaMg_3(Co_3)_4$
(3) $Mg_9(SiO_4)(OH, F)_2$
(4) Al_2O_3

FIGURE 9: XRD graph of AGF1109 flux.

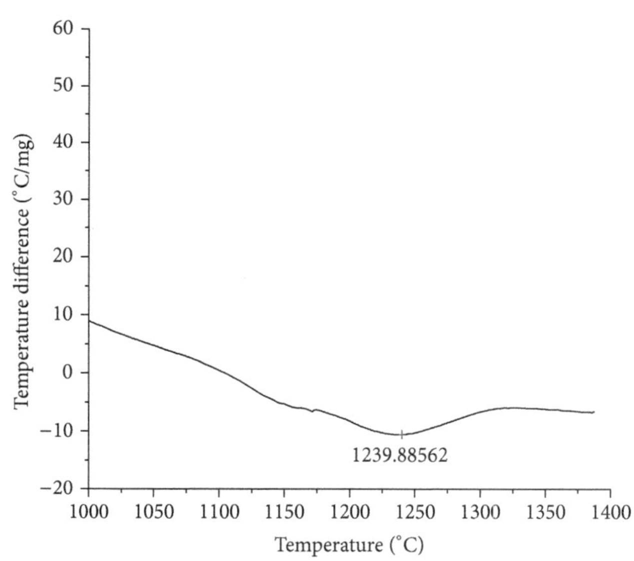

FIGURE 11: DTA graph of AGF1101 flux.

3. Result and Discussion

3.1. XRD and DTA Analysis of Developed Agglomerated Fluxes. The XRD results of developed agglomerated fluxes are shown in Figures 1, 2, 3, 4, 5, 6, 7, 8, 9, and 10 and by these graphs, it has been observed that a number of crystalline phases are formed due to chemical interaction among ions with potassium silicate binder. The oxides like MnO, MgO, SiO_2, Al_2O_3, and CaO were found in compound form, which means they do not react with one another. The names of different crystalline phases observed in XRD analysis along with their crystal structure were given in Table 2. The quantity of different oxides shown in Table 3 was based on XRD peaks with high intensity [13]. The development of different types of silicates and oxides is mainly due to mixing of potassium silicate during the agglomeration process.

The curves of differential thermal analysis (DTA) are shown in Figures 11, 12, 13, 14, 15, 16, 17, 18, 19, and 20 and it shows that most of the oxides were stable oxides up to 1000°C. The ranges of DTA curves were found between 1000

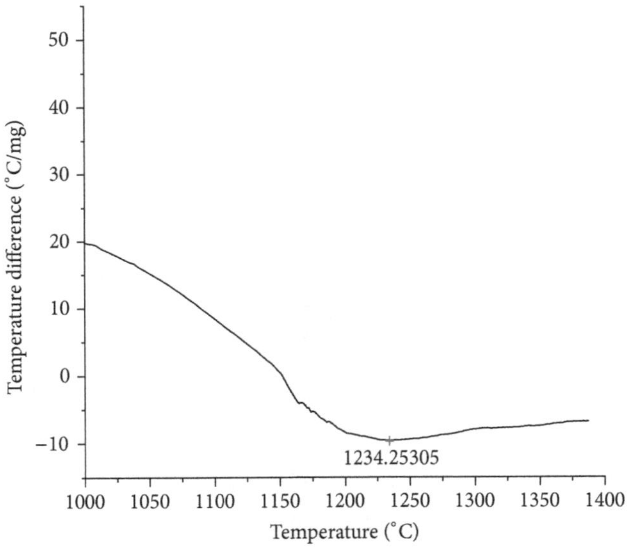

Figure 12: DTA graph of AGF1102 flux.

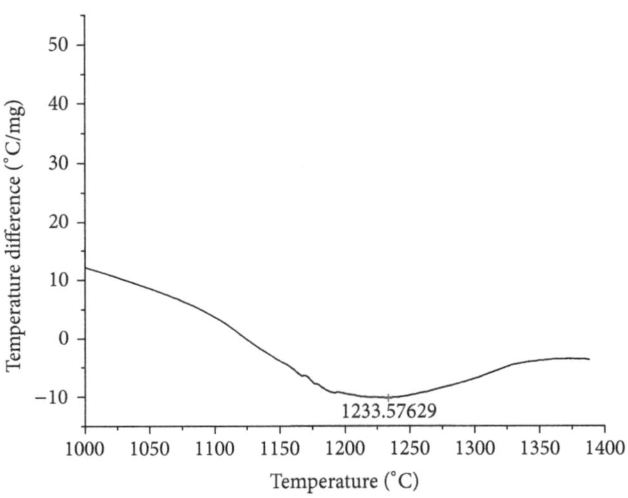

Figure 13: DTA graph of AGF1103 flux.

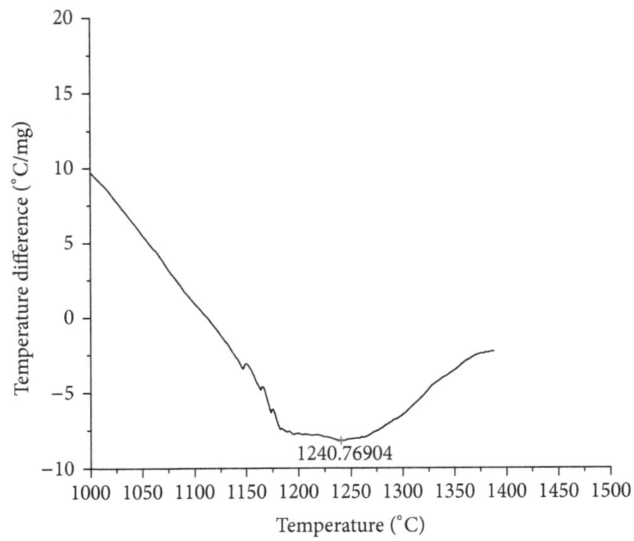

Figure 14: DTA graph of AGF1104 flux.

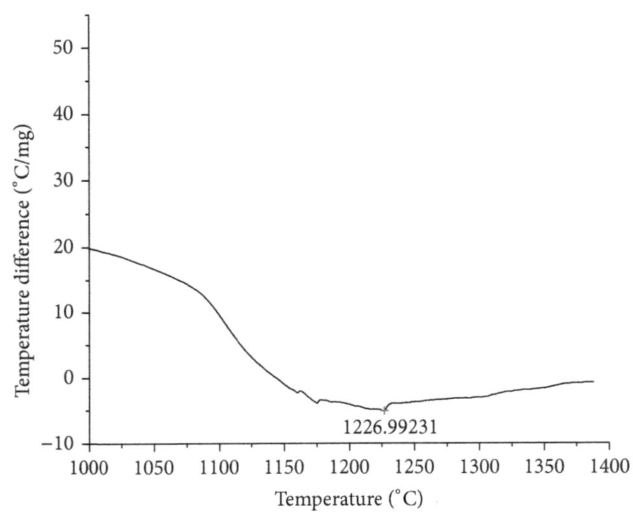

Figure 15: DTA graph of AGF1105 flux.

to 1400°C. With a few exceptions, most of the reactions were endothermic in nature and the endothermic peaks were observed at different temperatures for different fluxes given in Table 4. The endothermic peaks were different for different fluxes because of their different chemical compositions. These endothermic peaks were comparable with the melting reaction of different crystalline phases for different fluxes. In some DTA curves, a zigzag behavior was observed due to gassing or blistering of the glass formation [14]. The anions generally present in the fluxes are O^{2-} and F^{1-}. The cations from the flux like Ca^{2+}, Mg^{2+}, Al^{3+}, Fe^{2+}, Fe^{3+}, Si^{4+}, Mn^{2+}, Mn^{3+}, and Mn^{4+} can react with oxygen and tetrahedral unit $SiO_4{}^{4-}$ from the silicates formed in the fluxes. Since, calcium and magnesium ions have the largest negative heat of formation energy ΔG_f, therefore, they react quickly with oxygen in the welding arch. There were changes in oxidation number in some cationic constituents like Fe^{2+}, Fe^{3+}, Mn^{2+}, and Mn^{4+}. In most of the fluxes, calcium was present which increases the stability of electric arc during welding. The viscosity of fluxes increases by the presence of corundum and SiO_2 which are present in most of the fluxes. In most of the fluxes, it has been observed that the MnO and SiO_2 are jointly present and increase the thermomechanical behavior of weld joint.

4. Conclusion

The study identified the quantity of different ions present in the weld pool, and it also helps to know the different types of oxides and probable crystalline phases in developed

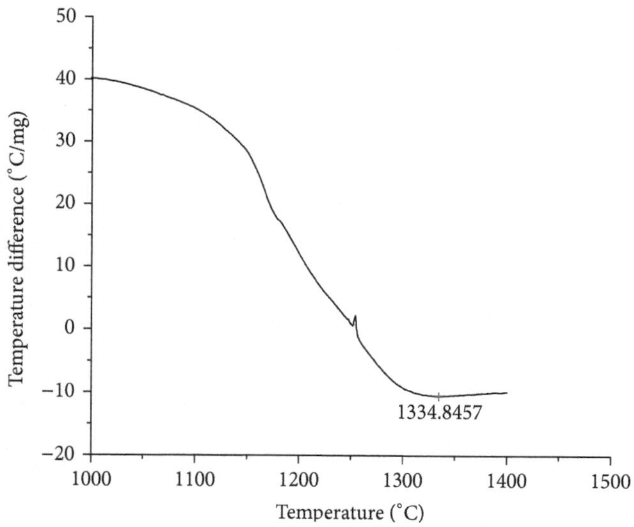

Figure 16: DTA graph of AGF1106 flux.

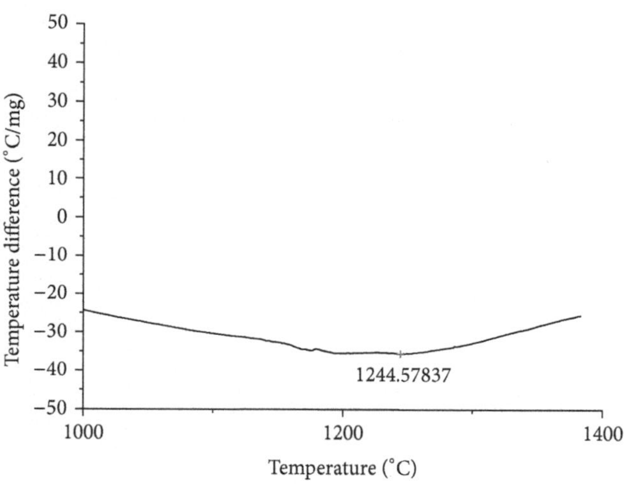

Figure 17: DTA graph of AGF1107 flux.

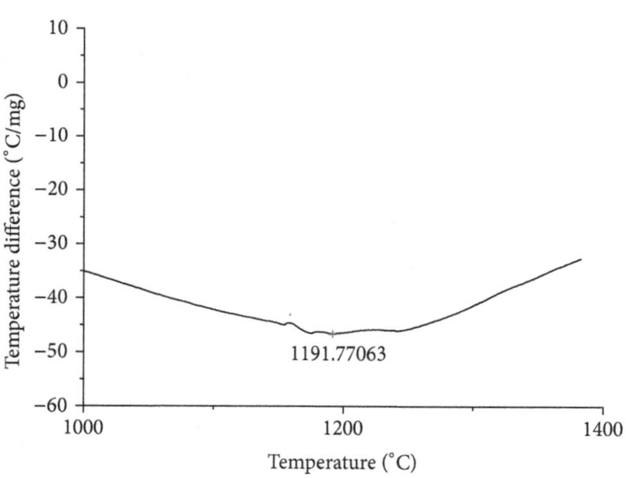

Figure 18: DTA graph of AGF1108 flux.

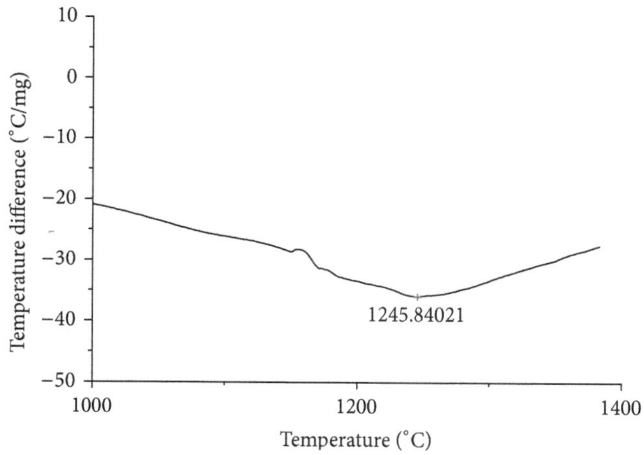

Figure 19: DTA graph of AGF1109 flux.

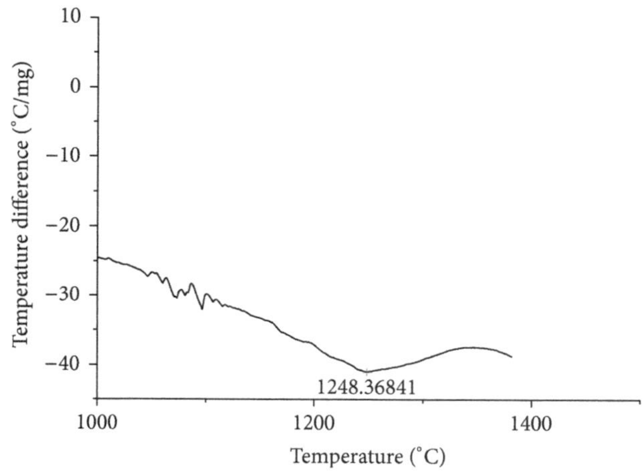

Figure 20: DTA graph of AGF1110 flux.

agglomerated fluxes during submerged arc welding. Such types of studies are very much helpful to know the effect of compositions of fluxes on thermomechanical behavior of weld metal.

References

[1] J. E. Indacochea, M. Blander, and S. Shah, "Submerged arc welding: evidence for electrochemical effects on the weld pool," *Welding Journal*, vol. 68, no. 3, pp. 77–79, 1989.

[2] C. E. Jackson, "Submerged-arc welding, fluxes and relations among process variables," in *Metals Hand Book*, pp. 73–77, ASM, Metals Park, Ohio, USA, 1982.

[3] G. G. Wittstock, "Selecting submerged arc fluxes for carbon and low alloy steels," *Welding Journal*, vol. 55, pp. 733–741, 1976.

[4] A. M. Paniagua-Mercado, V. M. López-Hirata, and M. L. Saucedo Muñoz, "Influence of the chemical composition of flux on the microstructure and tensile properties of submerged-arc welds," *Journal of Materials Processing Technology*, vol. 169, no. 3, pp. 346–351, 2005.

[5] P. Kanjilal, T. K. Pal, and S. K. Majumdar, "Combined effect of flux and welding parameters on chemical composition and mechanical properties of submerged arc weld metal," *Journal*

of Materials Processing Technology, vol. 171, no. 2, pp. 223–231, 2006.

[6] S. S. Singer, *Industrial Ceramics*, Chapman & Hall, London, UK, 1963.

[7] A. W. Allen, "Optical microscopy in ceramic engineering," in *Proceedings of the 3rd Berkeley International Materials Conference*, Berkeley, Calif, USA, June 1966.

[8] R. H. Redwine and M. A. Conrad, "Microstructures developed in crystallized glass ceramics," in *Proceedings of the 3rd Berkeley International Materials Conference*, Berkeley, Calif, USA, July 1961.

[9] A. M. Paniagua-Mercado, P. Estrada-Diaz, and V. M. López-Hirata, "Chemical and structural characterization of the crystalline phases in agglomerated fluxes for submerged-arc welding," *Journal of Materials Processing Technology*, vol. 141, no. 1, pp. 93–100, 2003.

[10] A. Muan and E. F. Osborn, *Phase Equilibria among Oxides in Steel Making*, Addison-Wesley, Reading, Mass, USA, 1965.

[11] D. C. Montgomery, *Design and Analysis of Experiments*, Wiley, New Delhi, India, 5th edition, 2007.

[12] A. Kumar, H. Singh, and S. Maheshwari, "Modeling and analysis by response surface methodology of hardness for submerged arc welded joints using developed agglomerated fluxes," *Indian Journal of Engineering & Materials Sciences*, vol. 19, pp. 379–385, 2012.

[13] B. D. Cullity, *Elements of X-Ray Diffraction*, Addison-Wesley, Reading, Mass, USA, 1978.

[14] P. K. Gordon, "Microstructure of complex ceramics," in *Proceedings of the 3rd Berkeley International Materials Conference*, Berkeley, Calif, USA, July 1966.

Mechanical Response of Al-1.09Mg_2Si Alloy under Varying Mould and Thermal Ageing Conditions

O. I. Sekunowo, G. I. Lawal, and S. O. Adeosun

Department of Metallurgical and Materials Engineering, Faculty of Engineering, University of Lagos, Lagos 101017, Nigeria

Correspondence should be addressed to O. I. Sekunowo, olatundeisrael@yahoo.co.uk

Academic Editor: Gerhard Sauthoff

Samples of the 6063 (Al-1.09Mg_2Si) alloy ingot were melted in a crucible furnace and cast in metal and sand moulds, respectively. Standard tensile, hardness, and microstructural test specimens were prepared from cast samples, solution treated at 520°C, soaked for 6 hrs, and immediately quenched at ambient temperature in a trough containing water to assume a supersaturated structure. The quenched specimens were then thermally aged at 175°C for 3–7 hrs. Results show that at different ageing time, varied fractions of precipitates and intermetallics evolved in the specimens' matrices which affect the resulting mechanical properties. The metal mould specimens aged for four hours (MTA-4) exhibited superior ultimate tensile strength of 247.8 MPa; microhardness, 68.5 HV; elongation, 28.2% . It is concluded that the extent of improvement in mechanical properties depends on the fractions, coherence, and distribution of precipitates along with the type of intermetallics developed in the alloy during ageing process.

1. Introduction

Casting is one of the most versatile methods of producing structural aluminium alloy components. However, the rather large preponderance of defects in cast aluminium components often limits their performance and adversely impacts their commercial values. Hence there is the need for a novel processing approach to improve the cast microstructure for enhanced performance. Generally, the poor mechanical properties of cast aluminium alloys can be improved through either alloy addition or various forms of heat treatment [1]. During the heat treatment of cast aluminium alloys, an advantage is made of the characteristic decrease in solubility at low temperature of magnesium (Mg) and silicon (Si) which are the main alloying elements in Al-Mg-Si alloy. Further, Keist [2] confirms that the appreciable decrease in concentration of the alloying elements at room temperature is the fundamental phenomenon that provides the basis for increasing substantially the hardness and strength of aluminium alloys through isothermal treatment.

Similarly, Siddiqui et al. [3] have shown that improved ductility can be achieved by process annealing at 415°C, soaked between two and three hours coupled with a cooling rate 30°C per hour.

Generally, strength improvement of most 6063 aluminium alloys can be effected in a three-pronged approach comprising solution heat treatment, quenching, and precipitation of solute atoms [4]. However, the greatest challenge usually encountered with this approach is effective control of the second-phase precipitates. Lumley et al. [5] proposed a method known as interrupted ageing for effective control and exploitation of secondary precipitation to improve the mechanical properties of aluminium alloys. This method promotes nucleation of finely dispersed precipitates in the final microstructure. Consequently, it was possible to achieve an average increase in 0.2% proof stress of 10–15% coupled with improved fracture toughness. This development underscores the effect of ageing conditions on the mechanical behaviours of the 6063 (Al-Mg-Si) alloy.

Gao et al. [6] investigated the influence of ageing conditions on the microstructure and tensile strength of 6063 aluminum alloy. Both temperature and soaking time were varied in the range 165°C–288°C at 25 and 64 hours, respectively. The results show that the higher the ageing temperature, the shorter time it takes to reach peak strength. Precipitation of secondary phase structures and its distribution within the matrix also play a major role in determining changes in mechanical properties of aluminium alloy [7]. It is therefore reasonable to leverage this possibility of microstructural manipulations for desirable property changes to achieve improved mechanical characteristics.

It has also been established that factors that determine the mechanical behaviours of a 6063 aluminium alloy component are the type, amount, and distribution of second-phase particles such as Mg_2Si, Al-Fe, Al-Fe-Si, and Al-Fe-Mn-Si dissolved in solution [8]. However, according to Tsao et al. [9], the phases that formed eventually depend mainly on the cooling rate and the Fe/Si ratio in the alloy. The combination of manganese with Fe, Si, and Al also forms α-$Al_x(Fe, Mn)_ySi_z$ phase that acts as nucleation sites for Mg_2Si and other crystals, which eventually influences the alloys behaviour [10]. Manganese as a common addition in 6XXX alloys increases strength as finely precipitated intermetallics modify the shape of plate-like iron phases which reduces their tendency to embrittlement [11].

The profound ability of 6063 aluminium alloys strengthening through precipitation hardening for industrial applications remains a major breakthrough in the field of materials processing [12]. Hence, precipitation hardened aluminium alloys are a commercially important group of materials because their mechanical properties can be modified by heat treatment.

However, the imperative of achieving a sufficient fraction of stable precipitates with appropriate morphology and their homogeneous dispersion within the matrix has been a major challenge. This underscores the need to develop a processing method that is capable of inducing in the alloy such desirable microstructural features that confer improved mechanical properties. Consequently, this work examines the mechanical response of 6063 (Al-1.09Mg_2Si) alloy processed through varying moulds and thermal ageing conditions.

2. Experimental Procedure

2.1. Material, Specimen Preparation, and Heat Treatment. Three kilogrammes (3 kg) of 6063 aluminium alloy ingot used for this study were obtained from the Nigerian Aluminium Extrusion Company (NIGALEX) Lagos, Nigeria, and its chemical composition is presented in Table 1. The ingot was sized into smaller pieces, charged into a pretreated low-pour-fuel-oil (LPFO) fired crucible furnace, and allowed to melt at slightly above 660°C. Metal and sand moulds of dimension, 15 × 146 mm, respectively, were prepared using a solid cylindrical wooden pattern of D 15 mm and 350 mm long. Four (4) samples each were cast in the metal and sand moulds, allowed to cool, and removed afterwards. With the aid of a hacksaw (manual) at ambient temperature, samples from both moulds were milled into twelve pieces and machined to standard tensile and microhardness test specimens. These were grouped into two and identified as presented in Table 2. The specimens were then solution heat treated in a muffle furnace model L40/11B170 at 520°C and soaked for six hours. They were immediately quenched in a trough containing water that was maintained at ambient temperature (36–39°C). The specimens were then thermally aged at 175°C for 3–7 hours.

2.2. Microstructure Analysis. Both the as-cast and thermally aged specimens prepared from metal and sand mould cast samples were ground in succession using emery paper of grits in the range of 220 to 600 microns. Each specimen surface was thereafter polished using aluminium-diamond paste under running water until a mirror-like surface was obtained. Microstructural examination of the specimens was carried out after their polished surfaces had been etched for 20 s using 1 g of sodium hydroxide (NaOH) in 100 mL of water. The specimen microstructural features were then viewed under an optical metallurgical microscope at a magnification of ×200, and the micrographs are presented in Figures 1(a)–1(f) and Figures 2(a)–2(f).

2.3. Mechanical Properties Evaluation. Standard tensile test specimens prepared from the as-cast, and thermally aged (MTA-i and STA-i) samples were subjected to tensile test using an Instron electromechanical machine model 3369. The microhardness values of both test specimens were also evaluated using a Leco Vicker micro hardness tester model LMT700AT under 100 g test load and at a dwelling time of 10 seconds. The specimen microhardness values (average) were obtained in three readings, namely, at the centre and adjacent vertices of the specimen surfaces.

3. Results and Discussion

Results of the ultimate tensile strength (UTS), ductility, and microhardness responses by test specimens are illustrated in Figures 3–9, while the microstructures induced in the test specimens are shown in Figures 1 and 2.

3.1. Microstructure. The micrographs in Figure 1 show the as-cast microstructure of the aluminium alloy (Figure 1(a)) displaying a near monolithic alpha aluminium matrix interspersed with low fraction of plate-like Mg_2Si precipitates, microsegregations, and inclusions. Given the relatively fast cooling of cast in a metal mould, microsegregation is normally not a common occurrence but for the presence of inclusions which aid its formation. The microstructure in Figure 1(a) is therefore typical of 6063 aluminium alloy ingot structure in its untreated condition.

On subjecting the specimens to a combined process of solution heat treatment and thermal ageing for varying periods of time, the specimens experienced various remarkable microstructural transformations. Figures 1(b)–1(f) reveal microstructures of the solution heat treated and thermally aged metal mould cast specimens. The micrograph in Figure 2(b) shows very low fractions of Mg_2Si precipitates developed after three hours of ageing. Moreover, the precipitates are not homogeneously dispersed within the alpha aluminium matrix. As the ageing period progressed up to four hours (Figure 2(c)), decomposition of the as-quenched structure continued and culminated in the formation of

Table 1: Chemical composition of as-received 6063 (Al-1.09Mg_2Si) alloy ingot.

Element	Mg	Si	Mn	Cu	Zn	Ti	Fe	Ca	Cr	Al
Composition (wt%)	0.482	0.441	0.123	0.001	0.004	0.012	0.199	0.001	0.001	98.736

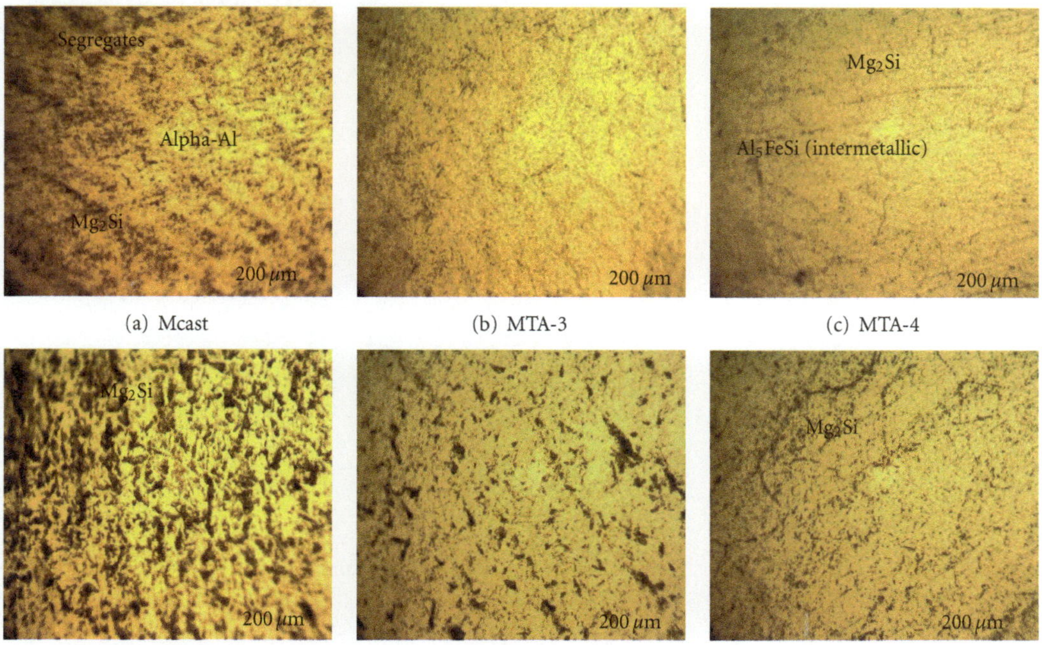

(a) Mcast (b) MTA-3 (c) MTA-4

(d) MTA-5 (e) MTA-6 (f) MTA-7

Figure 1: Micrographs of metal mould cast specimens: (a) as-cast, (b) solution treated and aged for 3 hrs, (c) solution treated and aged for 4 hrs, (d) solution treated and aged for 5 hrs, (e) solution treated and aged for 6 hrs and (f) solution treated and aged for 7 hrs.

Table 2: Specimens identification system.

Notation	Description
Mcast	Cast in metal mould
Scast	Cast in sand mould
STA-*i*	Cast in sand mould, solution treated and thermally aged for "*i*" time (where "*i*" is from 3–7 hours.)
MTA-*i*	Cast in metal mould, solution and thermally aged for "*i*" time (where "*i*" is from 3–7 hours.)

substantial fractions of fine needle-like (fibrous) Mg_2Si precipitates along with other intermetallics which include Al_5FeSi and $Al_{12}(FeMn)_3Si$ in consonance with the alloy composition (see Table 1). The precipitates are coherent and homogeneously dispersed within the alpha aluminium matrix. Coarse Mg_2Si precipitates and other intermetallics are seen in Figure 2(d). The growth of the crystals may be attributed to the period (five hours) within the Mg_2Si precipitates suffered coarsening due to an overaging treatment. This feature is absent in Figures 2(e) and 2(f) where dissolution of precipitates and clustering prevailed at six and seven hours of ageing, respectively. This agrees well with the work of Tsao et al. [9], which shows that the rate of microstructure transformation in Al-Mg-Si alloy that precipitates from metastable to stable phase is influenced by time.

Hence, in the present study, it is observed that the amount, size, and morphology of precipitates change with time. However, this is a descriptive analysis of the test specimens' microstructures as observed under an optical microscope. Further, the issue of precipitate stability during ageing is of great concern in 6063 aluminium alloys. It is established [11] that the presence of iron (Fe) in excess of certain concentrations has the tendency of exerting preferential stability of the precipitates. Consequently, the presence of Fe, which is about 0.199 wt% (see Table 1), in the Al-Mg-Si alloy under investigation is capable of skewing the amount of stable precipitates in favour of Mg_2Si.

Micrographs of the sand mould cast specimens are shown in Figures 2(a)–2(f). It is observed in Figure 2(a) that the fractions of coarse Mg_2Si precipitates are quite low, while the amount of microsegregates appears rather high and large in size compared with similar segregates in Figure 1(b). This can be attributed to the disparity in mould conditions as it affects the cast solidification rate. Figure 2(b) shows a predominantly alpha aluminium matrix with very low fractions of second-phase precipitates. However, when ageing time increased to four hours, plate-like crystals of Mg_2Si are strongly precipitated and evenly dispersed within the matrix (Figure 2(c)).

The fibrous Mg_2Si crystals that precipitated after five hours of ageing appear clustered along the grain boundaries and incoherent with the parent aluminium matrix

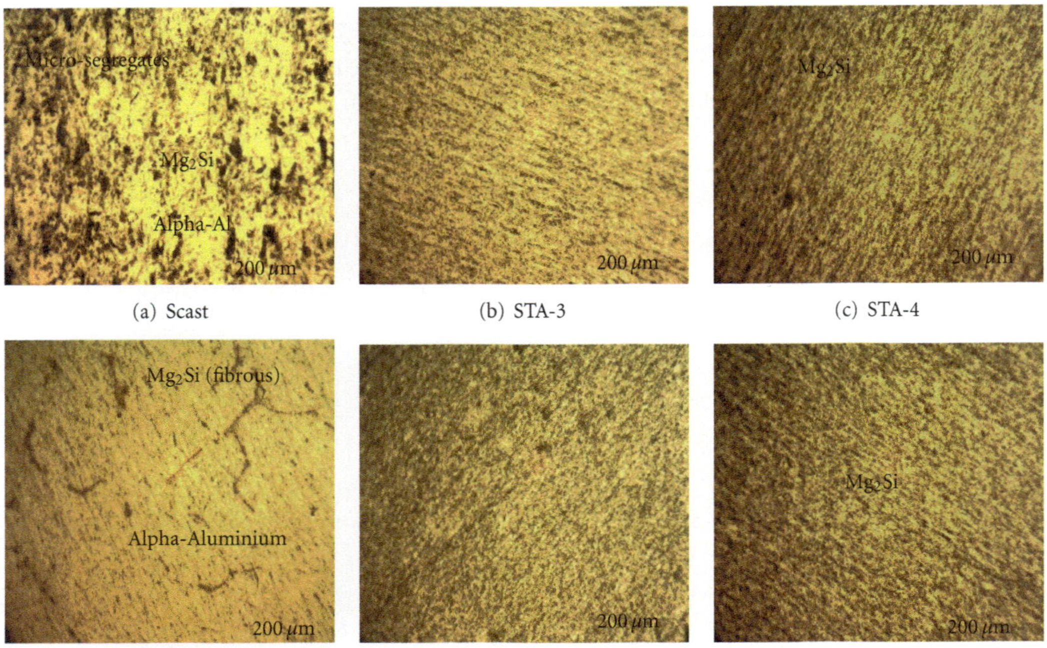

(a) Scast (b) STA-3 (c) STA-4
(d) STA-5 (e) STA-6 (f) STA-7

Figure 2: Micrographs of sand mould cast specimens: (a) as-cast, (b) solution treated and aged for 3 hrs, (c) solution treated and aged for 4 hrs, (d) solution treated and aged for 5 hrs, (e) solution treated and aged for 6 hrs and (f) solution treated and aged for 7 hrs.

(Figure 2(d)). This feature is absent in Figure 2(e) due to the presence of strongly precipitated second phase. However, the process of structure reversion beyond six hours of ageing (Figure 2(f)) is a stop gap for the nucleation of new crystals from the old unstable phase as a precursor to coarsening of Mg_2Si precipitates.

3.2. Ultimate Tensile Strength. Figure 3 illustrates the ultimate tensile strength (UTS) behaviour of both the as-cast specimens (Mcast and Scast) compared with the solution heat treated and thermally aged for three hours (MTA-3 and STA-3) specimens. It is evident from Figure 3 that the MTA-3 specimen exhibited the highest UTS of 193.8 MPa, while the Scast has the least UTS value 72.4 MPa. The huge difference in tensile values is attributable to the presence of the reinforcing fractions of Mg_2Si precipitates in the MTA-3 microstructure (Figure 1(b)) compared with the monolithic alpha aluminium structure of the Scast specimen (Figure 2(a)). However, both the Scast and STA-3 specimens demonstrated comparable UTS values of 72.4 MPa and 81.1 MPa, respectively, due to their similar microstructural features.

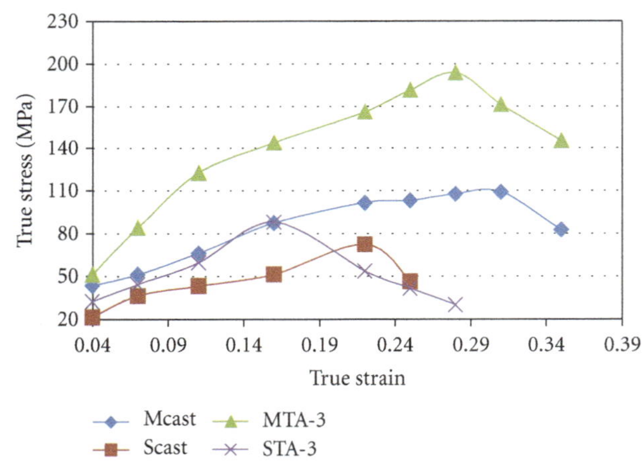

Figure 3: Variation of true stress-strain of as-cast metal and sand mould specimens after three hours of ageing.

The results of ageing time on the 6063 aluminium alloy demonstrate marked effect with regard to the MTA-4 specimen UTS value of 247.8 MPa as observed in Figure 4. This is predicated on the stable fine needle-like Mg_2Si precipitates coupled with the presence of hard intermetallics induced by the presence of iron in the alloy composition (see Table 1 and Figure 1(c)).

Needle-like Mg_2Si precipitates have the ability to form effective network of resistance to dislocations motion within the structure, hence the marked increase in UTS compared with 107.3 MPa of STA-4 due to its structure which is grossly deficient in fine needle-like precipitates (Figure 2(c)). These observations perfectly support the findings of [11] in relation to the effect of the combined presence of stable and hard intermetallics on tensile response of aged 6063 aluminium alloy. It is also evident in Figures 5, 6, and 7 that the thermally aged specimen UTS values decrease drastically under varying mould conditions as the ageing time increases from 5–7 hours in the following order: MTA-5, 148.2 MPa (STA-5, 82.8 MPa), MTA-6, 106.1 MPa (STA-6, 73.5 MPa), and MTA-7, 93.1 MPa (STA-7, 61.5 MPa).

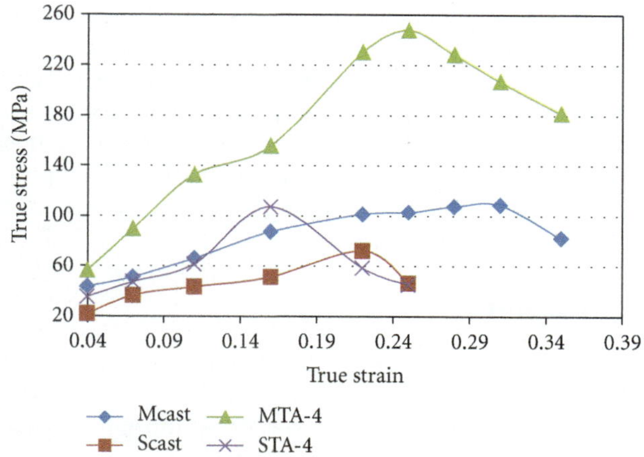

FIGURE 4: Variation of true stress-strain of as-cast metal and sand mould specimens after four hours of ageing.

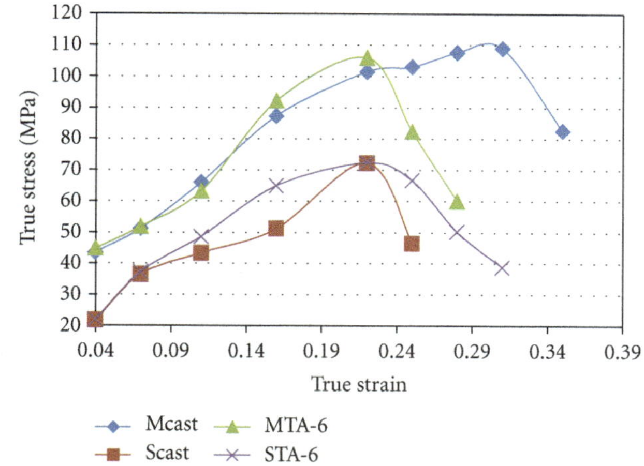

FIGURE 6: Variation of true stress-strain of as-cast metal and sand mould specimens after six hours of ageing.

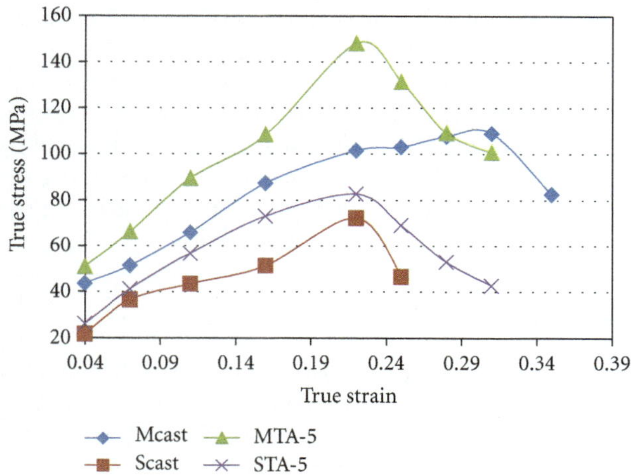

FIGURE 5: Variation of true stress-strain of as-cast metal and sand mould specimens after five hours of ageing.

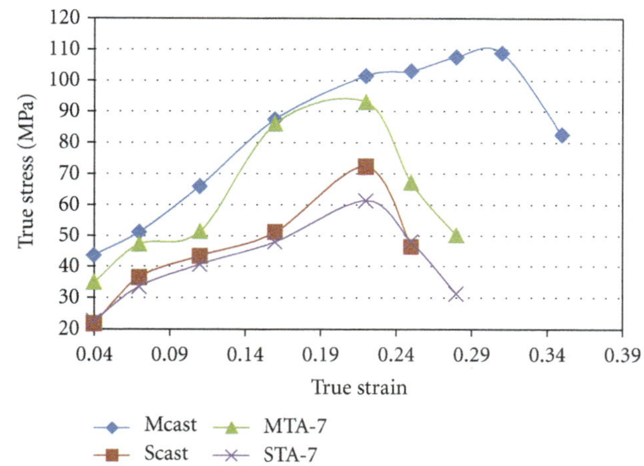

FIGURE 7: Variation of true stress-strain of as-cast metal and sand mould specimens after seven hours of ageing.

The progressive decrease in UTS is due to the reduction in weight fraction of reinforcing Mg_2Si crystals and Al_5FeSi and $Al_{12}(FeMn)_3Si$ intermetallics. As ageing time increases from 5–7 hours, the precipitates induced become unstable and incoherent in particular; the intermetallics formed increased in fraction and exhibit such characteristics that make them fracture easily under tensile force. This further validates the earlier work of Chennakesava and Zitoun [13] in which it is reported that the UTS of aged 6063 aluminium alloy decreases with larger fractions of the presence of iron forming intermetallics in its matrix.

Similar trends of progressive reduction in UTS values were demonstrated by the STA-5, 6, 7 specimens which correspond to the extent of precipitate coarsening and their heterogeneous dispersion within the matrix. As shown in Figure 7, both the as-cast and thermally aged for seven hours specimens exhibited close UTS values, namely, Mcast, 109.1 MPa, Scast, 72.1 MPa, MTA-7, 93.1 MPa, STA-7, and 61.5 MPa. In each of these instances, the influence of mould conditions coupled with varying ageing time appears to have been subdued. This is attributable to the structural similarity of the specimens as shown in Figures 1(d)–1(f) and 2(d)–2(f) occasioned by the few coarse and incoherent precipitates induced within their matrices.

3.3. Ductility. Presented in Figure 8 are the results of the alloy ductility behaviour at different mould conditions under varying ageing time. Generally, the ductility response of both the metal and sand cast specimens exhibits a similar trend at varying ageing time.

This is attributable to the progressive reduction with ageing time, the weight fractions of reinforcing Mg_2Si precipitates, and their replacement with coarse and incoherent ones along with iron forming intermetallics (Al_5FeSi and $Al_{12}(FeMn)_3Si$) which enhance ductility (see Figures 1(d)–1(f) and Figures 2(d)–2(f)).

However, compared with the sand cast specimen ductility which is in the range of 15.9–24.5%, the metal cast specimens show better ductility in the range of 19.4–35.7%. It is normally required that cast structural components possess

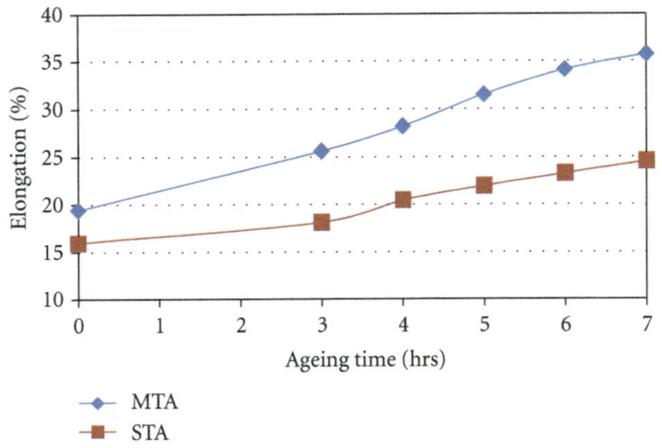

FIGURE 8: Variation of specimens' ductility with ageing time.

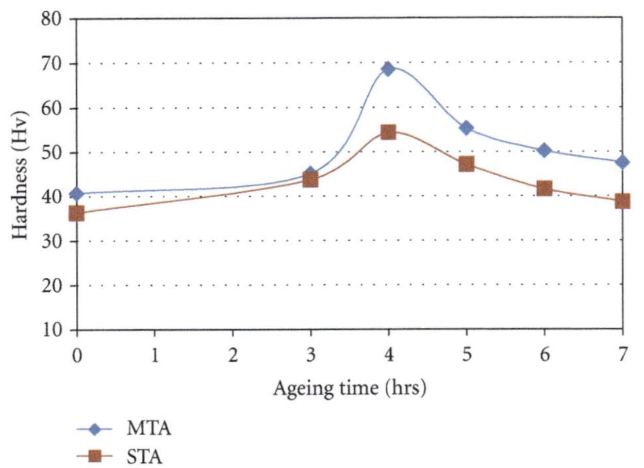

FIGURE 9: Variation of specimens' microhardness with ageing time.

certain level of ductility for enhanced formability. This can only be achieved through innovative processing such as friction stir casting of AA6063-T_6 employed by Karthikeyan and Senthil Kumar [14] that exhibited an increase of 133% ductility in relation to the parent material. The 28.2% ductility exhibited by the metal cast and thermally aged for four hours (MTA-4) specimen compared well with 14% ductility of conventional AA6063-T_6. This makes the alloy relatively suitable for structural applications.

3.4. Microhardness. The effects of both varying mould conditions and ageing time on the microhardness developed in the test specimens are shown in Figure 9. In both conditions, the microhardness values of the as-cast specimens in metal and sand moulds are the least, Mcast: 40.7 HV and Scast: 36.2 HV. This is predicated on the low fraction of fine plate-like crystals of Mg_2Si crystals and microsegregates within the matrix of the Mcast structure (see Figure 1(a)). Similarly, the Scast low hardness is due to the scanty coarse plate-like Mg_2Si crystals and large fraction of microsegregates in its structure (see Figure 2(a)). However, the precipitation of fine needle-like Mg_2Si crystals and coherent iron forming intermetallics from the onset of ageing gave rise to an increase in hardness until the peak value of 68.5 HV and 54.3 HV for the MTA-4 and STA-4 specimens, respectively. It is also observed in Figure 9 that a decline in microhardness sets in as ageing time increases beyond four hours due to precipitate coarsening and the presence of more $Al_{12}(FeMn)_3Si$ intermetallics that are incoherent with the alpha aluminium matrix. This type of structure often induces porosity in the alloy microstructure which impairs microhardness.

4. Conclusion

In this study, the fractions, morphology, and dispersion of Mg_2Si and the type of intermetallics (Al_5FeSi and $Al_{12}(FeMn)_3Si$) developed in test specimens through thermal ageing significantly influenced their mechanical properties. Further, the ageing time impacted both the morphology and stability or otherwise of the precipitates. The highest UTS of 247.8 MPa compared with 195 MPa of conventional AA6063-T_6, 28.2% ductility against 14% of standard AA6063-T_6, and a modest microhardness of 68.5 HV were all obtained through the MTA-4 treatment. This is the only process route that demonstrated improved balanced mechanical properties in the metal cast and thermally aged Al-1.09Mg_2Si alloy. The mechanical characteristics developed in the alloy through the MTA-4 process route have demonstrated the alloy suitability for structural applications coupled with the ability to withstand excessive wear in service.

References

[1] R. A. Siddiqui, H. A. Abdullah, and K. R. Al-Belushi, "Influence of aging parameters on the mechanical properties of 6063 aluminium alloy," *Journal of Materials Processing Technology*, vol. 102, no. 1, pp. 234–240, 2000.

[2] J. Keist, "The development of a fluidized bed process for the heat treatment of aluminum alloys," *Journal of Materials*, vol. 57, no. 4, pp. 34–39, 2005.

[3] R. A. Siddiqui, S. A. Abdul-Wahab, and T. Pervez, "Effect of aging time and aging temperature on fatigue and fracture behavior of 6063 aluminum alloy under seawater influence," *Materials and Design*, vol. 29, no. 1, pp. 70–79, 2008.

[4] W. Callister, *Materials Science and Engineering: An Introduction*, John Wiley and Sons, Clementi Loop, Singapore, 6th edition, 2006.

[5] R. N. Lumley, I. J. Polmear, and A. J. Morton, "Control of secondary precipitation to improve the performance of aluminium alloys," *Materials Science Forum*, vol. 396–402, no. 2, pp. 893–898, 2002.

[6] R. Q. Gao, K. Stiller, V. Hansen, A. Oskarsson, and F. Danoix, "Influence of aging conditions on the microstructure and tensile strength of aluminium alloy 6063," *Materials Science Forum*, vol. 396–402, no. 2, pp. 1211–1216, 2002.

[7] G. Al-Marahleh, "Effect of heat treatment on the distribution and volume fraction of Mg_2Si in structural aluminum alloy 6063," *Metal Science and Heat Treatment*, vol. 48, no. 5-6, pp. 205–209, 2006.

[8] M. Cai, D. P. Field, and G. W. Lorimer, "A systematic comparison of static and dynamic ageing of two Al-Mg-Si alloys," *Materials Science and Engineering A*, vol. 373, no. 1-2, pp. 65–71, 2004.

[9] C. S. Tsao, C. Y. Chen, U. S. Jeng, and T. Y. Kuo, "Precipitation kinetics and transformation of metastable phases in Al-Mg-Si alloys," *Acta Materialia*, vol. 54, no. 17, pp. 4621–4631, 2006.

[10] G. Sha, K. O'Reilly, B. Cantor, J. Worth, and R. Hamerton, "Growth related metastable phase selection in a 6xxx series wrought Al alloy," *Materials Science and Engineering A*, vol. 304-306, no. 1-2, pp. 612–616, 2001.

[11] A. K. Gupta, D. J. Lloyd, and S. A. Court, "Precipitation hardening in Al-Mg-Si alloys with and without excess Si," *Materials Science and Engineering A*, vol. 316, no. 1-2, pp. 11–17, 2001.

[12] T. Nahoto, C. Fumyochi, S. Tomohisa, and F. Kouichiro, "Precipitation behavior of Al-Mg-Si tenary alloys," *Materials Science Forum*, vol. 217–222, pp. 815–820, 1996.

[13] A. Chennakesava and E. Zitoun, "Tensile behaviour of 6063/Al_2O_3 particulate metal matrix composite fabricated by investment casting process," *International Journal of Applied Engineering Research*, vol. 1, no. 3, pp. 542–552, 2010.

[14] L. Karthikeyan and V. S. Senthil Kumar, "Relationship between process parameters and mechanical properties of friction stir processed AA6063-T_6 aluminum alloy," *Materials and Design*, vol. 32, no. 5, pp. 3085–3091, 2011.

23

Friction Stir Welding of Al 5052 with Al 6061 Alloys

N. T. Kumbhar[1] and K. Bhanumurthy[2]

[1] *Materials Science Division, Bhabha Atomic Research Centre, Trombay, Mumbai 400 085, India*
[2] *Scientific Information Resource Division, Bhabha Atomic Research Centre, Trombay, Mumbai 400 085, India*

Correspondence should be addressed to K. Bhanumurthy, aditya@barc.gov.in

Academic Editor: Stan A. David

Friction stir welding (FSW), a solid-state joining technique, is being extensively used in similar as well as dissimilar joining of Al, Mg, Cu, Ti, and their alloys. In the present study, friction stir welding of two aluminium alloys—AA6061 and AA5052—was carried out at various combinations of tool rotation speeds and tool traverse speeds. The transverse cross-section of the weld was used for optical as well as electron microscopy observations. The microstructural studies were used to get an indication of the extent of material mixing both at the macro- and microscales. It was observed that, at the interface region, both materials exhibited similar texture despite the nonrigorous mixing of the materials in the nugget. The extent of interdiffusion of alloying elements at the interface was studied using electron probe microanalysis. The tensile testing evaluation of these specimens showed good mechanical properties. The interdiffusion of alloying elements and development of similar orientations in the nugget could have contributed to the better tensile properties of the friction-stir-welded AA5052-AA6061 specimens.

1. Introduction

The joining of dissimilar materials is vital, since it is intended to obtain a product having a combination of desirable properties of both parent materials. The difficulties in joining materials with vastly differing properties using conventional fusion welding methods are well known. The selection of an appropriate filler material (having intermediate properties) is critical. Otherwise, the possible formation of complex, brittle, intermetallic compounds can degrade the quality of the weld, resulting in inferior welds. Similarly, it is not straight forward to directly join these materials using solid-state joining methods—given the compatibility issues of physical properties of the materials as well as formation of intermetallic compounds. Hence suitable interlayer which prevents the formation of intermetallic compounds is often employed in such cases.

Friction stir welding (FSW) has become an obvious choice for welding of "difficult to weld" aluminium alloys (5xxx series) or high-strength aluminium alloys (2xxx and 7xxx series). It has also facilitated welding of dissimilar aluminium alloys with same ease. The dissimilar welding of aluminium alloys has attracted more attention, since it offers an insight into many phenomena (e.g., material flow) which were not clear during the friction stir welding of similar Al alloys. Many topics like variation of microhardness, material flow, material location, temperature distribution, residual stresses, and so forth, across the interface of the abutting materials and their consequent effect on the mechanical properties are of interest to many researchers [1–3].

Lee et al. [1] studied the dissimilar lap joint FSW of AA5052-H112 and AA6061-T6 plates having thickness of 1 mm and 2 mm, respectively, with various tool rotation speeds and tool traverse speeds according to the fixed location of each material on top or bottom sheet. They found that the interface morphologies were characterized by interface pull-up and pull-down in the advancing side and retreating side. The thickness of the thinner AA5052 sheet lessened due to the vertical movement of the materials. It was identified that the amount of vertical transport increased and consequently the thickness of AA5052 decreased when the heat input was increased either by increasing the tool rotation speeds or by decreasing the tool traverse speeds. Joint strengths mainly depended on the interface morphology and vertical movement of material.

Table 1: Chemical composition of alloying elements in the aluminium alloys (in wt.%).

	Mg	Si	Mn	Cr	Ni	Fe	Ca	Cu	Zn	Ti	Al
AA5052	2.30	0.15	0.02	0.10	—	0.20	—	0.02	0.01	0.01	Bal
AA6061	0.92	0.60	0.06	—	0.18	0.33	0.2	—	0.03	0.02	Bal

Table 2: Mechanical properties of aluminium alloys AA5052 and AA6061.

Material	UTS (MPa)	YS (MPa)	Elongation (%)	Hardness (VHN)
AA5052	262	214	10	80
AA6061	310	276	12	107

Park and Kim [2] investigated the effect of tool rotation speed and tool traverse speed on the stirring action and friction heat during FSW experiments on dissimilar Al alloys—AA5052-O and AA6061-T6. They used a range of process parameters to determine the mechanical strength of weld nugget of the dissimilar materials. They concluded that the optimum conditions were a traveling speed of 61 mm/min and rotation speed of 1600 rpm. Their observations of the weld surface finish and plastic flow behaviour showed that the stirring effect increased and number of defects decreased when the traverse speed was decreased.

Park et al. [3] investigated the effect of material locations on the properties of dissimilar FSW joints of AA5052-H32 and AA6061-T6. They showed that the material mixing patterns in FSW joints vary depending on the location of base materials. For the given aluminium alloys, the materials were more properly mixed when AA5052-H32 was in the advancing side and AA6061-T6 was in the retreating side than the case of AA6061-T6 in the advancing side and AA5052-H32 aluminium alloy in the retreating side. It was found that for both combinations of material arrangement, AA5052-H32 showed the lowest value of microhardness in the heat-affected zone (HAZ), which clearly explained the reasons for the fracture of tensile test specimens at the 5052-H32 side.

The present study describes the microstructural development and resulting mechanical properties associated with the dissimilar friction stir welding of Al alloys—AA5052 and AA6061.

2. Experimental

In the present friction stir welding studies, two commercial aluminium alloys AA5052 and AA6061 were used—former being a partially recrystallized solid solution-strengthened aluminium alloy and latter being a precipitate-hardened aluminium alloy. The AA6061 was in the solutionized condition (solution heat treated at 530°C for duration of 0.5 h, followed by water quenching), whereas the AA5052 was in the as-received state. The chemical composition and mechanical properties of these alloys are listed in Tables 1 and 2, respectively.

The FSW experiments were carried out using a dedicated, locally designed and fabricated friction stir welding equipment at the Mechanical Engineering. Deptment, Indian Institute of Science, Bangalore. During the experiments, a data logger was used to collect the data (consisting of normal load, traverse load, spindle torque, etc.) from load cells attached to the equipment. The dimensions of the workpieces used were 300 mm × 50 mm × 5 mm. Prior to welding, one side of the workpieces was machined in the transverse long section plane (length-thickness plane) using a milling equipment. This facilitated a good contact of the mating surfaces with each other when arranged in a butt configuration. The workpieces were cleaned thoroughly by acetone to remove any dirt, organic material, and fine particles left after the machining process. A commercial high speed steel (HSS) tool, having a cylindrical geometry with 4.8 mm pin length and 6 mm pin diameter and having 25 mm shoulder diameter was used. The tool tilt was kept constant at 3° for all welding trials. Several FSW trials were carried out at 1120 and 1400 rpm and for various traverse speeds ranging from 60 mm/min, 80 mm/min and 100 mm/min. Depending on the combination of tool rotation speed and tool traverse speed used, the specimens were assigned identities according to the nomenclature (R“tool rotation speed”F“tool traverse speed” where R“tool rotation speed” refers to the tool rotation speed in rpm and F“tool traverse speed” refers to the feed or the tool travel rate in mm/min; e.g., the process parameter combination of 1120 rpm and 100 mm/min is designated as R1120F100). The qualification of defect-free joints was done using X-ray radiography and dye penetration tests. It was observed that for R1400F080 specimen (welded at tool rotation speed of 1400 rpm and tool traverse speed of 80 mm/min) defect-free joints were obtained all along the interface.

The transverse cross-section of the weld was used for optical as well as electron microscopy observations. This transverse section was carefully prepared using standard metallographic techniques to obtain a mirror-like surface finish. For optical microscopy, the polished specimens were chemically etched with an etchant to reveal the microstructure. The etchant used was a diluted solution of 5 mL HF + 10 mL H_2SO_4 in 85 mL water. A microhardness profile was obtained in the midthickness region across this structure using a load of 50 g and a 10 s dwell time. A Cameca SX100 electron probe microanalyzer (EPMA) equipped with three wavelength dispersive spectrometers was used at an operating voltage of 20 kV and regulated beam current of 20 nA for determining the chemical composition. For electron backscattered diffraction (EBSD), the specimens were electropolished at −40°C and 11 V in a solution containing 20 vol pct perchloric acid and 80 vol pct ethyl alcohol. The EBSD measurements were obtained on a Fei quanta-200 HV SEM with TSL-OIM package. The details of the conventions

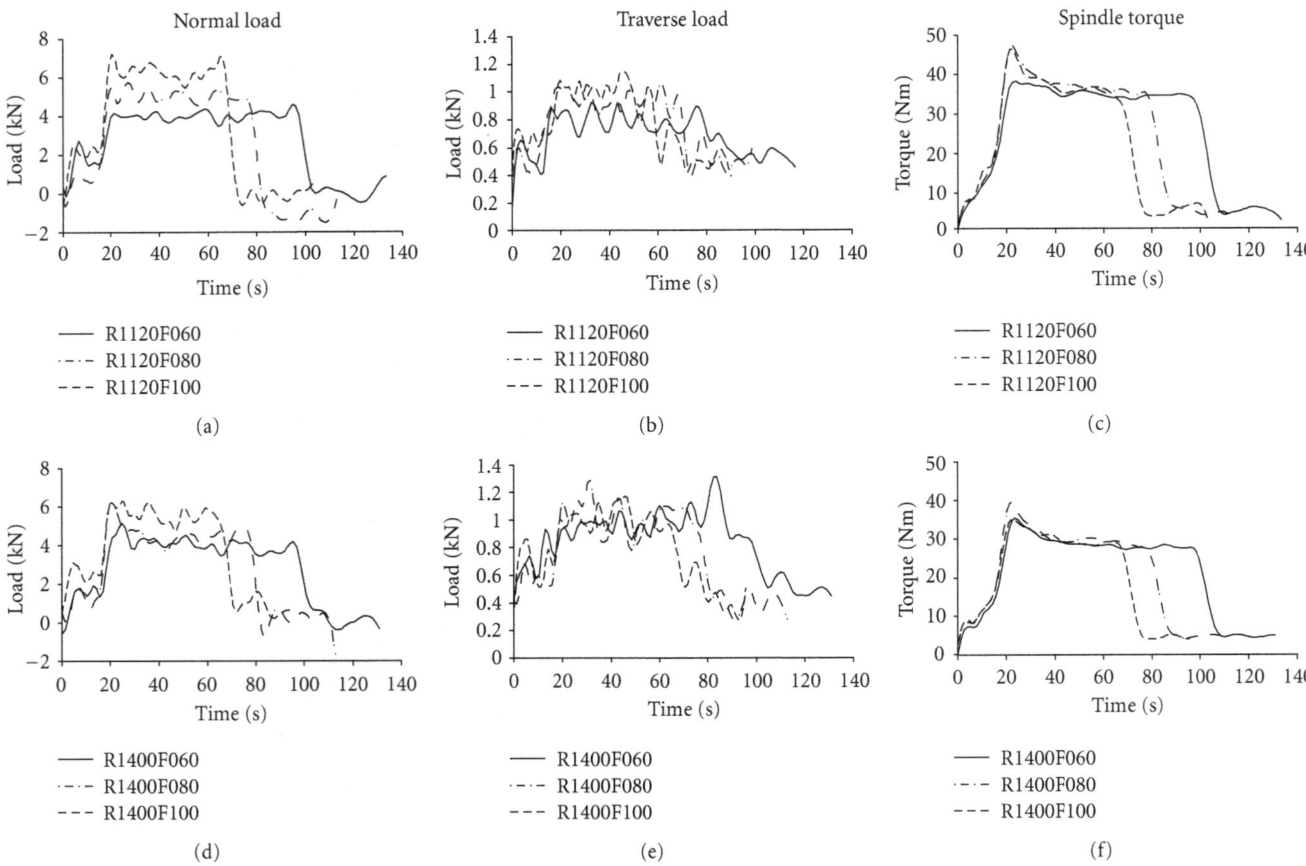

Figure 1: Variation of normal load, traverse load, and spindle torque with respect to time during friction stir welding of AA6061 and AA5052 experiments at various combinations of process parameters.

adopted for EBSD data analysis and various forms of representation/analysis of microtexture data are described in another work published by the authors [4].

Standard tensile specimen having gauge length of 25 mm and gauge width of 5 mm was cut using electrodischarge machine (EDM) from the welded plates by keeping the tensile axis perpendicular to the welding direction. The gauge length of the tensile specimen extended from the stir zone into the parent material. The tensile specimens were tested by using screw-driven Instron machine at a strain rate of $10^{-4}\,s^{-1}$.

3. Results and Discussion

3.1. Process Parameters Optimization. The variation of normal load, traverse load, and spindle torque with respect to time during the experimental trials for FSW of AA5052-AA6061 is shown in Figure 1. A detailed description of the dynamics of variation of load (both normal and traverse) and torque during the FSW of AA5052 is given in Section 4.1 of Chapter 4 of [5]. From the friction stir welding experiments for dissimilar combination of AA5052 and AA6061, it was noticed that the normal load experienced by the tool varied in the range 3.5–7 kN at rotation speed of 1120 rpm. Whereas at a higher rotation speed of 1400 rpm, the normal load was found to decrease and was in the range 3.5–6 kN. It was found that during FSW AA5052-AA6061 trials, the normal load was less at higher rotation speed of 1400 rpm. However no conclusive statement can be made for the normal load at rotation speed of 1120 rpm. The traverse load was in the range 0.6–1.2 kN and 0.7–1.3 kN for rotation speeds of 1120 and 1400 rpm, respectively. The spindle torque decreases with an increase in the rotation speed. Spindle torque values in the traversing phase were in the range of 33–38 Nm at 1120 rpm and 25–30 Nm at 1400 rpm. Further, it was observed that for a particular rotation speed, the spindle torque was not affected with the variation in traverse speeds (60, 80, and 100 mm/min).

3.2. Microstructural Characterization. Figure 2 shows a motif of the optical images of the cross-section of a FSW AA5052-AA6061 R1400F080 specimen. The different regions of the dissimilar friction stir weld are marked in the figure. It can be noticed that the interface between AA5052 and AA6061, which initially was linear prior to welding, now has a nonlinear, wavy, and distorted appearance. The interface appears to be serrated throughout the thickness of the weld. This interface can be considered to be an imperfection and was termed as "joint-line remnant" by Threadgill [6]. It must be noted that the presence of joint-line remnant does not affect the tensile properties of the welds [6]. However, with the

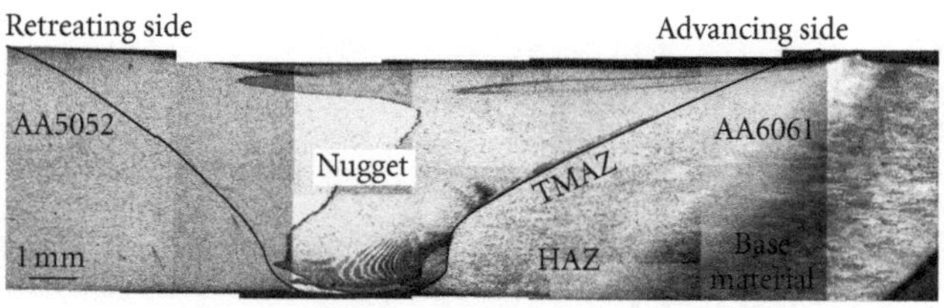

Figure 2: Motif of optical images of the transverse cross-section of FSW AA5052-AA6061 R1400F080 specimen showing the various regions.

present combination of process parameters (R1400F080) used, macroscopic mixing of the materials (AA5052 and AA6061) was not observed in the nugget. Several researchers [2, 7–11] have studied the formation of complex intercalation structures consisting of swirl-like features and intermingled dissimilar lamellae during the intermixing occurring in FSW of dissimilar materials. Park and Kim [2] have shown various micrographs to illustrate the degree of material mixing taking place for the same combination of aluminium alloys (AA5052 and AA6061) using a constant rotation speed of 1250 rpm and various traverse speeds (ranging from 15 mm/min to 507 mm/min). A primary reason for the observed effective mixing being the use of a threaded tool for the FSW trials. On the contrary, Colligan [12] pointed out that during the FSW process, not all the material influenced by the tool pin was stirred: much of the material movement was governed by extrusion.

In the present investigation, an unthreaded tool was used. Hence significant amount of intermixing among the two alloys was not seen. Lorrain et al. [10] reported that the material flow with an unthreaded tool had the same features as the material flow using threaded tools: deposition of the material at the advancing side (AS) in the upper part of the weld and at the retreating side (RS) in the lower part of the weld. In the present case, the mixing of the materials was not as effective as observed by Park et al. [3]. Park et al. [3] observed that the mixing of the materials was more effective when the AA5052 was on the advancing side for the AA5052-AA6061 FSW combination. In the present study, after etching, a swirl zone could be noticed at the bottom of the nugget zone confined to the AA6061 regime. This was in agreement with the results of Ouyang and Kovacevic [7], who reported vortex-like features for FSW AA6061. The advancing side interface at the AA6061 side was clearly seen, but the retreating side interface at the AA5052 side was difficult to demarcate.

In Figure 2 a distinct HAZ can be recognized on the AA6061-side, whereas no such characteristic region is distinguishable on the AA5052 at the retreating side. AA5052 being a solid solution-strengthened alloy, no effect of heat could be observed on it near the HAZ. AA6061 is a precipitation-hardened alloy and the effect of the heat during the deformation process was clearly reflected in the microstructure, in the form of HAZ which appeared brighter than the adjoining nugget and base material (BM).

Figure 3(a) shows the low magnification optical micrograph of the R1400F080 specimen. A secondary electron image of a region in the nugget denoted by the rectangle "X" is shown in Figure 3(b). Line AB in the midthickness region shows the location of microhardness profile across the cross-section of friction stir welded AA5052-AA6061 R1400F080 specimen. Similarly line PQ in Figure 3(b) shows the location of the line profile analysis for Al Kα, Mg Kα and Si Kα, across the AA5052-AA6061 interface shown in Figure 3(c). It was noticed that the width of interdiffusion was approximately 80 μm, for a relatively small duration of annealing (approx a few minutes). In this figure the occasional spikes in the Si Kα curve indicate the presence of Si-rich second-phase particles. Thus it was concluded that despite the insufficient intermixing of materials (AA5052 and AA6061) caused by the stirring action of the tool, it did not result in a simple mechanical interlocking of the aluminium alloys. It is observed from the EPMA profiles in Figure 3(c) that there is substantial interdiffusion of the alloying elements.

Figure 4(a) shows the inverse pole figure map of a region at the interface of AA5052 and AA6061 in the nugget of FSW AA5052-AA6061 R1400F080 specimen. The region shows refined grains in the nugget for both the aluminium alloys. This region shown was partitioned into two regions A and B. Region A comprised of fine grains of AA6061 while region B comprised of a refined microstructure of AA5052. The average grain diameters for regions A and B were 11 and 20 μm, respectively. From the chart of the {111} pole figures of these regions shown in Figure 4(b), it could be seen that the orientation distribution was similar for both the aluminium alloys. Most of the {111} poles were aligned in the ND direction (i.e., parallel to the welding direction).

3.3. Mechanical Properties Evaluation

3.3.1. Microhardness. Figure 5 shows the variation of microhardness across the transverse cross-section of FSW AA5052-AA6061 R1400F080 specimen in the midthickness region along line AB (shown in Figure 3(a)). In the figure, the filled and unfilled symbols denote the microhardness values corresponding to the regions/domains of AA5052 and AA6061 alloys, respectively. An abrupt transition across the AA5052-AA6061 interface in the nugget was observed as one proceeds from the AA5052 towards AA6061. It must be noted that the microindentation in the nugget was performed at intervals of 250 μm. From the EPMA profiles shown in Figure 3(c), it was observed that the width of the transition region was approximately 80 μm. Hence any possible smooth change (transition) in the microhardness was not observed using the microindentation technique. The microhardness values remained nearly constant in the nugget and the adjoining HAZ for both the aluminium alloys (56–61 VHN in AA5052 and 86–91 VHN in AA6061). Beyond the HAZ into the base material region, there was a smooth transition of the microhardness to the parent material microhardness values—decreasing from the higher hardness in the nugget at the AA6061 side and increasing from lower hardness in the nugget at the AA5052 side.

In an earlier work [4] reported by the authors on FSW of AA5052, it was observed that there was a slight increase

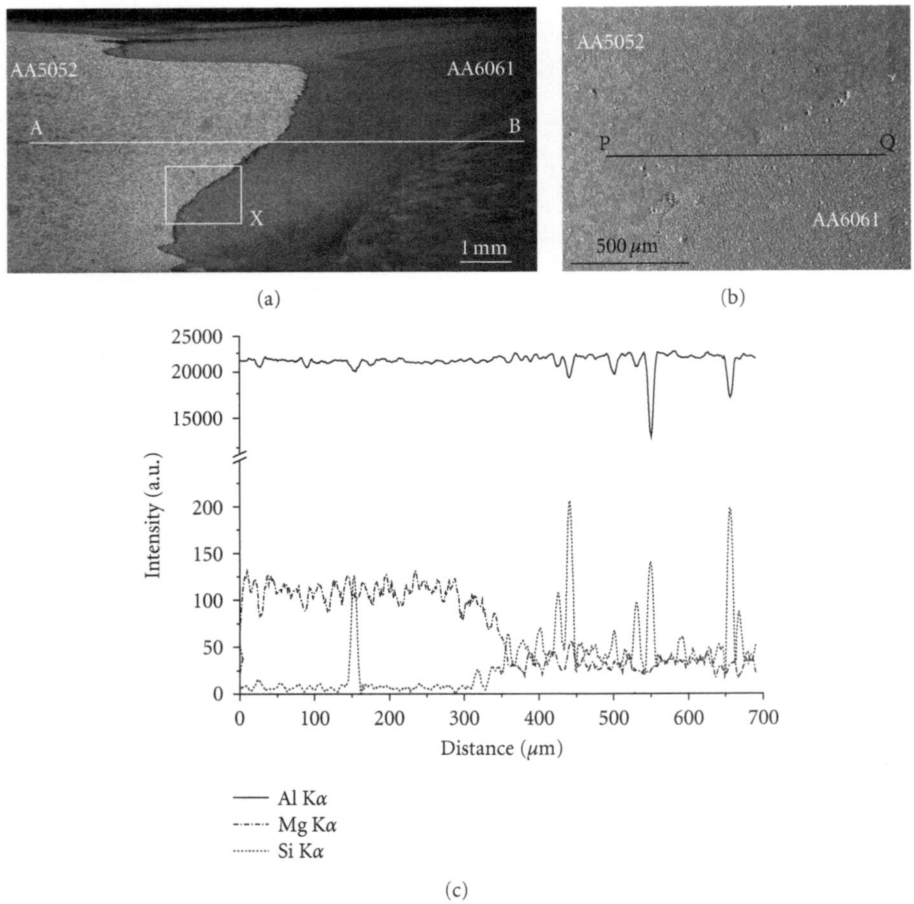

Figure 3: (a) Low-magnification optical micrograph of as-welded AA5052-AA6061 R1400F080 specimen. (b) Secondary electron image of region X in (a) corresponding to the interface of AA5052 and AA6061 in the nugget. (c) Intensity profiles of Al Kα, Mg Kα and Si Kα, X-ray lines along line PQ shown in (b).

in the microhardness at the nugget-TMAZ interface on the advancing side. No such distinct rise in the microhardness was noticed at the nugget-TMAZ interface on the advancing side for the present work on FSW of AA5052-AA6061.

3.3.2. Tensile Testing. Figure 6 shows the results of the tensile tests of FSW AA5052-AA6061 specimens. On the abscissa, the process parameters have been arranged in the decreasing order of the weld pitch (ratio of traverse speed to rotation speed expressed in mm/rev) implying that the heat input increases as one advances from the left to right. The process parameters on the left result in lower heat input whereas the process parameters on the right have higher heat input. It was noticed that the R1120F100 specimen with the lowest heat input resulted in inferior tensile properties compared to the other specimen. The rest of the specimen gave good indication of strength (~225 MPa) where the ultimate tensile strengths were comparable and the values of the yield strengths were in the range of 135–150 MPa. It was observed that the process parameters with high heat input resulted in good ductility (~7%). Overall, it was observed that the use of higher rotation speeds (1400 rpm) was essential to provide good heat input and the welds thus obtained had good tensile strength. The results for the tensile testing of FSW AA5052-AA6061 specimen were found to be consistent with the results obtained by Park et al. [3] for the same alloy system. According to their results, for tool rotation and traverse speeds of 2000 rpm and 100 mm/min, respectively, the ultimate tensile strength and yield strength were ~220 MPa and 110 MPa, respectively. However, the % elongation was found to be higher (10–17%) compared to the present study. However, the results in the present study are comparable with the work of Leitao et al. [13] on the dissimilar friction stir welding of AA5182 and AA6016, in which the % elongation was obtained as 8%, and the UTS and YS were 200–220 MPa and 115–150 MPa, respectively.

4. Conclusions

(1) Friction stir welding of dissimilar materials AA5052 and AA6061 was successfully performed. It was observed that at higher rotation speeds, the normal load and spindle torque requirement decreased.

(2) The microstructural studies suggested that there was no rigorous mixing of both materials in the nugget. There was an abrupt change in the microhardness

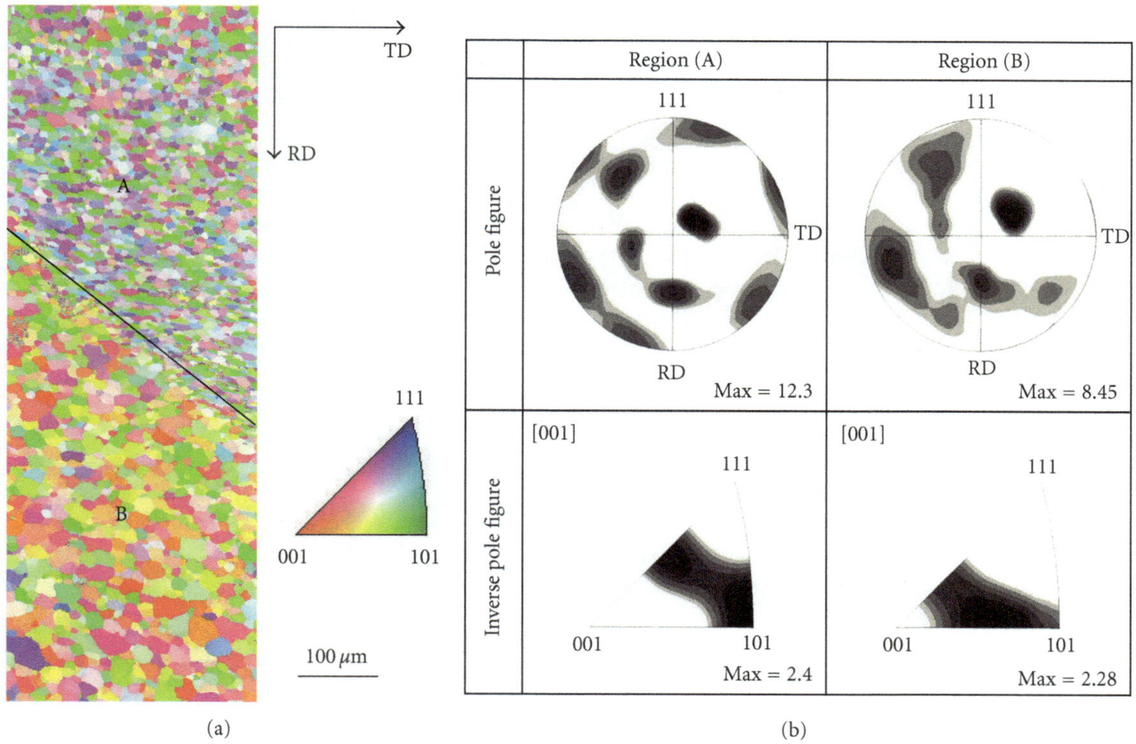

FIGURE 4: (a) Inverse pole figure map of the region across the interface of AA5052 and AA6061 in the nugget of FSW AA5052-AA6061 specimen. (b) Chart showing the pole figure and inverse pole figures for the various regions marked in (a).

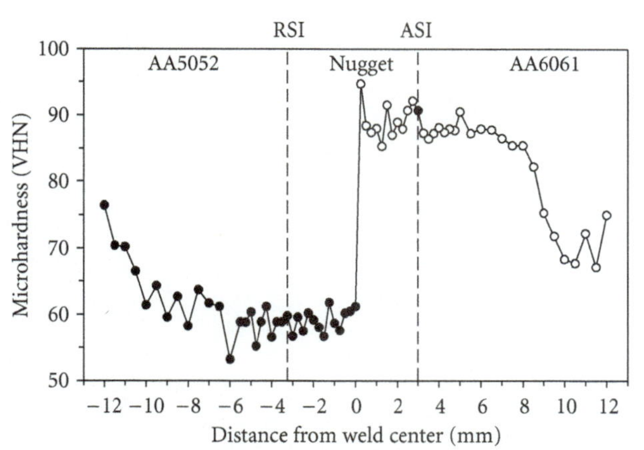

FIGURE 5: Microindentation hardness distribution across the transverse cross-section of FSW AA5052-AA6061 R1400F080 specimen.

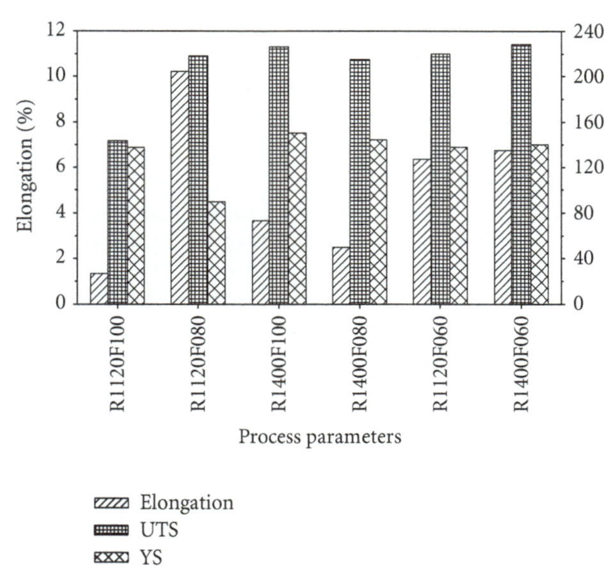

FIGURE 6: The % elongation, ultimate tensile strength (UTS), and yield strength (YS) for various combinations of process parameters for friction-stir-welded AA5052-AA6061 specimens.

across the interface in the nugget. However, electron probe microanalysis results stated that there was bonding at the atomic scale due to substantial interdiffusion of alloying elements at the interface of both the alloys in the nugget. Further, orientation imaging microscopy at the interfacial region suggested that despite the nonrigorous mixing and materials holding on to their domain, both the materials exhibited similar texture.

(3) Thus the interdiffusion of alloying elements and attaining of similar orientations in the nugget could have contributed to the good mechanical properties of the friction stir welded AA5052-AA6061 specimens. The tensile properties of the FSW AA5052-AA6061 specimens were better than the properties of the softest of the similar friction-stir-welded systems (i.e., FSW AA6061).

Acknowledgments

The authors would like to acknowledge Professor S. V. Kailas, Deptartment of Mechanical Engineering, Indian Institute of Science, Bangalore, for extending his support to carry out the friction stir welding experiments. The authors acknowledge the National Facility of Texture and OIM (a DST-IRPHA) facility for undertaking the microtexture studies. The first author acknowledges the fellowship provided by the Department of Atomic Energy, India, under the Mumbai University—BARC collaborative research scheme. The authors like to thank Dr. A. K. Suri, Director, Materials Group, BARC, and Dr. G. K. Dey, Head, Materials Science Division, BARC, for their keen interest in this work.

References

[1] C. Y. Lee, W. B. Lee, J. W. Kim, D. H. Choi, Y. M. Yeon, and S. B. Jung, "Lap joint properties of FSWed dissimilar formed 5052 Al and 6061 Al alloys with different thickness," *Journal of Materials Science*, vol. 43, no. 9, pp. 3296–3304, 2008.

[2] J. C. Park and S. J. Kim, "The effect of traveling and rotation speeds on mechanical properties during friction stir welding of dissimilar Al alloys," *Defect and Diffusion Forum*, vol. 297-301, pp. 590–595, 2010.

[3] S. K. Park, S. T. Hong, J. H. Park, K. Y. Park, Y. J. Kwon, and H. J. Son, "Effect of material locations on properties of friction stir welding joints of dissimilar aluminium alloys," *Science and Technology of Welding and Joining*, vol. 15, no. 4, pp. 331–336, 2010.

[4] N. T. Kumbhar, S. K. Sahoo, I. Samajdar, G. K. Dey, and K. Bhanumurthy, "Microstructure and microtextural studies of friction stir welded aluminium alloy 5052," *Materials and Design*, vol. 32, no. 3, pp. 1657–1666, 2011.

[5] N. T. Kumbhar, *Solid state methods for joining aluminium alloys*, Ph.D. thesis, 2011.

[6] P. L. Threadgill, "Terminology in friction stir welding," *Science and Technology of Welding and Joining*, vol. 12, no. 4, pp. 357–360, 2007.

[7] J. H. Ouyang and R. Kovacevic, "Material flow and microstructure in the friction stir butt welds of the same and dissimilar aluminum alloys," *Journal of Materials Engineering and Performance*, vol. 11, no. 1, pp. 51–63, 2002.

[8] P. Su, A. Gerlich, T. H. North, and G. J. Bendzsak, "Intermixing in dissimilar friction stir spot welds," *Metallurgical and Materials Transactions A*, vol. 38, no. 3, pp. 584–595, 2007.

[9] A. Gerlich, P. Su, M. Yamamoto, and T. H. North, "Material flow and intermixing during dissimilar friction stir welding," *Science and Technology of Welding and Joining*, vol. 13, no. 3, pp. 254–264, 2008.

[10] O. Lorrain, V. Favier, H. Zahrouni, and D. Lawrjaniec, "Understanding the material flow path of friction stir welding process using unthreaded tools," *Journal of Materials Processing Technology*, vol. 210, no. 4, pp. 603–609, 2010.

[11] J. C. McClure, Y. Li, R. Flores et al., "Intercalation vortices and related microstructural features in the friction-stir welding of dissimilar metals," *Materials Research Innovations*, vol. 2, no. 3, pp. 150–163, 1998.

[12] K. Colligan, "Material flow behavior during friction stir welding of aluminum," *Welding Journal*, vol. 78, no. 7, p. 229-s, 1999.

[13] C. Leitao, R. M. Leal, D. M. Rodrigues, A. Loureiro, and P. Vilaça, "Mechanical behaviour of similar and dissimilar AA-5182-H111 and AA6016-T4 thin friction stir welds," *Materials and Design*, vol. 30, no. 1, pp. 101–108, 2009.

Prediction of Microstructure Evolution in Hot Backward Extrusion of Ti-6Al-4V Alloy

Jong-Taek Yeom,[1] Jeoung Han Kim,[1] Jae-Keun Hong,[1] Nho-Kwang Park,[1] and Chong Soo Lee[2]

[1] *Special Alloys Group, Korea Institute of Materials Science, 531 Changwondaero, Changwon 641-831, Republic of Korea*
[2] *Department of Materials Science and Engineering, POSTECH, San 31, Hyojadong, Namgu, Pohang 790-784, Republic of Korea*

Correspondence should be addressed to Jong-Taek Yeom, yjt96@kims.re.kr

Academic Editor: Enrico Evangelista

Microstructure evolution of Ti-6Al-4V alloy during hot backward extrusion process was simulated with the combined approaches of finite element method (FEM) and microstructure prediction model. From experimental analysis, it can be found that the change of microstructure during hot forming process of titanium alloy has a close relation to α/β phase transformation and grain growth behaviour. A microstructure prediction model was established by considering the change of volume fractions and grain size of both phases varying with process variables and then implemented into the user-defined subroutine of FEM analysis. In order to demonstrate the reliability of the model, the volume fraction and grain size of primary α phase during the hot backward extrusion process of Ti-6Al-4V alloy were simulated. The simulation results were compared with the experimental ones.

1. Introduction

Computer simulation is an important and effective tool for the optimized forming processes and to reduce error which may be caused by trial and error method. Finite element (FE) analysis is widely used for the simulation of shape and microstructure changes in hot forming [1–3]. With the improvement in understanding of the relationship between microstructures and mechanical properties of structural components, the prediction of shape and microstructure changes has become a hot issue in the high-temperature forming.

The high-temperature forming processes for manufacturing titanium alloy products are normally conducted on the equiaxed structure in two-phase ($\alpha + \beta$) field [4]. The microstructure change is mainly indicated as the change of volume fraction and grain size of α and β phases. It may be due to phase transformation and grain growth during the high temperature forming. In the past, most of microstructure simulations have focused on the prediction of the single-phase alloys using the recrystallization and grain growth models [5]. However, only a limited work [6, 7] has been carried out on the simulation for microstructure evolution of Ti-6Al-4V alloy during hot forming processes. Therefore, the aim of this research is to establish the methodology for prediction of the microstructure evolution of Ti-6Al-4V components during high-temperature forming processes, and to develop a useful design tool for obtaining the optimum process condition in the high-temperature forming processes using finite element method (FEM) simulation.

2. Experimental Procedures

The material used in this work was a Ti-6Al-4V alloy billet annealed at 704°C for 2 h with a diameter of 100 mm. The chemical composition of the alloy is 6.32Al, 4.18V, 0.21Fe, 0.014C, 0.18O, 0.008N, 0.001Y, and balance Ti (in wt%). Initial microstructure of Ti-6Al-4V alloy was an equiaxed structure with the primary α grain size of about 16 μm as shown in Figure 1.

In order to investigate flow behavior and microstructure evolution of Ti-6Al-4V alloy in the α-β phase field,

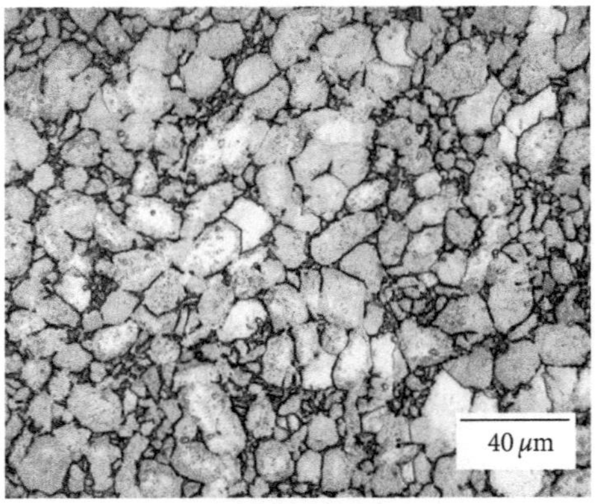

FIGURE 1: Typical microstructure of the annealed Ti-6Al-4V alloy billet.

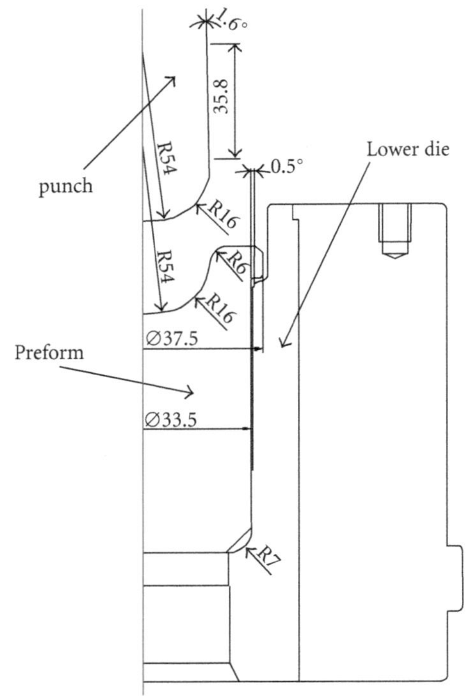

FIGURE 2: Design of the punch, lower die, and preform used in hot backward extrusion.

compression and heat treatment tests were carried out. Compression tests were performed in the temperature ranges between 850°C and 1000°C with 25°C intervals using the strain rate ranging from 10^{-3} to $10\,s^{-1}$. Heat treatment tests were carried out with the holding time of 30 min between 900 and 1015°C. To observe the actual microstructure at the test temperature, after the holding time of 30 min, the heat-treated samples were water-quenched. To analyze the microstructure evolution during cooling, the samples were heat-treated at the temperature of 900, 925, 950, and 975°C for 30 min. Following the heat treatment, the samples were cooled down using different cooling rates.

In order to verify the reliability of the established microstructure prediction model, Ti-6Al-4V alloy tube was manufactured by hot backward extrusion process using a 150 ton hydraulic press. Backward extrusion process for evaluating the reliability of the model was selected to predict microstructure variation during actual hot working considering friction, interface heat transfer, and process variables. The design of the dies (punch and lower die) and preform is represented in Figure 2. Oil-based graphite lubricant was spread into the punch and lower dies to reduce die friction. The preform of Ti-6Al-4V alloy was glass-coated and heated to an extrusion temperature of 970°C. The punch and die temperatures were selected at 500°C and 600°C, respectively. After the backward extrusion, the Ti-6Al-4V tube was water-quenched.

3. Results and Discussion

Figure 3 shows the change of microstructures with increasing the heating temperature obtained from isothermal heat treatment tests of Ti-6Al-4V alloy. It can be found that the volume fraction and grain size of α phase decrease with increasing the heating temperature, and especially the grain size of α phase markedly decreases from more than 950°C. The β phase grains remain constant up to 950°C and dramatically grow at 975°C. This is in agreement with the observations by Semiatin et al. [8].

Figure 4 shows the change of microstructure with increasing hold time obtained from isothermal heat treatment tests of Ti-6Al-4V alloy. It is noted that the α grain size does not change significantly with holding time whereas β grain size increases considerably. This is due to the high diffusivity in β phase.

In general, mechanical properties of Ti-6Al-4V alloy products forged at α-β region were directly affected by the volume fraction and grain size of equiaxed primary α phase. The microstructure prediction model reflecting microstructure evolution during heating and high temperature deformation is established in this work. To predict the volume fraction and grain size of primary α phase, the geometrical model for grain size change of α phase developed in previous work [2, 9, 10] was used. Assuming that the total numbers of α grains and β grain size are constant, the grain size of α phase (d) can be calculated by following equation:

$$d = d_0 \left(\frac{f_\alpha}{f_{\alpha 0}} \right)^{1/3} \quad (\mu\text{m}), \tag{1}$$

where d_0 and $f_{\alpha 0}$ are the initial grain size and volume fraction of primary α phase, respectively, and f_α is the present volume fraction of primary α phase. Meanwhile, the volume fractions of each phase during heating and soaking can be expressed as a function of initial volume fractions of each phase and temperature

$$\begin{gathered} f_\alpha = f_{\alpha,RT} + \left(1 - \exp\left(-k \cdot \left\langle T_\beta - T \right\rangle\right)\right) \\ f_\beta = 1 - f_\alpha, \end{gathered} \tag{2}$$

where f_{β_0} and f_β are initial volume fraction and present volume fraction of β phase, respectively. $f_{\alpha,RT}$ is the volume fraction of primary α phase at room temperature. The magnitude of $\langle T_\beta - T \rangle$ is zero when $T_\beta - T \le 0$, and it is

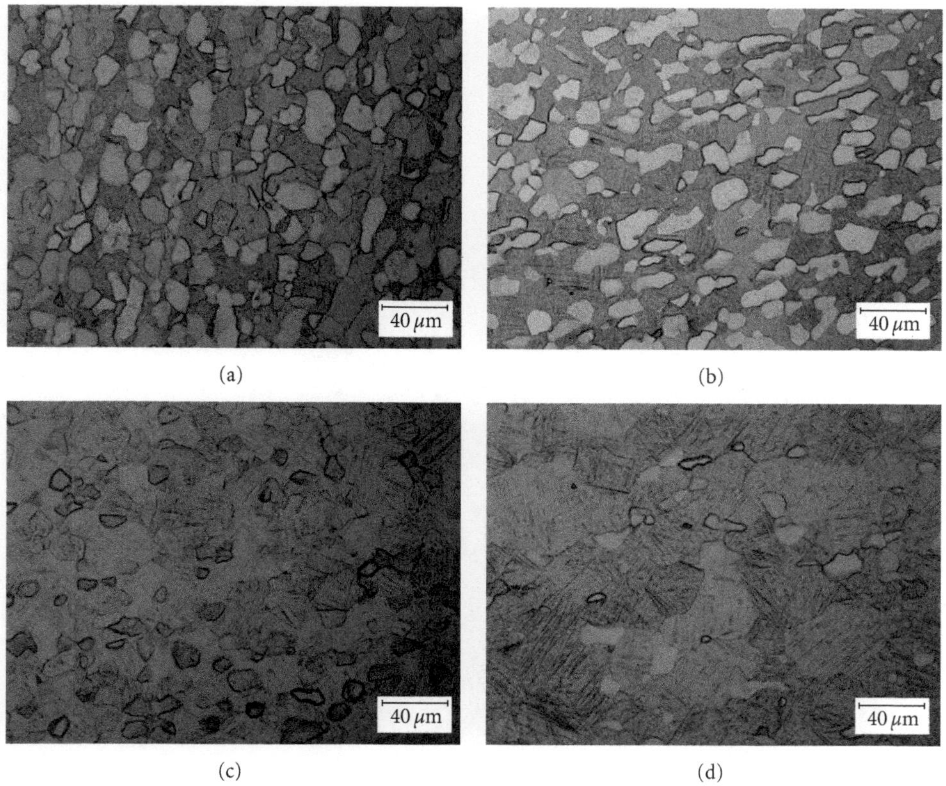

FIGURE 3: Microstructures of Ti-6Al-4V alloy samples water quenched after a 30 min heat treatment at (a) 900°C, (b) 925°C, (c) 950°C, and (d) 975°C.

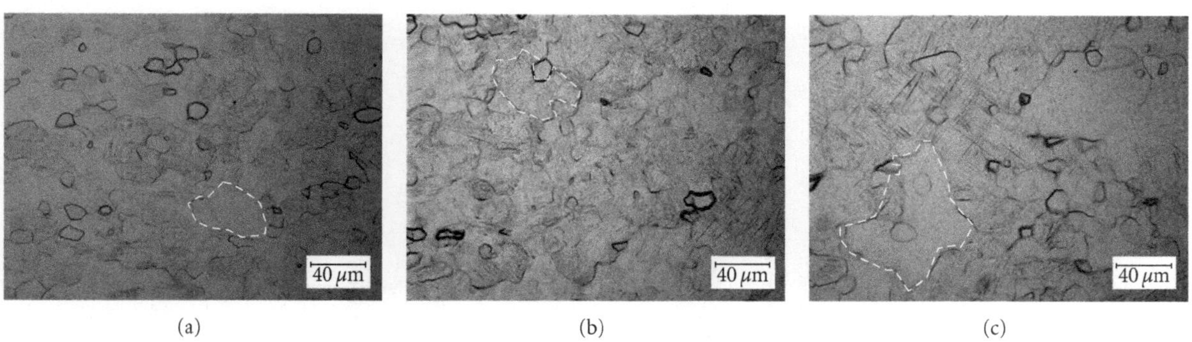

FIGURE 4: Microstructures of Ti-6Al-4V alloy samples water quenched after isothermal heat treatment at 950°C for (a) 10 min, (b) 1 h, and (c) 10 h. White dotted lines indicate β grains.

$T_\beta - T$ in other case. Through the isothermal heat treatment tests, the parameters $f_{\alpha,RT}$ and k in Ti-6Al-4V alloy with an equiaxed structure were determined by 0.942 and 8.17×10^{-3}, respectively.

In α-β Ti alloys, phase transformation ($\beta \rightarrow \alpha$) also occurs during cooling and forming. In order to express the volume fraction change of α phase during cooling and forming, the Avrami-type equation was applied

$$f_{\alpha,c} = f_{\alpha,h} + \left(f_{\alpha,RT} - f_{\alpha,h}\right) \cdot \left(1 - \exp\left(-f(T) \cdot t^{n}\right)\right)$$
$$f(T) = A \cdot \left(\frac{T_{\text{heat}} - T}{T_{\text{heat}} - T_e}\right)^{n'}, \tag{3}$$

where $f_{\alpha,h}$ are the volume fraction of α phase after heating and soaking. T_{heat} is heating or soaking temperature. T_e represents the finishing temperature of phase transformation and this value is assumed to be 298 K. The values of n, n', and A were 1, 6.67, and 0.0023, respectively.

A commercial FEM code, DEFORM 2D, was used to simulate the effect of process variables in hot backward extrusion process of Ti-6Al-4V alloy on the distribution of the internal state variables such as strain, strain rate, and temperature. The flow stress data obtained from compression tests were corrected for temperature increment calculated by the following equation:

$$\Delta T = \frac{0.9 \int_0^{\varepsilon} p\sigma d\varepsilon}{\rho C}, \tag{4}$$

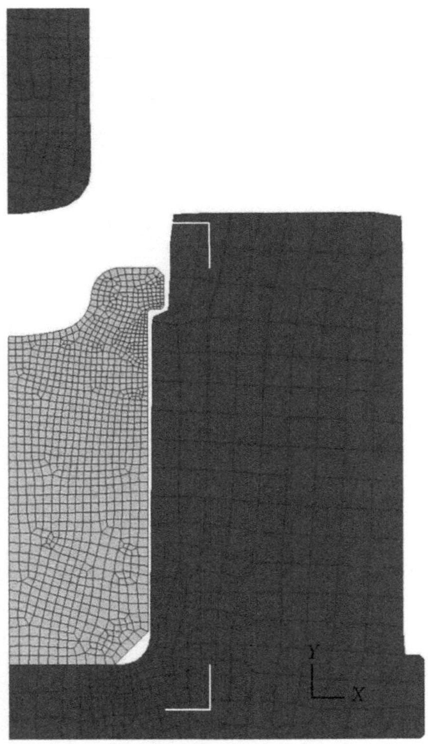

FIGURE 5: FEM modeling for hot backward extrusion process of Ti-6Al-4V alloy.

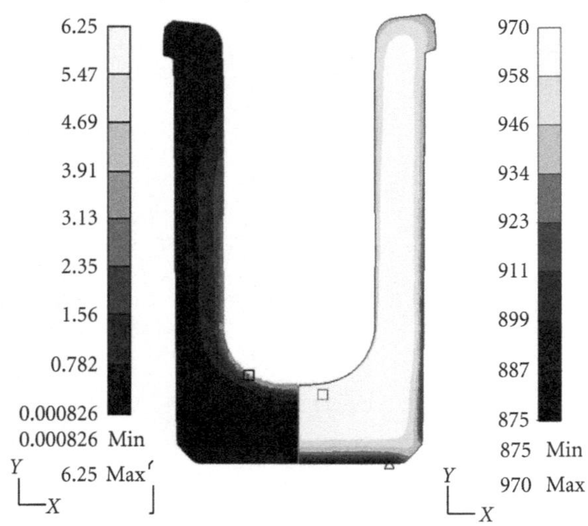

FIGURE 6: Temperature and effective strain contours for hot backward extrusion simulation of Ti-6Al-4V alloy.

where ε_p is plastic strain, ρ is density of material, and C is specific heat. Corrected flow stress data were directly used to simulate material behavior of Ti-6Al-4V alloy during hot backward extrusion.

Generally, the friction between dies and workpiece is expressed by the friction law of constant factor. Also, interface heat transfer between dies and workpiece greatly affects the temperature change of the workpiece during hot forming process. Based on the results of previous work [11], the friction coefficient and interface heat transfer coefficient were determined at 0.3 and 5.0 $kW/m^2{}^\circ C$, respectively. The element used in the simulation is brick elements, and automatic remeshing system was adopted during simulation. Figure 5 shows the 2D modeling for dies and initial preform shapes.

Figure 6 shows simulation results of strain and temperature distributions of hot backward extrusion process of Ti-6Al-4V alloy tube. The simulation results indicate that the highest strain region in the workpiece is found at inner wall, which is in contact with the punch during

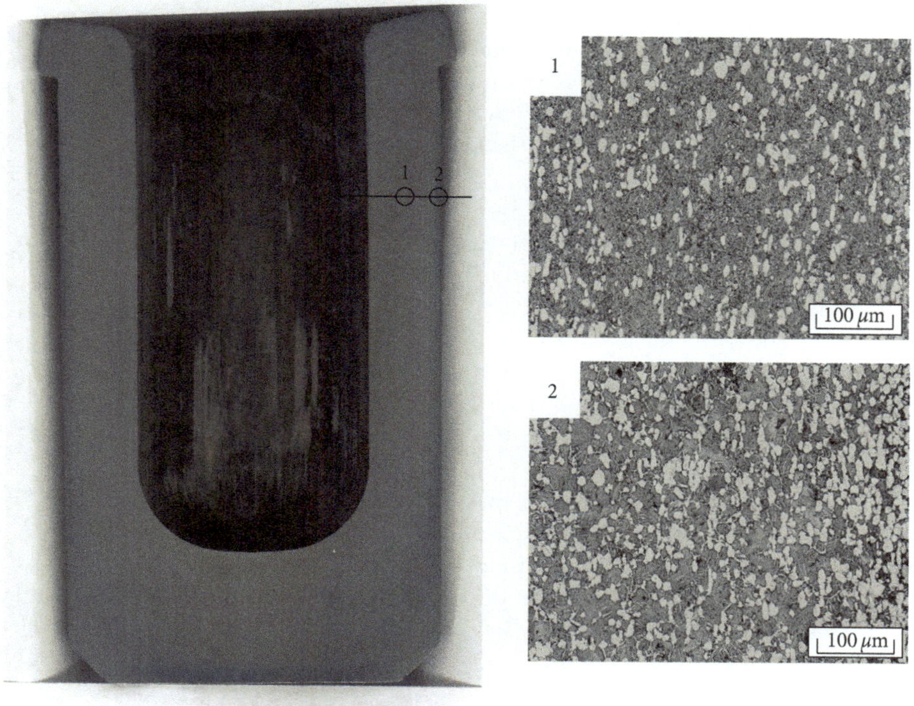

FIGURE 7: Microstructures observed at different locations for backward extruded Ti-6Al-4V alloy tube.

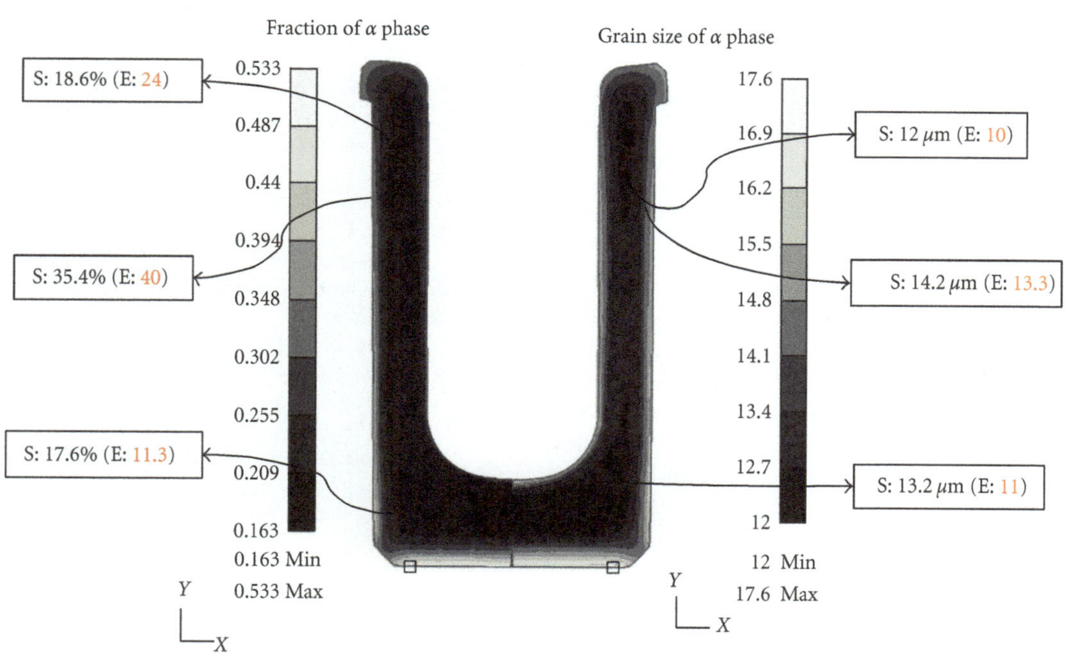

FIGURE 8: Contours of volume fraction and grain size of primary α phase for hot backward extrusion simulation of Ti-6Al-4V alloy. The symbols "S" and "E" in the bracket denote simulation and experimental results, respectively.

the forming process, and the temperature level at die contact area (or surface area) is lower than that at middle area due to heat transfer between dies and workpiece. Also, the lowest temperature region in the workpiece is found at the bottom, which is in constant contact with the lower die throughout the forming process from the moment the preform is transferred to the die.

Figure 7 shows the Ti-6Al-4V alloy tube manufactured by hot backward extrusion process and the microstructures observed at the cross-section of the tube. It can be found that extruded Ti-6Al-4V alloy tube is formed without the surface defects. Microstructures observed at most locations indicate a bimodal structure composed of equiaxed α and transformed beta phases. In the microstructural observation from surface to middle positions, it can be seen that the volume fraction of α phase at middle position is smaller than that of α phase at surface.

These microstructure changes during the extrusion process were closely related to the plastic strain and temperature varying at different positions. In this work, in order to predict the α grain size and the α/β phases volume fraction change, the decoupled approach between the microstructure model and FE analysis was applied. The model for predicting the volume fraction change and grain size of α phase was implemented into a post-user-defined subroutine of a commercial FE-code, DEFORM-2D.

Figure 8 shows the comparison between experimental data and simulation results. The experimental data measured from extruded tube were overlapped aside from simulation results. the values in the brackets especially indicate the experimental data. It was found that reasonable accuracy was obtained at most of region and the average deviation was below 15%. Only at a certain part of inner wall, 36% of error was noticed which is due to the contact of the inner wall with the punch occurring in the given microstructures. It is clearly demonstrated that simulation results of volume fraction and grain size of α phase at most locations for the extruded Ti-6Al-4V tube were relatively close to measured values. Consequently, it is suggested that the microstructure simulation module is very useful for hot forming process design of Ti-6Al-4V alloy.

4. Conclusions

In this work, the microstructure evolution of Ti-6Al-4V tube during hot extrusion process was predicted by decoupled approach of FE analysis and microstructure prediction model. In order to evaluate the volume fraction and grain size of α phase during the hot backward extrusion process, the microstructure prediction model of Ti-6Al-4V alloy was suggested and implemented into the post-user subroutine of FE code. The comparison of the microstructure simulation module with the actual microstructures of the backward extruded Ti-6Al-4V alloy tube has successfully validated the reliability of the present module in the prediction of the volume fraction and grain size of α phase.

Acknowledgments

This research was partly supported by a grant through the research program funded by the Agency for Defense Development, Korea, and also partly supported by a grant through the research program (UCN248-2865.C) funded by the Ministry of Knowledge Economy, Korea.

References

[1] X. Duan and T. Sheppard, "Simulation and control of microstructure evolution during hot extrusion of hard aluminium alloys," *Materials Science and Engineering A*, vol. 351, no. 1-2, pp. 282–292, 2003.

[2] J. T. Yeom, J. H. Kim, N. K. Park, S. S. Choi, and C. S. Lee, "Ring-rolling design for a large-scale ring product of Ti-6Al-4V alloy," *Journal of Materials Processing Technology*, vol. 187-188, pp. 747–751, 2007.

[3] J. H. Kim, N. S. Reddy, J. T. Yeom, J. K. Hong, C. S. Lee, and N. K. Park, "Microstructure prediction of two-phase titanium alloy during hot forging using artificial neural networks and fe simulation," *Metals and Materials International*, vol. 15, no. 3, pp. 427–437, 2009.

[4] S. L. Semiatin, V. Seetharaman, and I. Weiss, "The thermomechanical processing of alpha/beta titanium alloys," *JOM*, vol. 49, no. 6, pp. 33–68, 1997.

[5] C. M. Sellars and J. A. Whiteman, "Recrystallization and grain growth in hot rolling," *Metal Science*, vol. 13, no. 3-4, pp. 187–194, 1978.

[6] Z. M. Hu, J. W. Brooks, and T. A. Dean, "Experimental and theoretical analysis of deformation and microstructural evolution in the hot-die forging of titanium alloy aerofoil sections," *Journal of Materials Processing Technology*, vol. 88, no. 1, pp. 251–265, 1999.

[7] S. L. Semiatin, J. C. Soper, and I. M. Sukonnik, "Short-time beta grain growth kinetics for a conventional titanium alloy," *Acta Materialia*, vol. 44, no. 5, pp. 1979–1986, 1996.

[8] S. L. Semiatin, F. Montheillet, G. Shen, and J. J. Jonas, "Self-consistent modeling of the flow behavior of wrought alpha/beta titanium alloys under isothermal and nonisothermal hot-working conditions," *Metallurgical and Materials Transactions A*, vol. 33, no. 8, pp. 2719–2727, 2002.

[9] J. D. Miller and S. L. Semiatin, "Effect of the size distribution of alpha particles on microstructure evolution during heat treatment of an alpha/beta titanium alloy," *Metallurgical and Materials Transactions A*, vol. 36, no. 1, pp. 259–262, 2005.

[10] Y. H. Lee, C. S. Lee, T. J. Shin, S. M. Hwang, and S. M. Shim, "High Temperature Forming of Ti-6Al-4V Alloy Considering Microstructural Evolution," *Key Engineering Materials*, vol. 274–276, pp. 117–122, 2004.

[11] I. T. Yeom, N. K. Park, Y. H. Lee et al., "An improved process design for the hot backward extrusion of Ti-6Al-4V tubes using a finite element method and continuum instability criterion," *Proceedings of the Institution of Mechanical Engineers, Part B*, vol. 221, no. 2, pp. 255–265, 2007.

Progress in Wettability Study of Reactive Systems

Bolarinwa Komolafe and Mamoun Medraj

Department of Mechanical Engineering, Concordia University, 1455 de Maisonneuve Boulevard West, Montreal, QC, Canada H3G 1M8

Correspondence should be addressed to Mamoun Medraj; mmedraj@encs.concordia.ca

Academic Editor: Sunghak Lee

A review of various factors affecting wetting and spreading has been presented. Various methods for wettability studies and critical issues concerning contact angle measurement are discussed. Reactive wetting has been given a special attention in the paper and recent developments in this area are summarized and discussed. Although advanced understanding of nonreactive wetting is currently available in the literature, complete understanding of reactive wetting is still lacking due to the complexity of the phenomenon. To date, there is no single model that is generally applicable to all reactive wetting systems. Reaction between the liquid drop and substrate can take the form of dissolution, intermetallic formation, or combination of both processes. Factors affecting wettability have been summarized and discussed as well as the major available techniques to investigate wetting behaviour.

1. Introduction

Wettability study is an age-old field with the pioneering theoretical work being credited to Young and Laplace [1]. It is of great importance to many industrial practices such as soldering, coating, and composite manufacturing [2–15]. It gives an indication of the extent of intimacy in contact between a liquid and a solid, making it, therefore, an important behaviour to evaluate the success of an industrial operation like brazing [16].

The earliest approach to wettability study was to use physical theory of surface tension to explain and characterize the phenomenon [1, 17, 18]. This idea is well reflected in Young's equation, which is expressed as [19–22]

$$\gamma_{sv} - \gamma_{sl} = \gamma_{lv} \cos\theta, \tag{1}$$

where θ represents the contact angle measured and γ_{sv}, γ_{sl}, and γ_{lv} represent the solid-vapour, solid-liquid, and liquid-vapour interfacial tension, respectively.

Laplace, on the other hand, gave the mathematical description of the pressure difference between the two sides of a curved surface as [23]

$$\Delta P = \gamma_{lv} \left(\frac{1}{R_1} + \frac{1}{R_2} \right), \tag{2}$$

where R_1 and R_2 represent the two principal radii of curvature.

Dupre defined the work of adhesion between a solid and a liquid, yielding the famous Young-Dupre equation [24]. The significance of the Young-Dupre equation lies in the fact that it relates the work of adhesion to the measurable quantities, surface tension of the liquid, and the equilibrium contact angle, rather than to the inaccessible interfacial tensions involving the solid surface as [25, 26]

$$W_{sl} = \gamma_{lv} (1 + \cos\theta). \tag{3}$$

However, wetting is more complex than as described by (3) [27]. The derivations of Young's and Young-Dupre equations are made under the assumptions of spreading of nonreactive liquid on an ideal (chemically inert, smooth, homogenous, and rigid) solid. It also assumed that the contact angles are large enough to allow accurate measurement. However, these conditions are rarely met in practical situations [24].

1.1. Wetting and Spreading. With respect to the literature, there seems to be no generally accepted distinction between wetting and spreading. According to Kumar and Prabhu [24], wetting is essentially a surface phenomenon in which a liquid covers the surface of a solid when it is placed over such

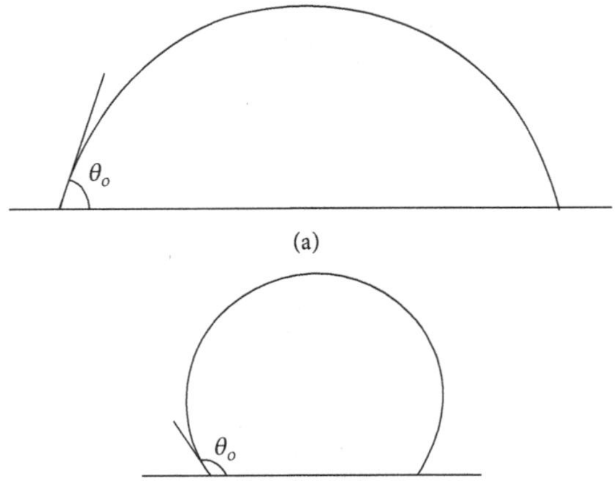

FIGURE 1: Profiles of (a) wetting (b) nonwetting sessile drops resting on horizontal surface [37].

a surface, while spreading is considered as a physical process by which the liquid wets the surface of the solid. Yin et al. [29] considered wetting to be a process confined to the contact line region only and spreading as a macroscopic process which involves the flow of a liquid mass over a solid surface in response to an external effect such as capillary force. However, Sharps et al. [30] stated that for spreading to occur, the apparent contact angle must be zero; otherwise, it is wetting or at best good wetting, depending on how close the angle is to zero. This view [30] represents the clearest distinction between wetting and spreading.

The degree of wetting of a substrate by a liquid is usually quantified by the equilibrium or static contact angle, θ, which the liquid metal drop forms on such a substrate [5, 7, 31–36]. The system is usually described as wetting or nonwetting if the contact angle is lesser or greater than 90° as shown in Figures 1(a) and 1(b), respectively [37].

If the liquid droplet wets the substrate with a finite equilibrium contact angle, θ_e, it is referred to as partial wetting; but when θ_e is equal to zero, it is called complete wetting [38, 39].

In the case of nonequilibrium situation whereby the surface energy of the solid is larger than the addition of the solid/liquid and liquid/vapour interfacial energies, then the difference is referred to as spreading coefficient, S, and there will be complete spreading [1]:

$$S = \gamma_{so} - \gamma_{sl} - \gamma, \quad (4)$$

where γ_{so} is the surface energy of a solid with a dry surface and γ is the same as liquid/vapour surface energy.

1.2. Critical Issues with Contact Angle and Its Measurement. In nonreactive wetting systems, the final contact angle is truly the equilibrium contact angle because of the lack of reactivity in the system. However, in systems undergoing reactive wetting, chemical equilibrium is hardly achieved or takes a long period of time to achieve, making the use of equilibrium contact angle to assess the degree of wetting lose its meaning in some circumstances or cases. For a system in which the reaction leads to the formation of solid phases either at the contact line or throughout the brazing droplet, this might lead to isothermal solidification, thus leading to a premature arrest of the spreading [32]. Also, in cases involving the dissolution of the substrate, the contact angle that is observed at the macroscopic level is the apparent contact angle [40]. The true contact angle can only be observed through metallographic inspection. Therefore, the use of contact angle to quantify the degree of wetting in such systems has to be approached with caution.

1.3. Experimental Methods for Wettability Study. Various techniques abound in the literature to evaluate spreading and measure the equilibrium contact angle. Conventional sessile drop method is probably the most used technique in wettability studies [7, 16, 30]. This is due to its simplicity and ease of use. The conventional sessile drop approach/methodology for studying wettability involves placing a solid particle (pure metal, alloy, etc.) on the substrate and then heating up the assembly until the particle melts and flows on the substrate. However, there are some serious concerns with this approach.

In the course of this process, for the case of an alloy, its composition may change due to evaporation and/or diffusion. Also, it can be very difficult to identify when the solid particle is fully molten, which can affect the reference time for wetting/spreading in the study [41]. Equally, the wettability study is usually carried out at a temperature above the melting point of the solid particle. Therefore, this process cannot be described as strictly isothermal because while the pure metal or alloy is melting, the temperature of the furnace is still increasing or yet to stabilize. This might explain some of the discrepancies in the spreading time reported in the literature.

An improved form of the conventional sessile drop method for wettability study is the "drop-transfer technique" [13, 41]. In this approach, the brazing alloy is heated up until it is completely molten on a plate or material with which the alloy is known to have little or no reactivity. Once the furnace temperature is stabilized, the molten alloy is then "transferred" to the substrate as shown in Figure 2 [41]. However, the drawback attached to this approach is that the drop might be driven at the beginning by the transfer inertia. Also, there might be interaction between the wetting process and the drop transfer mechanism.

Another method being employed in wettability study is the dispensed drop method [28, 42]. In this method, the droplet is extruded through a capillary introducer as shown in Figure 3 [42]. However, this method is not free from the problem of perturbation which occurs as the droplet initiates a contact with the substrate.

One of the impediments to comprehensive understanding of reactive wetting, especially in high temperature experiments, is the difficulty of having in situ and real-time monitoring of the interface between the solid and the spreading droplet; hence, quenching and subsequent microstructural analyses are usually the method used to study the interface [43]. However, a new drop experiment technique has been

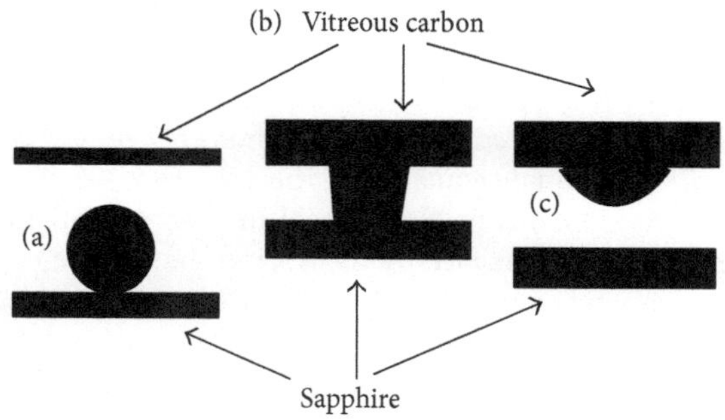

FIGURE 2: Shadow pictures of drop configuration in a transferred drop wetting experiment of Cu-Cr alloy on vitreous carbon: (a) initial configuration, (b) drop configuration upon initial contact, and (c) hanging drop after separation from lower substrate [41].

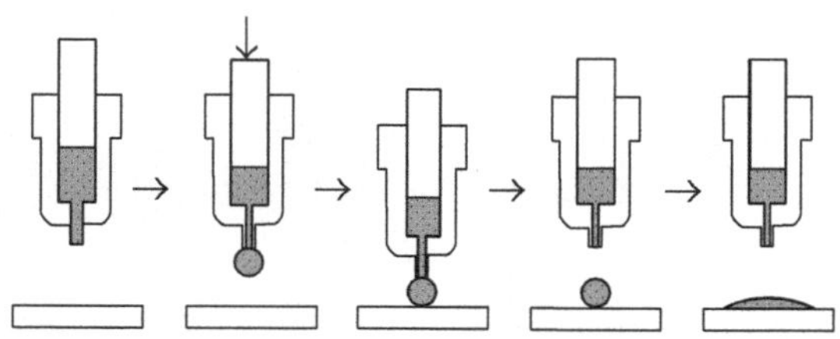

FIGURE 3: Schematic representation of dispense drop method [42].

introduced that allows for real-time monitoring and observation of both the solid-liquid and liquid-vapour interface evolution. This was achieved by performing high temperature reactive wetting experiments in Hele-Shaw cells [29, 44]. This technique was based on drop transfer technique in a very small gap geometry. The transparent cell walls made a real-time observation of the interfaces possible; a schematic of the experimental set-up is shown in Figure 4 [44]. The drop in this experiment is essentially 2-dimensional.

1.4. Contact Angle Hysteresis. For many surfaces, different values of contact angle can be observed or measured depending on whether the measurement is performed with a drop of increasing size (the advancing contact angle, θ_a) or of diminishing size (the receding contact angle, θ_r) [22, 25]. This phenomenon of the existence of more than one contact angle for a particular drop is referred to as contact angle hysteresis [24]. θ_a corresponds to the highest static value that θ can attain [20]. The difference between the advancing and the receding contact angle gives the measure of the hysteresis [45, 46]. Hysteresis plays a special role in practical capillary problems, especially where superhydrophobicity of a surface is required [47–49]. This phenomenon is mostly attributed to surface roughness and chemical heterogeneity of a geometrical surface, among other factors [20, 46, 50, 51]. A very detailed theoretical analysis on contact angle hysteresis has been done by Joanny and de Gennes [52]. Also, Shanahan [22] linked the second order force ripple occasionally observed during dynamic experiments using the Wilhemy plate technique to the phenomenon of contact angle hysteresis leading to the formation of protrusions on the wetting front. There are several other relevant and important

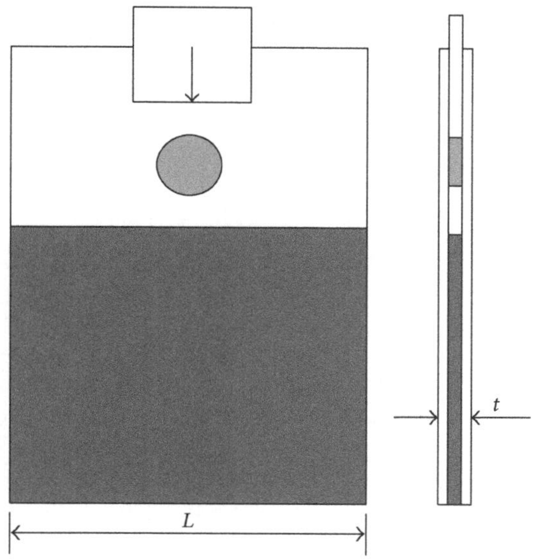

FIGURE 4: Schematic of the two-dimensional experimental set-up [44].

works that have been done on the study of contact angle hysteresis such as [53–58].

1.5. Factors Affecting Wetting. According to Shibata et al. [59], issues such as metal purity, gas-phase oxygen content, surface roughness, and substrate purity must all be controlled in order to accurately carry out a wettability study.

1.5.1. Substrate Surface Roughness. The influence of roughness of the solid surface on the contact angle obtained in practice was first discussed by Wenzel [60, 61]. Wenzel [62] argued that for a drop of water resting on a horizontal solid surface, the specific energy content of the solid interface will, in general, be different for the wetted area under the drop than for the dry area around it. He stated that for identically the same increase in the free liquid area at the upper surface of the drop, a greater amount of actual surface is wetted under it, when the surface of the solid is rough than when it is smooth. Consequently, for the process involving the rough surface, there is a greater net energy decrease to induce spreading, and the rough surface is wetted more rapidly. The same reasoning applies for a water-repellent surface, in which case the dry interface has the lower specific energy. The drop will then spontaneously assume a more spherical form. He proposed an equation that gives a relation between equilibrium contact angle and the apparent angle formed on a rough surface as [24, 61]

$$\cos\theta_w = r\cos\theta, \quad (5)$$

where θ is the equilibrium contact angle on a smooth surface, θ_w is the apparent contact angle on a rough surface (generally known as Wenzel angle), and r is the average roughness ratio, the factor by which roughness increases the solid-liquid interfacial area.

From this equation, it can easily be deduced that for any equilibrium contact angle less than 90°, the apparent contact angle will decrease with increase in roughness and vice versa.

Effect of Surface Roughness on Hysteresis. Kandlikar and Steinke [63] observed the interfacial behaviour of deionized (DI) water on copper and stainless steel to investigate the effect of surface roughness and surface temperature on advancing and receding contact angles. For the studied surface roughness range, always the equilibrium contact angle was found to be in between the two limits of the corresponding advancing and receding contact angles. In the case of copper, the highest advancing contact angle was obtained with the smoothest surface while equilibrium contact angle remained relatively constant with increase in surface roughness. For stainless steel, however, the equilibrium contact angle decreased with increase in surface roughness.

It is worth mentioning the fact that, in reactive wetting systems, the asperities and grooves associated with surface roughness may act as preferable sites for reaction and adsorption. Hence, the effect of surface roughness on wettability might be confounded with the effect of other factors [24].

Hysteresis of a Wenzel drop (a drop that wets the groove of a rough surface) on a rough hydrophobic surface made of polydimethylsiloxane (PDMS) was studied by He et al. [45]. The experiments were conducted to measure both the advancing and receding contact angles. The advancing contact angle for the Wenzel drop on the rough surface was found to lie between 142 and 143°. However, no conclusion could be drawn regarding the receding contact angle of a Wenzel drop because the apparent contact angle kept decreasing with decreasing droplet size down to the smallest drop volume that could be handled. They [45] predicted the advancing contact angle of the Wenzel drop using the Wenzel's formula in the form of

$$\cos\theta^{w}_{\text{adv}} = r\cos\theta^{f}_{\text{adv}}, \quad (6)$$

where f is the area fraction of the peaks of the square pillars on the horizontal surface, subscript "adv" denotes advancing angle, w denotes a wetted contact area, and r is the solid roughness, defined as the ratio of the actual solid-liquid contact area to the contact area projected on the horizontal plane.

Equation (6) gave the value of θ^{w}_{adv} as 149°.

1.5.2. Heterogeneity of the Surface. The equilibrium contact angles are influenced by, among other factors, any surface impurities that may be present on the liquid-vapor interface or the solid surface [64]. Many materials of practical interest are multiphase solids with heterogeneous surfaces [65].

Cassie extended Wenzel's treatment of rough surfaces to heterogeneous surfaces. He derived an important equation analogous to Wenzel's for such surfaces [24, 65, 66] as

$$\cos\phi = \sigma_1\cos\theta_1 + \sigma_2\cos\theta_2, \quad (7)$$

where ϕ is the contact angle on the heterogeneous surface, σ_1 is fraction of the surface having the intrinsic contact angle θ_1, and σ_2 is the fraction of the surface having the intrinsic angle θ_2.

Effect of Surface Heterogeneity on Contact Angle Hysteresis. He et al. [45] equally quantified the hysteresis of a Cassie drop (a drop that sits on the peak of the surface roughness) on a rough PDMS surface by measuring the advancing and receding contact angles. An advancing contact angle between 152 and 153° was observed experimentally for a Cassie drop on the rough substrate. The corresponding receding contact angle for the Cassie drop was found to be about 132°. They [45] predicted the advancing contact angle of the Cassie drop using Cassie formula of the form:

$$\cos\theta^{c}_{\text{adv}} = f\cos\theta^{f}_{\text{adv}} + f - 1, \quad (8)$$

where c represents composite contact, f is the area fraction of the peaks of the square pillars on the horizontal surface, and subscript "adv" denotes advancing angle.

Using (8), a value of 153.5° was obtained for θ^{c}_{adv}.

They [45] also estimated the receding contact angle of the Cassie drop using an equation of the form:

$$\cos\theta^{c}_{\text{rec}} = f\cos\theta^{f}_{\text{rec}} + f - 1, \quad (9)$$

where rec represents receding angle.

An estimated value of 143° was obtained for θ^{c}_{rec} from (9). They [45] equally estimated the receding contact angle of the Cassie drop using a model proposed by Patankar [67]. A value of 127.6° was obtained for θ^{c}_{rec}, which was much more close to the experimental value. It must be noted that in the model proposed by Patankar [67], the droplet was assumed to leave a thin film of liquid behind on the peaks of the pillars of the surface roughness as against leaving behind a dry surface as assumed in (9).

1.5.3. Atmosphere. The surface tension of liquid metals is very sensitive to impurities and, in particular, to the oxygen concentration in the liquid phase [68]. If the liquid filler metal is to wet and flow over the base metal, any intervening oxide layers must be removed. Three methods are used in practice to reduce and control the formation of oxide layers in wettability studies: fluxing with molten salts, heating in a reducing atmosphere, and heating in a vacuum [2, 69].

Takao et al. [70] investigated the effect of fluxes (ULF-300R and PO-Z-7) on the ability of Sn-3.5Ag to wet Cu substrate at 270°C. UFL-300R is a more activated flux containing halogen while PO-Z-7 has no halogen. For UFL-300R, a final contact angle (θ) and interfacial liquid-flux tension (γ_{lf}) of 38° and 0.406 N/m were obtained, respectively. While for PO-Z-7, the values of θ and γ_{lf}, were reported to be 43° and 0.470 N/m, respectively. The relative decrease of 5° in the contact angle was attributed to the relative 0.064 N/m decrease in γ_{lf} values of the two fluxes.

López and Kennedy [16] investigated the effect of a K-Al-F-based flux on the spreading of Al on TiC, at temperatures up to 900°C, in Ar and in air. They reported that, while obtuse contact angles were observed without flux, the flux facilitated rapid spreading to a perfect wetting condition, in both Ar and air.

In all cases involving the use of reducing atmosphere, the active reducing constituent is hydrogen or carbon monoxide [2]. For a few metals, hydrogen is satisfactory as a reducing atmosphere [71]. Practical limits are set to the use of hydrogen reduction: firstly in techniques for drying the gas and secondly in keeping it dry because it takes up moisture from the surfaces of the tubes through which it passes into the test chamber [69].

1.5.4. Temperature. The wetting behaviour of a liquid on a solid is sensitive to temperature changes as temperature affects a number of properties of liquid such as viscosity and surface tension [72].

Samarasekera and Munir [73] investigated the contact angles formed by liquid tin, indium, and gallium on zirconium and niobium over temperature range of 303–923 K. They observed three characteristic regions of the temperature dependence of contact angles. A steady-state region in which θ was relatively independent of temperature was preceded and followed by regions in which θ decreased rapidly with increasing temperature.

An experimental investigation into contact angle temperature dependence for water droplets on practical aluminum surfaces was carried out by Bernardin et al. [64]. They observed two distinct temperature-dependent regimes. In the lower temperature regime, below 120°C (surface temperature), a relatively constant contact angle of 90° was observed; while in the high temperature regime, above 120°C, the contact angle decreased in a fairly linear manner. It is worth noting that they [64] used a pressure vessel to increase the water temperature up to 170°C.

Feduska [74] studied the reactions between five high-temperature base metals and six high-temperature brazing alloys in three different atmospheres at nominal and superheat brazing temperature levels. He observed that the wettability of the substrates by the brazing alloys decreased as the brazing temperature was increased above the nominal level. This trend was attributed to an increase in reactivity between the brazing alloy elements and the base metal at the higher brazing temperature.

Also, contact angle of pure aluminum on the surface of polycrystalline alumina substrate was studied in the temperature range of 750–1100°C by Sangghaleh and Halali [75]. In this system, contact angle was observed to reduce with increasing temperature. In temperatures in excess of 1050°C, the condition of the system changed from nonwetting to wetting.

With reference to the works of López and Kennedy [16], the spreading process was seen to exhibit a temperature dependency. The spreading time was considered to follow an Arrhenius behaviour of the form:

$$\frac{1}{t_0} = K \exp\left(-\frac{E_a}{RT}\right), \tag{10}$$

where t_0 corresponds to the spreading time, with K being a preexponential constant.

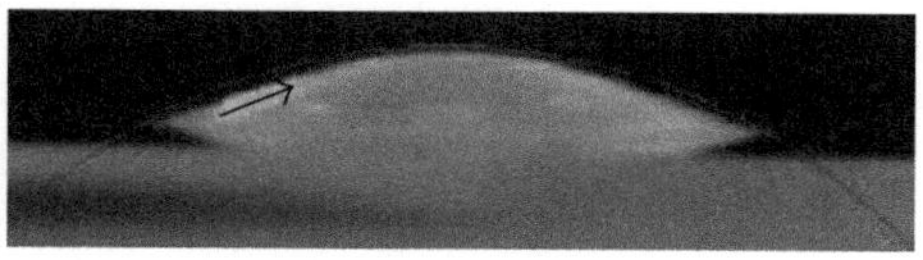

(a)

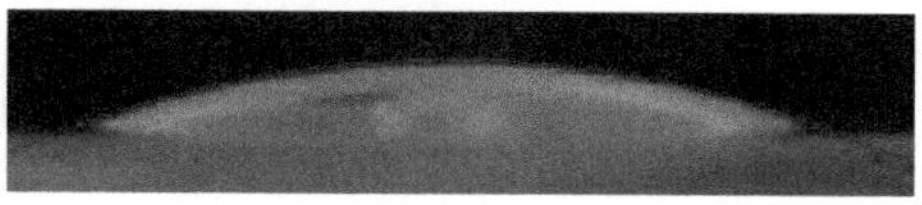

(b)

FIGURE 5: Spreading of (a) pure Cu and (b) saturated Cu-Si on Si substrates after 50 ms [29].

1.5.5. Alloying Elements. Yang et al. [76] investigated, in detail, the effects of Ni and Cr contents and wetting temperatures on the wettability and the wetting mechanisms of copper on tungsten substrate. The results showed that the wetting angles of Cu on the W substrate decreased with an increase in the content of Ni or Cr and also decrease with increase in the wetting temperatures.

Equally, wettability at a liquid Al-alloy/SiC interface was evaluated by the sessile drop method at 750°C by Candan [77]. The wetting angle, θ, of the sessile drop on SiC substrate was equally reported to decrease with the addition of Pb, Mg and Ca to pure Al. The reduction in θ obtained with the Al-Pb alloy drop was proportional to the reduction in surface tension, γ_{lv}, of Al; whereas for Al-Mg and Al-Ca alloys, the reduction in θ was greater than the reduction in γ_{lv} of Al. The reduction in contact angle was attributed to reactions that took place at the Al-alloy/SiC interface.

Yu et al. [78] studied the interfacial microstructure and wetting property of Sn-Zn-Cu solders with Cu substrate. They reported that wettability improved remarkably with increasing Cu content in the solder.

1.6. Marangoni Convention. In reactive wetting systems where there is alloying between the chemical species of the liquid and the solid substrate, there is a likely tendency for solute-capillary effect, if the difference in surface tension between the two chemical species is large [28, 29]. In the early stages of reactive regime in wetting where there is dissolution of the substrate into the liquid droplet, the concentration of the solute is usually larger at the contact line/solid-liquid interface than at the top of the droplet. This concentration gradient is likely to set up a solute-capillary flow. The presence of this effect was demonstrated in the spreading of pure Cu and saturated Cu-Si drops on pure Si substrate after 50 ms from the initiation of wetting as shown in Figure 5 [29]. This observation was related to the fact that the surface tension of pure Cu (1280 mNm^{-1}) is higher than that of saturated Cu-Si solution (840 mNm^{-1}) at 1100°C. The effect of early dissolution of Si into the drop at the contact line reduces the surface tension at the contact line relative to that at the top of the drop. Therefore, spreading will be retarded at this stage. On the other hand, spreading will be accelerated if the surface tension of the solute is higher than that of the liquid [28].

1.7. Nonreactive Wetting. When a liquid spreads on a dense and nonporous solid substrate without any reaction, it is referred to as a nonreactive or an inert wetting [79]. For this kind of system, the interface remains flat and unchanged materially [30]. The system is in chemical equilibrium and the degree of wetting is determined by the Young's equation for a flat and nondeformable solid [19]. Also, the spreading time for a metallic millimeter sized droplet in a nonreactive system in which the equilibrium contact angle value is higher than 20° is reported to be less than 10^{-1} s [28, 80]. For the case of complete wetting of a solid surface by a liquid, wetting dynamics are reportedly dominated by a precursor film which spreads ahead of the main body of fluid [1, 38]. A precursor film can either be a bulk liquid phase identical to the drop phase or an adsorption layer [51].

Contact line motion is normally correlated with the dynamic contact angle to describe the kinetics of a wetting system using both theoretical (hydrodynamic, molecular kinetics, and combined hydrodynamic-molecular kinetics) and empirical approach [6].

For droplet spreading where the effect of gravitation is minimal, the macroscopic shape of the droplet is found to be close to a spherical cap [1, 81]. In the case of complete wetting, at late times of spreading, the shape of the drop is spherical and the flow dynamics can be described by Tanner's law [82, 83]:

$$\frac{r}{R} \sim \left(\frac{\gamma t}{\eta R}\right)^{1/10}, \tag{11}$$

where r is the instantaneous radius of the wetted area, R is the initial radius of the drop at the start of spreading, t is time, η is viscosity, and γ represents the liquid surface tension.

Tanner's law does not apply to early stages of spreading where the shape of the drop is completely different (high curvature at the contact line) from that at late time [15]. At short time of spreading, Biance et al. [81] reported the presence of an inertial regime in the spreading drop of water on glass plate. The duration of this regime was found to be in the order of 1 ms for a millimetric water drop and the wetting rate followed a square root growth of the form:

$$r^2(t) \sim \sqrt{\frac{\gamma R}{\rho}} t. \tag{12}$$

The characteristic time or duration of this inertial regime was found to be mainly dependent on the drop size.

In partial wetting experiment of different water-glycerol mixtures on solid surfaces with various degree of wettability, early-time inertially dominated wetting was also reported [39]. The spreading dynamics was fit with a power law in the form of:

$$\frac{r}{R} = C\left(\frac{t}{\tau}\right)^{\alpha}, \quad \text{where } \tau \equiv \sqrt{\frac{\rho R^3}{\gamma}}, \tag{13}$$

where C is a prefactor and α is an exponent.

They [39] found that for low equilibrium angles ($\theta_e <$ 10°), $\alpha \approx 0.5$, and the spreading dynamics is similar to that of complete wetting. However, both α and C were found to decrease with increase in contact angle.

In the short time dynamics of viscous drop spreading investigated by Eddi et al. [15], however, spreading did not follow a pure power-law growth. The spreading velocity was, instead, found to decrease logarithmically in time and the dynamics was reported to be identical to that of coalescing viscous drop. The initial stage of wetting was also found to be independent of substrate wettability. The effective exponent for coalescence was derived as

$$\alpha = \frac{\ln(r/R)}{\ln(r/R) - 1}. \tag{14}$$

Lavi and Marmur [6] investigated the partial spreading of organic liquids on coated silicon wafer. An exponential power law was suggested for the kinetics in systems with partial wetting as

$$\frac{A}{A_f} = 1 - \exp\left(-\frac{K}{A_f}\tau^n\right), \tag{15}$$

where A is the normalized and dimensionless drop base area, A_f is the final, equilibrium value of the normalized wetted (solid-liquid) area, τ is the dimensionless time, and K and n are empirical constants.

Another empirical model for describing the mobility relations of partially wetting systems is [43]

$$\frac{\cos\theta_0 - \cos\theta_D}{\cos\theta_0 - 1} = \tanh\left(4.96\text{Ca}^{0.702}\right), \tag{16}$$

where Ca is the capillary number (ratio of viscous force to surface tension force), while θ_0 and θ_D represent the static and dynamic contact angles, respectively. There are other relevant works that have been done on nonreactive wetting [54, 84–94].

2. Reactive Wetting

This type of wetting is characterized with dissolution of the substrate and/or formation of intermediate compounds at the interface between the liquid alloy and the substrate [29]. The wetting kinetics, in this case, is controlled not by viscous dissipation but by the rate of interfacial reaction and the spreading time is reported to be in the range of 101–104 s [41, 42, 80, 95–97]. A spreading time of 0.06 s was reported for eutectic Ag-Cu liquid droplet on Cu substrate by Sharps et al. [30]. Reactive wetting is very important in metal-metal joining processes such as brazing and soldering [43].

The main difference between wetting in purely dissolutive systems and wetting in compound forming systems is the formation of intermetallic compounds in the latter, which tend to arrest the spreading at the triple line [44]. Interfacial reactions have often been suggested to lead to the reduction of the corresponding interfacial energy and attempts have been made to relate the degree of wetting with the free energy of interfacial reactions by replacing the γ_{sl} term in (1) by [98]

$$\gamma_{sl}(t) = \gamma_{sl} + \Delta G_r(t), \tag{17}$$

where $\Delta G_r(t)$ corresponds to the change in Gibbs energy released per unit area by the reaction in the "immediate vicinity of the solid/liquid interface" [98].

2.1. Purely Dissolutive Reactive Wetting. This kind of wetting is synonymous with high degree of dissolution of the substrate in the molten alloy, leading to a nonplanar solid-liquid interface [43]. It occurs not only in metal/metal systems but also in metal/ceramic systems.

Protsenko et al. [40] studied the wetting of single crystal silicon wafers (0 0 1) by pure copper and silicon-saturated CuSi alloy, using the dispensed drop technique at 1100°C under high vacuum. This was done to investigate both dissolutive (nonequilibrium) and nonreactive (equilibrium) wetting for the same system. The spreading time for the nonreactive CuSi/Si was observed to take place in a few milliseconds with an average final visible contact angle of 19±3°. In the Cu/Si system, wetting occurred in two different stages. In the first stage (nonreactive), a visible contact angle close to 40° (angle formed by pure Cu on nearly unreacted Si) was observed after a few milliseconds. The second stage (dissolutive) led to the formation of a visible contact angle of 5° on a nonplanar interface within 1-2 s. This second reactive or dissolutive stage was subsequently broken down into two steps. The first step involved the decrease of the contact angle from 40° to 19° which is formed on Si by Si-saturated copper. This angle decrease took place rapidly in about 40 ± 10 ms. At this point, the solid Si was reportedly in equilibrium with the liquid at the solid-liquid interface but not in equilibrium with the bulk liquid. The implication of this is that dissolution rate at the solid-liquid interface was faster than the transport rate of the Si atoms from the interface to the bulk liquid. The second reactive step led to the decrease of the visible contact angle from 19° to 5° couple with the formation of a crater at the interface. The variation of the drop base diameter and visible contact angle with time, on a logarithmic scale, is shown in Figure 6 [40]. It is important to know that Protsenko et al. [99] obtained similar result on (1 1 1) Si wafers.

Recently, an interesting study was made into the wetting of Ag-Cu alloys on Cu substrates at 900°C, using dispensed drop technique [28]. Equilibrium and nonequilibrium systems were considered by using Cu-saturated and non-Cu-saturated liquid. The plotted variation of the drop base diameter, d, and contact angle, θ, with time on a logarithmic scale is shown in Figure 7 [28].

For both the equilibrium and nonequilibrium systems, the spreading curves were nearly overlapping up to 20–30 ms. Based on this observation, the early stage of wetting was described as nonreactive, suggesting the absence of dissolution and its effects in the nonequilibrium system at this stage. However, after 30 ms, the spreading of the non-Cu-saturated drop became faster than the Cu-saturated drop. This observed faster spreading in the nonequilibrium system was attributed to the dissolution of Cu into the Ag-Cu alloy. In both cases, the spreading time exceeded 100 s. This observed long spreading time in the equilibrium Ag-Cu/Cu system is unusual, considering the lack of any significant reaction between the solid and liquid phases. The long spreading time observed in the equilibrium system was attributed to the very low equilibrium contact angle, about 1°, of the system.

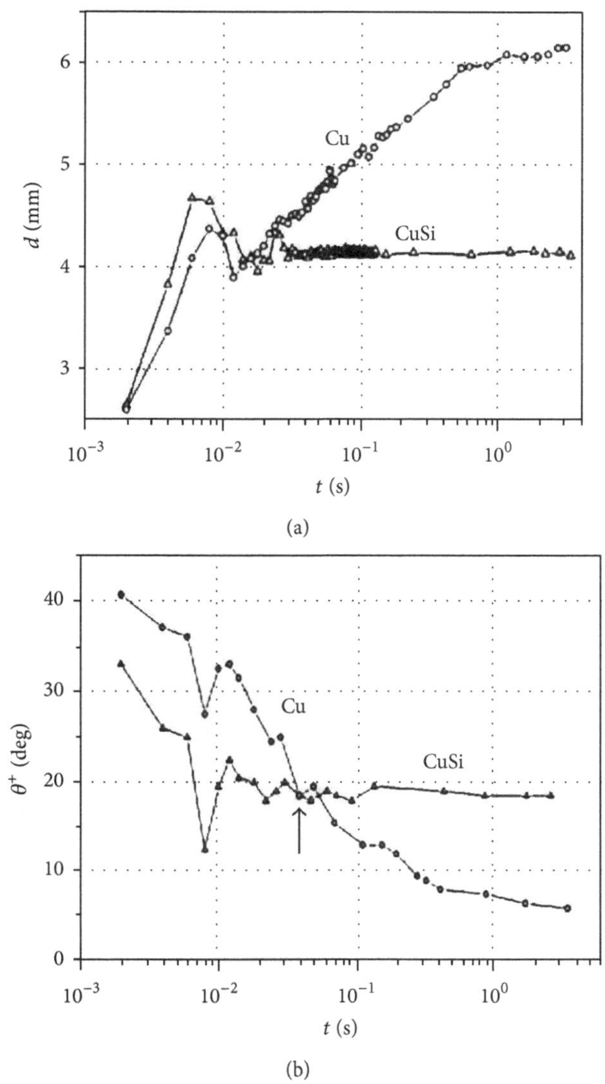

FIGURE 6: Drop base radius (a) and visible contact angle (b) as a function of time plotted on a logarithmic scale for Cu and Cu presaturated in Si on a (0 0 1) Si surface at 1100°C. The arrow in the $\theta^+(t)$ diagram indicates the time when the visible contact angle equalled the value of 19° observed with Si presaturated Cu (CuSi) [40].

Three distinct spreading stages were observed in the non-Cu-saturated system as shown in Table 1 [28]. The highlight of this is the presence of nonreactive spreading both before and after the dissolutive stage. Schematic representation of the spreading stages for both nonequilibrium and equilibrium Ag-Cu/Cu systems are shown in Figure 8 [28].

2.2. Compound Forming Systems. One way of improving the wetting of a substrate by a metal is to alloy such a metal with chemical species that react with the substrate to form a reaction product that is more wettable than the original substrate [41, 42, 76, 97]. One of the earliest efforts

Table 1: Spreading regimes in nonequilibrium Ag-Cu/Cu system [28].

Time interval (s)	Drop base diameter (d/d_{dr}) evolution	Spreading regime	Average spreading rate (m/s)
0–0.03	0–2.6	Nonreactive	10^{-1}
0.03–1	2.6–4.6	Dissolutive	10^{-3}
1–100	4.6–7.3	Diffusion into the solid	10^{-5}

$T = 900^\circ$C.

towards understanding this type of reactive wetting was made by Aksay et al. [19]. Using a thermodynamic approach, they concluded that, under chemical nonequilibrium, mass transfer across the interface leads to a transient decrease in the corresponding interfacial energy.

Landry and Eustathopoulos [95] made a significant contribution towards understanding this reactive wetting by relating the experimentally obtained wetting radius, $R(t)$, curves for Al/C_v with the mechanism of growth of the interfacial reaction product (Al_4C_3) at 1100 K. The main feature of the $R(t)$ curve was the presence of a nearly linear spreading region/regime during which the liquid/solid interface was completely covered by a layer of (Al_4C_3). The wetting kinetics was suggested to be controlled by the kinetics of lateral growth of Al_4C_3 at the triple line.

Also in the experimental study of wetting of Cu-Si alloys on carbon substrate [98], linear spreading was observed. A 3-step change in contact angle and base radius were noticed. Spreading was described as rapid in the first stage before slowing down significantly in the second stage (linear spreading regime). The drop radius and contact angle remained steady in the third stage. The presence of a continuous layer of silicon carbide was noticed at the solid/liquid interface.

The concept of diffusion-limited reactive wetting was introduced in 1999 by Voitovitch et al. [41] in the study of the spreading kinetics of Cu-Cr alloys on carbon. A dense reaction layer, Cr_7C_3, was formed at the solid/liquid interface, with the thickness depending on Cr concentration and the time of holding. The observed triple line velocity was not constant with time, it was found to be a function of the contact angle. A final contact angle of 40° was observed, which was reported to be nearly equal to the contact angle of Cu-Cr alloy on a substrate of chromium carbide. The diffusion of chromium to the triple line was concluded to be the rate limiting factor in the spreading kinetics of Cu-Cr/C system.

2.2.1. Nonreactive Stage in Reactive Wetting. Different stages have been observed in different reactive systems [7, 28, 40, 42, 80, 98, 100]. However, there has been a common observation among researchers of the existence of nonreactive stage at the onset of this wetting which is characterised with fast spreading with no reaction with the substrate, thereby leaving the solid/liquid interface in its original flat state. However, until recently, there was no physical proof of this due to lack of in situ and real-time observation of the solid/liquid interface.

Yin et al. [29] performed experimental investigation into dissolutive wetting of Bi-Sn system in Hele-Shaw cells. The obtained results, as shown in Figure 9 [29], tend to support the presence of this nonreactive stage at the onset of reactive wetting. At the early stage of wetting (less than 100 ms), no change in the solid/liquid interface was observed.

2.3. Cold Wetting. Application of hysteresis phenomenon and reactive wetting to liquid droplets spreading on solid surfaces is well documented in the literature [101–105]. A surface having primarily polar groups will have a strong affinity for water and a low contact angle will be formed on it by water. Such a surface is referred to as being hydrophilic. On the other hand, if the surface is made up of nonpolar groups, a large contact angle will be formed on it by water and such a surface is termed hydrophobic [25]. The idea of free and rapid motion of liquid droplets on solid surface is essentially based on the ability of a liquid to modify a hydrophilic surface by some reactive mechanism and render if hydrophobic. The droplet would then be out of equilibrium between the hydrophilic surface it meets at the front and the hydrophobic trail it leaves behind, thereby leading to a free running droplet [102].

Chaudhury and Whitesides [105] equated the unbalanced force experienced by a drop placed on a surface with gradient in surface tension, as shown in Figure 10 [105], to the unbalanced Young's force (dF_Y), as given in the following [105]:

$$dF_Y = \left[(\gamma_{SV} - \gamma_{SL})_A - (\gamma_{SV} - \gamma_{SL})_B\right] dx, \tag{18}$$

where dx is the thickness of the section of the drop.

In terms of contact angles at points A and B, the unbalanced Young's force was represented as

$$dF_Y = \gamma_{LV} \left(\cos\theta_A - \cos\theta_B\right) dx. \tag{19}$$

They [105] stated that the presence of gradient in surface tension only on a solid surface is not sufficient to ensure motion of liquid drops, the surface must also have low hysteresis in contact angles and be free of defects that can pin the edge of the drop. By producing gradients in chemical compositions and surface tension on solid surfaces through silanization reaction, they [105] were able to cause 1-2 μL drops of water to move up a 15° slope along the direction of increasing surface free energy with average velocities of 1 to 2 mm/s.

Bain et al. [101] put to test the concept of reactive flow of liquid droplet on solid surface by experimenting with a drop of Decane containing a fluorinated fatty acid, $CF_3(CF_2)_6CO_2H$ (PFOA), on an ordinary glass microscope slide inclined at an angle of 43° to the horizontal. The drop was observed to flow up the slide as shown in Figure 11 [101].

The chemistry occurring at the surface of the glass was identified to be the key behind this behaviour. The PFOA was reported to adsorb from Decane to form an oriented

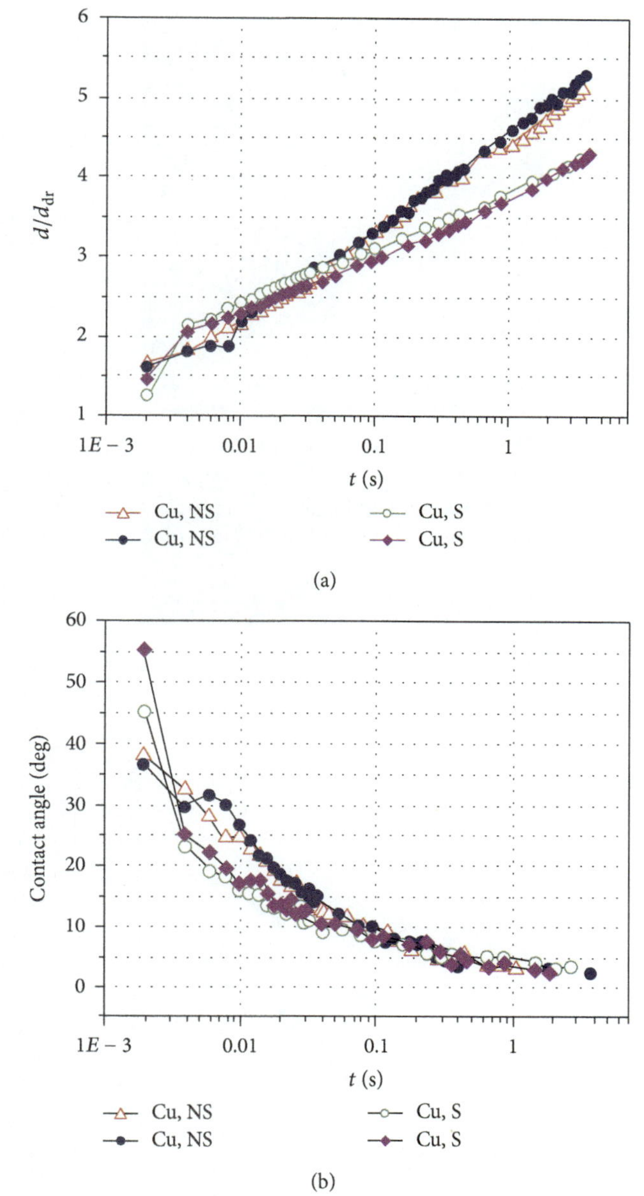

FIGURE 7: Normalised drop base diameter (a) and contact angle (b) as a function of logarithm of time for two saturated (S) and two nonsaturated (NS) Ag-Cu drops spreading on monocrystalline Cu at $t < 4$ s [28].

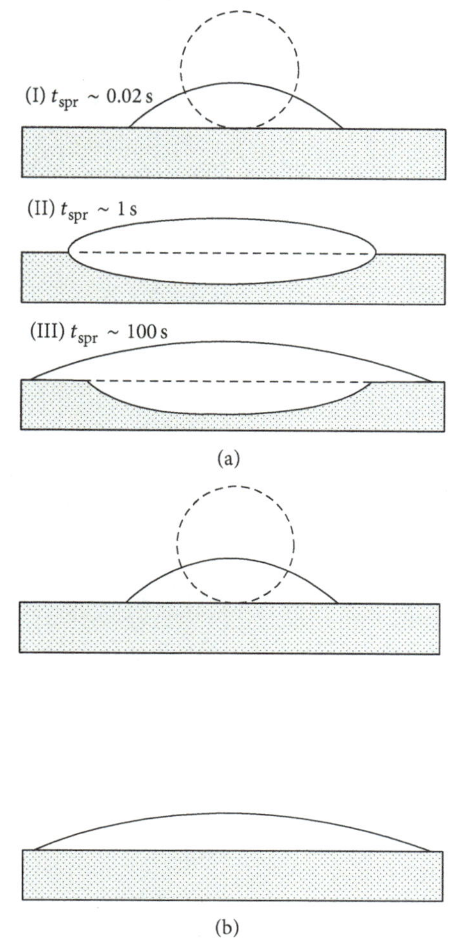

FIGURE 8: Schematic representation of spreading stages in (a) nonequilibrium and (b) equilibrium Ag-Cu/Cu systems [28].

monolayer attached to the glass through the carboxylic acid group. The outer surface of the monolayer allegedly consists of oil-repelling CF_2 and CF_3 groups. While Decane beads up on a fluorinated surface because of its lower surface energy; it spreads on a clean surface of glass.

The idea of reactive wetting in free-running droplet was also established by dos Santos and Ondarçuhu [102]. They experimented with the flow of droplets of n-alkanes (n-octane and n-dodecane) containing 1H,1H,2H,2H-perfluorodecyltrichlorosilane on glass surface. They reported that the silane molecules can form dense grafted monolayers on silicon or glass, thereby rendering the surface hydrophobic. Two types of droplets were used for the experiment, based on the capillary length ($\kappa^{-1} = (\gamma/\rho g)^{1/2}$; the capillary length: γ is the surface tension of the liquid and ρ is its density). After dropping the droplet on the glass surface, motion was initiated by pushing it with a pipette. The droplet was observed to move on the substrate and the self-supporting movement ended only when the droplet had no more hydrophilic surface available. The movement was equally observed on an inclined substrate with the droplet running uphill. For droplets with length smaller than the capillary length, they were observed to move with constant velocity V and constant length L. When the silane concentration of the small droplets was changed, the velocity initially increased with increase in concentration before reaching a saturation level at high concentrations, as shown in Figure 12 [102].

There are other important contributions to the field of liquid cold wetting [106, 107].

2.4. Modelling of Kinetics of Reactive Systems. Due to the lack of thorough understanding and complexity of reactive wetting systems, there is still no single model that is generally applicable to all reactive systems.

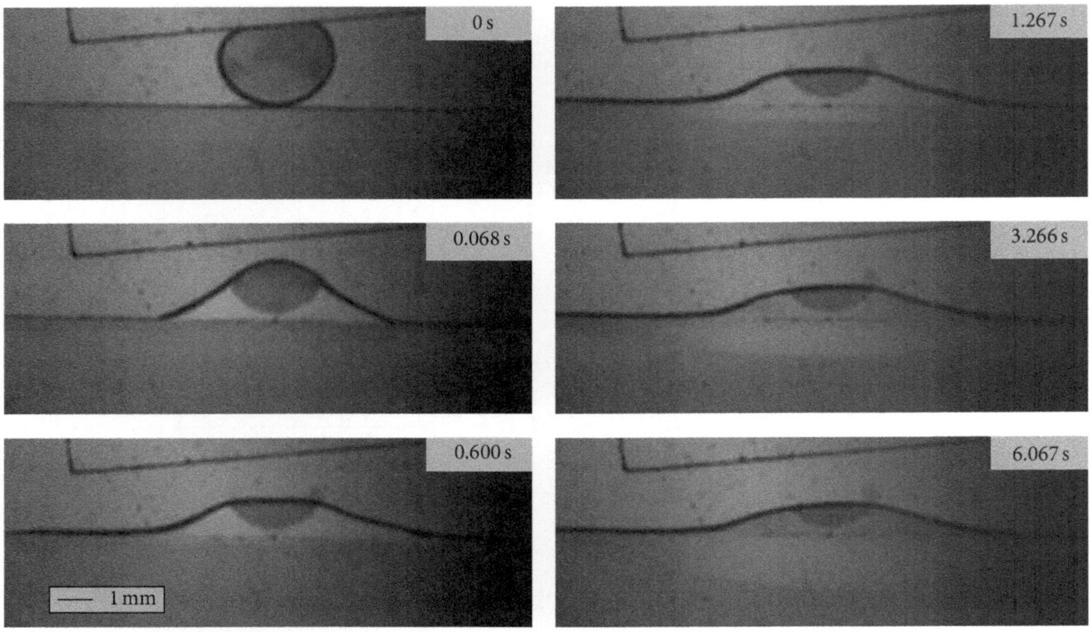

FIGURE 9: Images of a pure Sn drop wetting pure Bi in a Hele-Shaw cell [29].

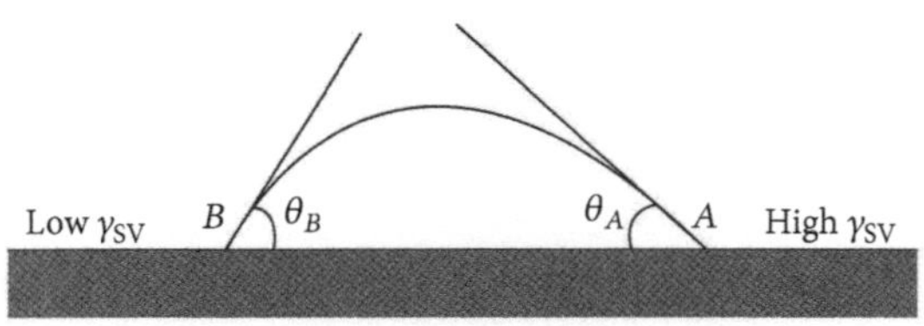

FIGURE 10: Idealized diagram of a thin cross section of a liquid drop on a gradient surface [105].

In the Cu/Si system studied by Protsenko et al. [40], the wetting rate was modelled by an equation of the form:

$$U = \frac{4}{3\theta_Y^2} \cdot \frac{D}{\delta} \left(\theta^+ - \theta_F^+\right), \tag{20}$$

where θ_Y, θ^+, θ_F^+, and D/δ represent Young's contact angle, instantaneous visible contact angle, final visible contact angle, and mass transport parameter, respectively.

When the transport of Si in the liquid was assumed to be purely diffusion based, the experimental data fit the equation well but with a much higher than usual value for the diffusion coefficient, indicating that the diffusion coefficient lost its physical meaning and became an adjustable parameter.

Also, Kozlova et al. [28] reported that the spreading kinetics of saturated Ag-Cu alloy on Cu followed the relation:

$$R^n \sim At, \tag{21}$$

where R represents the droplet base radius, t is the time, and n and A are empirical constants.

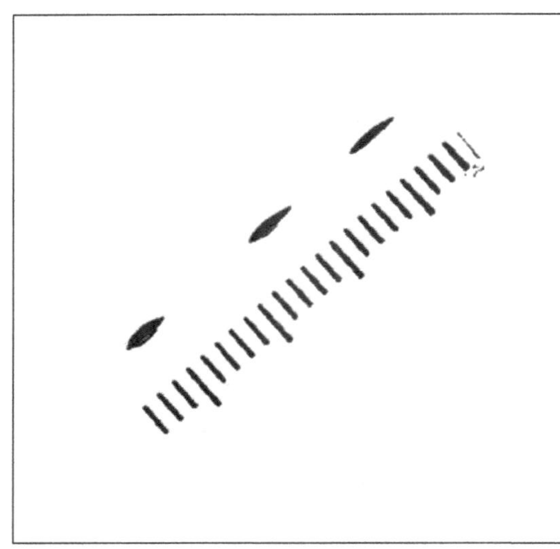

FIGURE 11: Photograph of a 2 μL drop of Decane containing 1.5 mM pentadecafluorooctanoic acid (PFOA) flowing up a glass microscope slide inclined at an angle of 43° from the horizontal. Three photographs taken at 0.45 s intervals have been superimposed in the same image with the first photograph being on the left [101].

The diffusion-limited spreading kinetics of Cu-Cr/C_v was equally described with an equation of the form [41]:

$$\frac{dR}{dt} = \frac{2DF_{(t)}}{en_v} \left(C_0 - C_e\right)\theta, \tag{22}$$

where D is the diffusion coefficient in the liquid phase, n_v is the number of moles of reactive solute per unit volume of the reaction product, e is the reaction thickness at the triple line, $F_{(t)}$ is the dynamic spreading force, C_o is the bulk drop

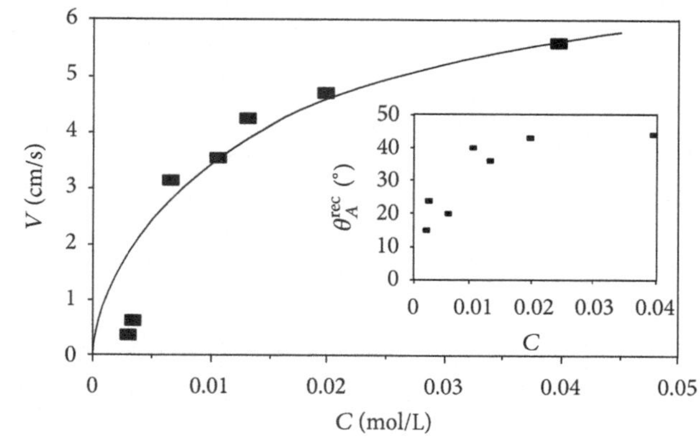

FIGURE 12: Plot of the velocity of octane droplets of length $L = 3.5$ mm as a function of the concentration in silane. Inset: receding contact angle of pure octane measured on the hydrophobic trail left by the same droplets [102].

concentration, and C_e is the concentration of the reactive solute at the triple line.

However, the experimental results obtained for this system, Cu-Cr/C_v, disagreed with this model. The intercept of the line described by (22) should be at $\theta = 0$. But for this system, it was at finite angle of 40°. Equation (22) was based on the assumption that all solute flux is consumed by the reaction at the triple line, therefore neglecting any thickening of the reaction layer behind the triple line. This was cited as the reason for the disagreement between the model and the experimental results for this system [80]. On the other hand, experimental results obtained for CuSn-Ti/C_v system were successfully modelled by (3) [80].

For reaction-limited spreading, Dezellus et al. [42] suggested that instantaneous contact angle, $\theta(t)$, could be defined by the relation:

$$\cos\theta_F - \cos\theta = (\cos\theta_F - \cos\theta_0)\cdot\exp(-kt) \qquad (23)$$

with k being the kinetic constant.

For liquid droplet spreading in cold wetting systems, Raphael [108] showed that for an ideal surface, the droplet will reach a constant velocity given by

$$V = \frac{\gamma}{6l\eta}\tan\theta^*\left(\cos\theta_{Pe} - \cos\theta_{Ae}\right), \qquad (24)$$

where γ is the surface tension of the liquid, η is its viscosity, l is a logarithmic factor ($l = \ln(x_{max}/x_{min})$, the ratio of macroscopic and molecular lengths), θ_{Pe} is the equilibrium angle on the hydrophilic surface (P), θ_{Ae} is the equilibrium angle on the hydrophilic surface (A), and θ^* is the dynamic contact angle of the droplet given as

$$2\cos\theta^* = \cos\theta_{Ae} + \cos\theta_{Pe}. \qquad (25)$$

Dos Santos and Ondarçuhu [102] applied (24) and (25) to real surface by replacing θ_{Pe} and θ_{Ae} by θ_P^{adv} (advancing contact angle on P) and θ_A^{rec} (receding contact angle on A), respectively, and they succeeded in fitting the obtained experimental results.

For a droplet with size lesser than the capillary length, Brochard-Wyart and de Gennes [109] presented a model predicting the steady state velocity V as a function of the drop size L, while incorporating the reaction kinetics as

$$L = V\tau\ln\left(\frac{V_1}{V_1 - V}\right), \qquad V_1 = \frac{\gamma_1\theta^*}{6l}, \qquad (26)$$

where τ is the reaction time and γ_1 is a constant.

3. Conclusion

The level of understanding of inert wetting is quite good. Models that can successfully describe the process are available in the literature. Reactive wetting systems, on the other hand, are quite complex to study. The nature of reaction in these systems could take the form of dissolution, intermetallic formation, or combination of both processes. Microstructural changes in such systems can occur within a fraction of seconds. Equally, different stages are involved in reactive wetting. The early stage of reactive wetting is usually very fast and independent of any reaction. However, the kinetics of the latter stages is usually controlled by the nature of the reaction. It is difficult, if not impossible, to find a single model that will perfectly describe the kinetics of all the stages involved in any reactive system. Variation of contact angle, with respect to advancing and receding droplets, has been reflected upon. Reactive wetting, as it applies to cold systems, has also been reviewed.

The Hele-Shaw reactive wetting experiment has the potential to radically advance the state of knowledge in interfacial dynamics of reactive systems. Identification and clarification of the different wetting regimes in the Cu-Si/Si system is a great development considering the importance of this system to the electronic industry. Also, the identification of two nonreactive spreading regimes before and after the dissolutive spreading in Ag-Cu/Cu system is a new development. Equally, the effect of solutal Marangoni convection during the early stage of reactive regime has been clearly identified in the literature. The development of reaction and diffusion limited spreading models will also go a long way in shedding more light on compound-forming reactive wetting systems.

Conflict of Interests

The authors declare that there is no conflict of interests of any sort that might have influenced the discussions or conclusions of this paper.

References

[1] P. G. De Gennes, "Wetting: statics and dynamics," *Reviews of Modern Physics*, vol. 57, no. 3, pp. 827–863, 1985.

[2] D. R. Milner, "A survey of the scientific principles related to wetting and spreading," *British Welding Journal*, vol. 5, pp. 90–105, 1958.

[3] C. A. Leon, V. H. Lopez, E. Bedolla, and R. A. L. Drew, "Wettability of TiC by commercial aluminum alloys," *Journal of Materials Science*, vol. 37, no. 16, pp. 3509–3514, 2002.

[4] M. M. Schwartz, *Brazing*, ASM International, Materials Park, Ohio, USA, 1987.

[5] S. Amore, E. Ricci, G. Borzone, and R. Novakovic, "Wetting behaviour of lead-free Sn-based alloys on Cu and Ni substrates," *Materials Science and Engineering A*, vol. 495, no. 1-2, pp. 108–112, 2008.

[6] B. Lavi and A. Marmur, "The exponential power law: partial wetting kinetics and dynamic contact angles," *Colloids and Surfaces A*, vol. 250, no. 1–3, pp. 409–414, 2004.

[7] G. W. Liu, F. Valenza, M. L. Muolo, G. J. Qiao, and A. Passerone, "Wetting and interfacial behavior of Ni-Si alloy on different substrates," *Journal of Materials Science*, vol. 44, no. 22, pp. 5990–5997, 2009.

[8] A. Shafiei, P. Abachi, K. Dehghani, and K. Pourazarang, "On the formation of intermetallics during the furnace brazing of pure titanium to 304 stainless steel using Ag (30–50%)-Cu filler metals," *Materials and Manufacturing Processes*, vol. 25, no. 11, pp. 1333–1340, 2010.

[9] T. D. Blake and K. J. Ruschak, "A maximum speed of wetting," *Nature*, vol. 282, no. 5738, pp. 489–491, 1979.

[10] D. Chatain and W. C. Carter, "Spreading of metallic drops," *Nature Materials*, vol. 3, no. 12, pp. 843–845, 2004.

[11] E. Saiz, M. Benhassine, J. De Coninck, and A. P. Tomsia, "Early stages of dissolutive spreading," *Scripta Materialia*, vol. 62, no. 12, pp. 934–938, 2010.

[12] E. Saiz, A. P. Tomsia, N. Rauch et al., "Nonreactive spreading at high temperature: molten metals and oxides on molybdenum," *Physical Review E*, vol. 76, no. 4, Article ID 041602, 2007.

[13] E. Saiz and A. P. Tomsia, "Atomic dynamics and Marangoni films during liquid-metal spreading," *Nature Materials*, vol. 3, no. 12, pp. 903–909, 2004.

[14] T. Ondarçuhu and M. Veyssié, "Relaxation modes of the contact line of a liquid spreading on a surface," *Nature*, vol. 352, no. 6334, pp. 418–420, 1991.

[15] A. Eddi, K. G. Winkels, and J. H. Snoeijer, "Short time dynamics of viscous drop spreading," *Physics of Fluids*, vol. 25, no. 1, Article ID 013102, 2013.

[16] V. H. López and A. R. Kennedy, "Flux-assisted wetting and spreading of Al on TiC," *Journal of Colloid and Interface Science*, vol. 298, no. 1, pp. 356–362, 2006.

[17] D. Bonn, J. Eggers, J. Indekeu, and J. Meunier, "Wetting and spreading," *Reviews of Modern Physics*, vol. 81, no. 2, pp. 739–805, 2009.

[18] A. M. Cazabat, S. Gerdes, M. P. Valignat, and S. Villette, "Dynamics of wetting: from theory to experiment," *Interface Science*, vol. 5, no. 2-3, pp. 129–139, 1997.

[19] I. A. Aksay, C. E. Hoge, and J. A. Pask, "Wetting under chemical equilibrium and nonequilibrium conditions," *Journal of Physical Chemistry*, vol. 78, no. 12, pp. 1178–1183, 1974.

[20] A. M. Schwartz, "The dynamics of contact angle phenomena," *Advances in Colloid and Interface Science*, vol. 4, no. 4, pp. 349–374, 1975.

[21] C. Bergman, C. Girardeaux, C. Perrin-Pellegrino et al., "Wetting of decagonal $Al_{13}Co_4$ and cubic AlCo thin films by liquid Pb," *Philosophical Magazine*, vol. 86, no. 6–8, pp. 849–854, 2006.

[22] M. E. R. Shanahan, "Simple analysis of local wetting hysteresis on a Wilhelmy plate," *Surface and Interface Analysis*, vol. 17, no. 7, pp. 489–495, 1991.

[23] L. Leger and J. F. Joanny, "Liquid spreading," *Reports on Progress in Physics*, vol. 55, no. 4, pp. 431–486, 1992.

[24] G. Kumar and K. N. Prabhu, "Review of non-reactive and reactive wetting of liquids on surfaces," *Advances in Colloid and Interface Science*, vol. 133, no. 2, pp. 61–89, 2007.

[25] G. T. Barnes and I. R. Gentle, *Interfacial Science: An Introduction*, Oxford University Press, 2nd edition, 2011.

[26] A. B. D. Cassie, "Contact angles," *Discussions of the Faraday Society*, vol. 3, pp. 11–16, 1948.

[27] D. Quéré, "Rough ideas on wetting," *Physica A*, vol. 313, pp. 32–46, 2002.

[28] O. Kozlova, R. Voytovych, P. Protsenko, and N. Eustathopoulos, "Non-reactive versus dissolutive wetting of Ag-Cu alloys on Cu substrates," *Journal of Materials Science*, vol. 45, no. 8, pp. 2099–2105, 2010.

[29] L. Yin, B. T. Murray, S. Su et al., "Reactive wetting in metal-metal systems," *Journal of Physics Condensed Matter*, vol. 21, no. 46, Article ID 464130, 2009.

[30] P. R. Sharps, A. P. Tomsia, and J. A. Pask, "Wetting and spreading in the CuAg system," *Acta Metallurgica*, vol. 29, no. 5, pp. 855–865, 1981.

[31] J.-G. Li, "Wetting of ceramic materials by liquid silicon, aluminium and metallic melts containing titanium and other reactive elements: a review," *Ceramics International*, vol. 20, no. 6, pp. 391–412, 1994.

[32] J. C. Ambrose and M. G. Nicholas, "Wetting and spreading of nickel-phosphorus brazes: detailed real time observations of spreading on iron-chromium substrates," *Materials Science and Technology*, vol. 12, no. 1, pp. 72–80, 1996.

[33] F. Gnecco, E. Ricci, S. Amore et al., "Wetting behaviour and reactivity of lead free Au-In-Sn and Bi-In-Sn alloys on copper substrates," *International Journal of Adhesion and Adhesives*, vol. 27, no. 5, pp. 409–416, 2007.

[34] H. Xu, Z. Yuan, H. Matsuura, and F. Tsukihashi, "Hysteresis phenomenon and wetting characteristics of molten Sn-3.0 wt.%Ag-0.5 wt.%Cu on different tilting substrates," *Journal of Alloys and Compounds*, vol. 487, no. 1-2, pp. 121–125, 2009.

[35] P. R. Chidambaram, G. R. Edwards, and D. L. Olson, "A thermodynamic criterion to predict wettability at metal- alumina interfaces," *Metallurgical Transactions B*, vol. 23, no. 2, pp. 215–222, 1992.

[36] D. Chatain, "Anisotropy of wetting," *Annual Review of Materials Research*, vol. 38, pp. 45–70, 2008.

[37] S. J. Hitchcock, N. T. Carroll, and M. G. Nicholas, "Some effects of substrate roughness on wettability," *Journal of Materials Science*, vol. 16, no. 3, pp. 714–732, 1981.

[38] T. D. Blake and J. de Coninck, "The influence of solid-liquid interactions on dynamic wetting," *Advances in Colloid and Interface Science*, vol. 96, no. 1–3, pp. 21–36, 2002.

[39] J. C. Bird, S. Mandre, and H. A. Stone, "Short-time dynamics of partial wetting," *Physical Review Letters*, vol. 100, no. 23, Article ID 234501, 2008.

[40] P. Protsenko, J.-P. Garandet, R. Voytovych, and N. Eustathopoulos, "Thermodynamics and kinetics of dissolutive wetting of Si by liquid Cu," *Acta Materialia*, vol. 58, no. 20, pp. 6565–6574, 2010.

[41] R. Voitovitch, A. Mortensen, F. Hodaj, and N. Eustathopoulos, "Diffusion-limited reactive wetting: study of spreading kinetics of Cu-Cr alloys on carbon substrates," *Acta Materialia*, vol. 47, no. 4, pp. 1117–1128, 1999.

[42] O. Dezellus, F. Hodaj, and N. Eustathopoulos, "Progress in modelling of chemical-reaction limited wetting," *Journal of the European Ceramic Society*, vol. 23, no. 15, pp. 2797–2803, 2003.

[43] L. Yin, B. T. Murray, and T. J. Singler, "Dissolutive wetting in the Bi-Sn system," *Acta Materialia*, vol. 54, no. 13, pp. 3561–3574, 2006.

[44] L. Yin, A. Chauhan, and T. J. Singler, "Reactive wetting in metal/metal systems: dissolutive versus compound-forming systems," *Materials Science and Engineering A*, vol. 495, no. 1-2, pp. 80–89, 2008.

[45] B. He, J. Lee, and N. A. Patankar, "Contact angle hysteresis on rough hydrophobic surfaces," *Colloids and Surfaces A*, vol. 248, no. 1–3, pp. 101–104, 2004.

[46] H. Tavana and A. W. Neumann, "On the question of rate-dependence of contact angles," *Colloids and Surfaces A*, vol. 282-283, pp. 256–262, 2006.

[47] W. Li and A. Amirfazli, "A thermodynamic approach for determining the contact angle hysteresis for superhydrophobic surfaces," *Journal of Colloid and Interface Science*, vol. 292, no. 1, pp. 195–201, 2005.

[48] A. Nakajima, "Design of hydrophobic surfaces for liquid droplet control," *NPG Asia Materials*, vol. 3, no. 5, pp. 49–56, 2011.

[49] X. S. Cui and W. Li, "On the possibility of superhydrophobic behavior for hydrophilic materials," *Journal of Colloid and Interface Science*, vol. 347, no. 1, pp. 156–162, 2010.

[50] S. Vedantam and M. V. Panchagnula, "Constitutive modeling of contact angle hysteresis," *Journal of Colloid and Interface Science*, vol. 321, no. 2, pp. 393–400, 2008.

[51] W. D. Kaplan, D. Chatain, P. Wynblatt, and W. C. Carter, "A review of wetting versus adsorption, complexions, and related phenomena: the rosetta stone of wetting," *Journal of Materials Science*, vol. 48, no. 17, pp. 5681–5717, 2013.

[52] J. F. Joanny and P. G. De Gennes, "A model for contact angle hysteresis," *The Journal of Chemical Physics*, vol. 81, no. 1, pp. 552–562, 1984.

[53] M. Iwamatsu, "Contact angle hysteresis of cylindrical drops on chemically heterogeneous striped surfaces," *Journal of Colloid and Interface Science*, vol. 297, no. 2, pp. 772–777, 2006.

[54] B. B. Sauer and W. G. Kampert, "Influence of viscosity on forced and spontaneous spreading: wilhelmy fiber studies including practical methods for rapid viscosity measurement," *Journal of Colloid and Interface Science*, vol. 199, no. 1, pp. 28–37, 1998.

[55] B. He, N. A. Patankar, and J. Lee, "Multiple equilibrium droplet shapes and design criterion for rough hydrophobic surfaces," *Langmuir*, vol. 19, no. 12, pp. 4999–5003, 2003.

[56] H. Kamusewitz, W. Possart, and D. Paul, "The relation between Young's equilibrium contact angle and the hysteresis on rough paraffin wax surfaces," *Colloids and Surfaces A*, vol. 156, no. 1–3, pp. 271–279, 1999.

[57] J. Long, M. N. Hyder, R. Y. M. Huang, and P. Chen, "Thermodynamic modeling of contact angles on rough, heterogeneous surfaces," *Advances in Colloid and Interface Science*, vol. 118, no. 1-3, pp. 173–190, 2005.

[58] A. Marmur, "Wetting on hydrophobic rough surfaces: to be heterogeneous or not to be?" *Langmuir*, vol. 19, no. 20, pp. 8343–8348, 2003.

[59] H. Shibata, X. Jiang, M. Valdez, and A. W. Cramb, "The contact angle between liquid iron and a single-crystal magnesium oxide substrate at 1873 K," *Metallurgical and Materials Transactions B*, vol. 35, no. 1, pp. 179–180, 2004.

[60] G. L. J. Bailey and H. C. Watkins, "The flow of liquid metals on solid metal surfaces and its relation to soldering, brazing and hot-dip coating," *Journal of the Institute of Metals*, vol. 80, pp. 57–76, 1951.

[61] R. Shuttleworth and G. L. J. Bailey, "The spreading of a liquid over a rough solid," *Discussions of the Faraday Society*, vol. 3, pp. 16–22, 1948.

[62] R. N. Wenzel, "Resistance of solid surfaces to wetting by water," *Industrial & Engineering Chemistry*, vol. 28, pp. 988–994, 1936.

[63] S. G. Kandlikar and M. E. Steinke, "Contact angles and interface behavior during rapid evaporation of liquid on a heated surface," *International Journal of Heat and Mass Transfer*, vol. 45, no. 18, pp. 3771–3780, 2002.

[64] J. D. Bernardin, I. Mudawar, C. B. Walsh, and E. I. Franses, "Contact angle temperature dependence for water droplets on practical aluminum surfaces," *International Journal of Heat and Mass Transfer*, vol. 40, no. 5, pp. 1017–1033, 1997.

[65] N. Eustathopoulos, M. G. Nicholas, and B. Drevert, *Wettability at High Temperatures*, Pergamon Materials Series, Elsevier Science, 1st edition, 1999.

[66] R. E. Johnson Jr. and R. H. Dettre, "Contact angle hysteresis—III. Study of an idealized heterogeneous surface," *Journal of Physical Chemistry*, vol. 68, no. 7, pp. 1744–1750, 1964.

[67] N. A. Patankar, "On the modeling of hydrophobic contact angles on rough surfaces," *Langmuir*, vol. 19, no. 4, pp. 1249–1253, 2003.

[68] P. Costa, A. Passerone, and E. Ricci, "Oxygen fluxes and the measurement of the surface tension of liquid metals," *High Temperatures*, vol. 20, no. 1, pp. 59–72, 1988.

[69] D. R. Milner and R. L. Apps, *Introduction to Welding and Brazing*, Pergamon Press, London, UK, 1st edition, 1968.

[70] H. Takao, T. Tsukada, K. Yamada, M. Yamashita, and H. Hasegawa, "Development of wettability evaluation technique applying contact angle measurement during soldering," *R&D Review of Toyota CRDL*, vol. 39, no. 2, pp. 41–48, 2004.

[71] G. Humpston and D. M. Jacobson, *Principles of Soldering and Brazing*, ASM International, Materials Park, Ohio, USA, 3rd edition, 1996.

[72] Y. O. Kabova, V. V. Kuznetsov, and O. A. Kabov, "Temperature dependent viscosity and surface tension effects on deformations of non-isothermal falling liquid film," *International Journal of Heat and Mass Transfer*, vol. 55, no. 4, pp. 1271–1278, 2012.

[73] H. D. Samarasekera and Z. A. Munir, "An investigation of the contact angles between low melting metals and substrates of niobium and zirconium," *Journal of The Less-Common Metals*, vol. 64, no. 2, pp. 255–266, 1979.

[74] W. Feduska, "High-tempeerature brazing alloy-base metal wetting reactions," *Welding Journal*, vol. 38, no. 3, pp. 122–131, 1959.

[75] A. Sangghaleh and M. Halali, "An investigation on the wetting of polycrystalline alumina by aluminium," *Journal of Materials Processing Technology*, vol. 197, no. 1–3, pp. 156–160, 2008.

[76] X. H. Yang, P. Xiao, S. H. Liang, J. T. Zou, and Z. K. Fan, "Alloying effect of Ni and Cr on the wettability of copper on W substrate," *Acta Metallurgica Sinica*, vol. 21, no. 5, pp. 369–379, 2008.

[77] E. Candan, "Effect of alloying elements to aluminium on the wettability of AL/SiC system," *Turkish Journal of Engineering and Environmental Sciences*, vol. 26, no. 1, pp. 1–5, 2002.

[78] D. Q. Yu, H. P. Xie, and L. Wang, "Investigation of interfacial microstructure and wetting property of newly developed Sn-Zn-Cu solders with Cu substrate," *Journal of Alloys and Compounds*, vol. 385, no. 1-2, pp. 119–125, 2004.

[79] P. Bhagawath, K. N. Prabhu, and Satyanarayan, "Wetting behavior of reactive and non-reactive wetting of liquids on metallic substrates," *World Academy of Science, Engineering and Technology*, vol. 73, pp. 978–981, 2013.

[80] O. Dezellus and N. Eustathopoulos, "Fundamental issues of reactive wetting by liquid metals," *Journal of Materials Science*, vol. 45, no. 16, pp. 4256–4264, 2010.

[81] A. L. Biance, C. Clanet, and D. Quéré, "First steps in the spreading of a liquid droplet," *Physical Review E*, vol. 69, no. 1, Article ID 016301, 2004.

[82] L. H. Tanner, "The spreading of silicone oil drops on horizontal surfaces," *Journal of Physics D*, vol. 12, no. 9, pp. 1473–1484, 1979.

[83] S. Rafaï, D. Sarker, V. Bergeron, J. Meunier, and D. Bonn, "Superspreading: aqueous surfactant drops spreading on hydrophobic surfaces," *Langmuir*, vol. 18, no. 26, pp. 10486–10488, 2002.

[84] L. Chen, G. K. Auernhammer, and E. Bonaccurso, "Short time wetting dynamics on soft surfaces," *Soft Matter*, vol. 7, no. 19, pp. 9084–9089, 2011.

[85] J. Eggers, J. R. Lister, and H. A. Stone, "Coalescence of liquid drops," *Journal of Fluid Mechanics*, vol. 401, pp. 293–310, 1999.

[86] L. Duchemin, J. Eggers, and C. Josserand, "Inviscid coalescence of drops," *Journal of Fluid Mechanics*, no. 487, pp. 167–178, 2003.

[87] K. G. Winkels, J. H. Weijs, A. Eddi, and J. H. Snoeijer, "Initial spreading of low-viscosity drops on partially wetting surfaces," *Physical Review E*, Article ID 055301, 2012.

[88] L. Courbin, J. C. Bird, M. Reyssat, and H. A. Stone, "Dynamics of wetting: from inertial spreading to viscous imbibition," *Journal of Physics Condensed Matter*, vol. 21, no. 46, Article ID 464127, 2009.

[89] J. D. Paulsen, J. C. Burton, and S. R. Nagel, "Viscous to inertial crossover in liquid drop coalescence," *Physical Review Letters*, vol. 106, no. 11, Article ID 114501, 2011.

[90] S. Case and S. Nagel, "Coalescence in low-viscosity liquids," *Physical Review Letters*, vol. 100, no. 8, Article ID 084503, 2008.

[91] M. Wu, T. Cubaud, and C.-M. Ho, "Scaling law in liquid drop coalescence driven by surface tension," *Physics of Fluids*, vol. 16, no. 7, pp. L51–L54, 2004.

[92] A. Carlson, M. Do-Quang, and G. Amberg, "Dissipation in rapid dynamic wetting," *Journal of Fluid Mechanics*, vol. 682, pp. 213–240, 2011.

[93] C. Redon, F. Brochard-Wyart, H. Hervet, and F. Rondelez, "Spreading of "heavy" droplets—II. Experiments," *Journal of Colloid And Interface Science*, vol. 149, no. 2, pp. 580–591, 1992.

[94] R. Holdich, V. M. Starov, P. Prokopovich et al., "Spreading of liquid drops from a liquid source," *Colloids and Surfaces A*, vol. 282-283, pp. 247–255, 2006.

[95] K. Landry and N. Eustathopoulos, "Dynamics of wetting in reactive metal/ceramic systems: linear spreading," *Acta Materialia*, vol. 44, no. 10, pp. 3923–3932, 1996.

[96] N. Eustathopoulos, "Progress in understanding and modeling reactive wetting of metals on ceramics," *Current Opinion in Solid State and Materials Science*, vol. 9, no. 4-5, pp. 152–160, 2005.

[97] A. Mortensen, B. Drevet, and N. Eustathopoulos, "Kinetics of diffusion-limited spreading of sessile drops in reactive wetting," *Scripta Materialia*, vol. 36, no. 6, pp. 645–651, 1997.

[98] K. Landry, C. Rado, and N. Eustathopoulos, "Influence of interfacial reaction rates on the wetting driving force in metal/ceramic systems," *Metallurgical and Materials Transactions A*, vol. 27, no. 10, pp. 3181–3186, 1996.

[99] P. Protsenko, O. Kozlova, R. Voytovych, and N. Eustathopoulos, "Dissolutive wetting of Si by molten Cu," *Journal of Materials Science*, vol. 43, no. 16, pp. 5669–5671, 2008.

[100] J. C. Ambrose, M. G. Nicholas, and A. M. Stoneham, "Dynamics of braze spreading," *Acta Metallurgica et Materialia*, vol. 40, no. 10, pp. 2483–2488, 1992.

[101] C. D. Bain, G. D. Burnett-Hall, and R. R. Montgomerie, "Rapid motion of liquid drops," *Nature*, vol. 372, no. 6505, pp. 414–415, 1994.

[102] F. D. dos Santos and T. Ondarçuhu, "Free-running droplets," *Physical Review Letters*, vol. 75, no. 16, pp. 2972–2975, 1995.

[103] M. E. R. Shanahan and P.-G. de Gennes, "Start-up of a reactive droplet," *Comptes Rendus l'Académie des Sciences*, vol. 324, no. 4, pp. 261–268, 1997.

[104] C. D. Bain and G. M. Whitesides, "A study by contact angle of the acid-base behavior of monolayers containing ω-mercaptocarboxylic acids adsorbed on gold: an example of reactive spreading," *Langmuir*, vol. 5, no. 6, pp. 1370–1378, 1989.

[105] M. K. Chaudhury and G. M. Whitesides, "How to make water run uphill," *Science*, vol. 256, no. 5063, pp. 1539–1544, 1992.

[106] F. Brochard, "Motions of droplets on solid surfaces induced by chemical or thermal gradients," *Langmuir*, vol. 5, no. 2, pp. 432–438, 1989.

[107] P.-G. de Gennes, "Running droplets in a random medium," *Comptes Rendus de l'Academie de Sciences*, vol. 327, no. 1, pp. 147–154, 1999.

[108] E. Raphael, "Etalement de gouttes sur une surface bigarrée," *Comptes Rendus de l'Academie de Sciences*, vol. 306, 1988.

[109] F. Brochard-Wyart and P.-G. De Gennes, "Spontaneous motion of a reactive droplet," *Comptes Rendus de l'Academie de Sciences*, vol. 321, no. 7, pp. 285–288, 1995.

Some Thiophene Derivatives as Corrosion Inhibitors for Carbon Steel in Hydrochloric Acid

A. S. Fouda,[1] A. A. Attia,[2] and A. A. Negm[3]

[1] *Department of Chemistry, Faculty of Science, Mansoura University, El-Mansoura, Egypt*
[2] *Department of Chemistry, Faculty of Science, Zagazig University, Zagazig, Egypt*
[3] *East Delta Electricity Production Company, Damietta, Egypt*

Correspondence should be addressed to A. S. Fouda; asfouda@mans.edu.eg

Academic Editor: Elena V. Pereloma

Corrosion inhibitive performance of some thiophene derivatives during the acidic corrosion of carbon steel surface in 1 M HCl was investigated by chemical technique (weight loss) and electrochemical techniques (potentiodynamic polarization, electrochemical frequency modulation, and electrochemical impedance spectroscopy). The effect of temperature on the corrosion rate was investigated by the weight loss method, and some thermodynamic parameters for corrosion and adsorption processes were determined and discussed. The results show that the inhibition efficiency increased with the increase in inhibitor concentration and temperature. The adsorption of thiophene derivatives on the carbon steel surface obeys Langmuir adsorption isotherm. The obtained results indicated that the investigated compounds are chemically adsorbed on the steel surface. Potentiodynamic polarization studies showed that these compounds are mixed-type inhibitors and the results obtained from the techniques are in good agreement.

1. Introduction

Carbon steel has remarkable economic and attractive materials for engineering applications owing to its low cost, easy availability, and high mechanical strength. The use of inhibitors is one of the most practical methods to protect metals against acid attack. The most widely used acid inhibitors are organic compounds containing oxygen, nitrogen, and/or sulfur [1–5]. Acid inhibitors have many important roles in the industrial field such as a component in pretreatment composition, cleaning solutions, and in acidification of oil wells. Heteroatoms as sulfur, nitrogen, and oxygen as well as aromatic rings in their structure are the major adsorption centers. Some works have studied the influence of organic compounds containing nitrogen on the corrosion resistance of steel in acidic media [6–18] most organic inhibitors decrease corrosion rate by adsorption on the metal surface. To be effective, an inhibitor must also displace water from the metals surface to block active corrosion sites and interact with the anodic or cathodic reaction sites to retard the oxidation and/or reduction of corrosion reaction. So, the inhibition efficiency of organic compounds depends on the structure of the inhibitor, the characteristics of the environment, and so forth. Several organic compounds have been evaluated as corrosion inhibitors for carbon steel in acidic media, so the aim of this work is to study the inhibition properties of some thiophene derivatives for the corrosion of 1018 carbon steel in 1 M HCl. The main techniques were employed: chemical (weight loss) and electrochemical ones (potentiodynamic, electrochemical frequency modulation, EFM, and electrochemical impedance spectroscopy, EIS). These techniques were performed in 1 M HCl without and with the presence of the investigated compounds in the concentration range (0 to 12×10^{-5} M).

2. Methods

2.1. Materials. Rectangular specimens with dimensions $2 \times 2 \times 0.2$ cm of carbon steel that is chemical composition can be seen from Table 1. It was used for weight loss measurements. For electrochemical tests, the exposed surface area of metal was 1.00 cm^2.

Table 1: Chemical composition (weight %) of the carbon steel.

Element	C	Mn	P	Si	Fe
Weight (%)	0.2	0.35	0.024	0.003	Rest

2.2. Inhibitors. Thiophene derivatives were prepared as reported before [19]. Table 2 shows the structures, names, molecular weights, and molecular formulas of these compounds.

2.3. Solutions. Hydrochloric acid (37%), ethyl alcohol, and acetone were BDH grades and purchased from Algamhoria Co. (Egypt). Bidistilled water was used for preparing test solutions for all measurements.

2.4. Procedures Used for Corrosion Measurements

2.4.1. Weight Loss Tests. Medium carbon steel specimens were mechanically abraded up to 1200 grades emery paper and degreased with acetone, rinsed with bidistilled water two times, and finally dried between filter paper. After weighting accurately, the specimens were immersed in 100 mL of 1 M HCl with and without different concentrations of thiophene derivatives. After different immersion times (30, 60, 90, 120, 150, and 180 min) the carbon steel samples were taken out, washed with bidistilled water, dried, and weighted accurately. The experiments were done in triplicate and the average value of the weight loss was taken. The average weight loss of the three parallel carbon steel sheets could be obtained at required temperature. The inhibition efficiency (%IE) and the degree of surface coverage (θ) of the investigated inhibitors on the corrosion of carbon steel were calculated as follows [20]:

$$\mathrm{IE\%} = \theta \times 100 = \left[\frac{(W_o - W)}{W_o}\right] \times 100, \quad (1)$$

where W_o and W are the values of the average weight loss in the absence and presence of the inhibitor, respectively.

2.4.2. Potentiodynamic Polarization Measurements. Polarization experiments were carried out in a conventional three-electrode cell with platinum gauze as the auxiliary electrode and a saturated calomel electrode (SCE) coupled to a fine Luggin capillary as reference electrode. The working electrode was in the form of a square cut from copper sheet of equal composition embedded in epoxy resin of polytetrafluoroethylene so that the flat surface area was 1 cm^2. Prior to each measurement, the electrode surface was pretreated in the same manner as the weight loss experiments. Before measurements, the electrode was immersed in solution for 30 min. until a steady state was reached. The potential was started from −600 to +400 mV versus open circuit potential (Eocp). All experiments were carried out in freshly prepared solutions at 25°C and results were always repeated at least three times to check the reproducibility.

2.4.3. Electrochemical Impedance Spectroscopy Measurements. Impedance measurements were carried out using AC signals of 5 mV peak to peak amplitude at the open circuit potential in the frequency range of 100 kHz to 0.1 Hz. All impedance data were fitted to appropriate equivalent circuit using the Gamry Echem Analyst software.

2.4.4. Electrochemical Frequency Modulation Technique. EFM experiments were performed with applying potential perturbation signal with amplitude 10 mV with two sine waves of 2 and 5 Hz. The choice for the frequencies of 2 and 5 Hz was based on three arguments [21–23]. The larger peaks were used to calculate the corrosion current density ($i_{corr.}$), the Tafel slopes (β_c and β_a), and the causality factors CF-2 and CF-3 [24].

All electrochemical experiments were carried out using Gamry instrument PCI300/4 Potentiostat/Galvanostat/Zra analyzer, DC105 corrosion software, EIS300 electrochemical impedance spectroscopy software, EFM140 electrochemical frequency modulation software, and Echem Analyst 5.5 for results plotting, graphing, data fitting, and calculating.

3. Results and Discussion

3.1. Weight Loss Measurements. Weight loss of carbon steel was determined, at various time intervals, in the absence and presence of different concentrations of thiophene derivatives compounds (A, B, and C). The obtained weight loss time curves are represented in Figure 1 for inhibitor (A), the most effective one. Similar curves were obtained for other inhibitors (not shown). The inhibition efficiency of corrosion was found to be dependent on the inhibitor concentration. The curves obtained in the presence of inhibitors fall significantly below that of free acid. In all cases, the increase in the inhibitor concentration was accompanied by a decrease in weight loss and an increase in the percentage inhibition. These results lead to the conclusion that the compounds under investigation are fairly efficient as inhibitors for carbon steel dissolution in hydrochloric acid solution. Also, the degree of surface coverage (θ) by the inhibitor, calculated from (1), would increase by increasing the inhibitor concentration. In order to get a comparative view, the variation of the percentage inhibition (%IE) of the three inhibitors with their molar concentrations was calculated according to (2). The values obtained are summarized in Table 3. Careful inspection of these results showed that, at the same inhibitor concentration, the order of inhibition efficiencies is as follows: A > B > C.

3.1.1. Adsorption Isotherm. One of the most convenient ways of expressing adsorption quantitatively is by deriving the adsorption isotherm that characterizes the metal/inhibitor/environment system. Various adsorption isotherms were applied to fit θ values, but the best fit was found to obey Langmuir adsorption isotherm that are represented in Figures 2, 3, and 4 for inhibitors (A), (B), and (C), respectively; Langmuir adsorption isotherm may be expressed by the following:

$$\frac{C}{\theta} = \left(\frac{1}{K}\right) + C, \quad (2)$$

Table 2: Chemical structure of inhibitors.

Inhibitor	Structures	Names	Mol.Wt., Mol. formulas
A	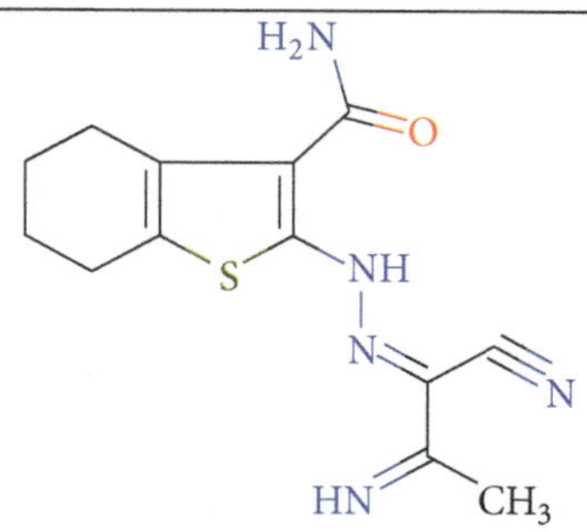	2-[(2Z)-2-(1-cyano-2-iminopropylidene)hydrazinyl]-4,5,6,7-tetrahydro-1-benzothiophene-3-carboxamide	289.35 $C_{13}H_{15}N_5OS$
B	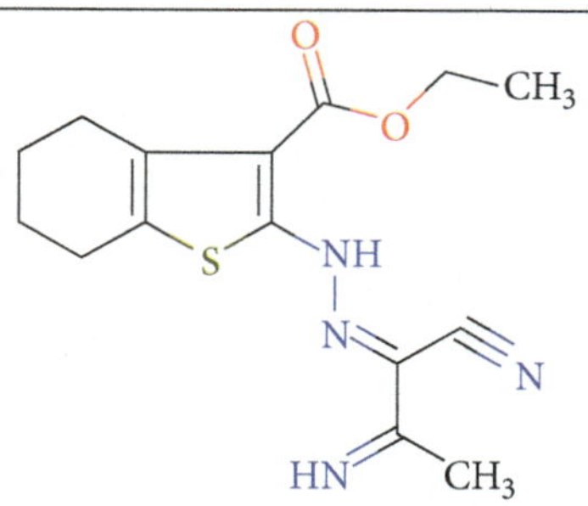	ethyl 2-[(2Z)-2-(1-cyano-2-iminopropylidene)hydrazinyl]-4,5,6,7-tetrahydro-1-benzothiophene-3-carboxylate	318.39 $C_{15}H_{18}N_4O_2S$
C	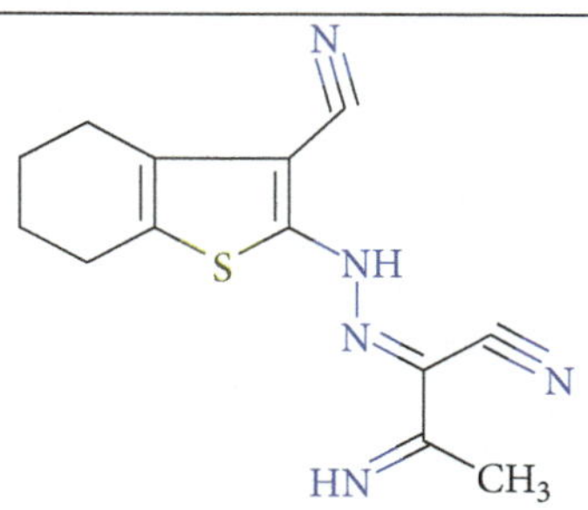	2-[(2Z)-2-(1-cyano-2-iminopropylidene)hydrazinyl]-4,5,6,7-tetrahydro-1-benzothiophene-3-carbonitrile	271.34082 $C_{13}H_{13}N_5S$

where C is the concentration (mol L^{-1}) of the inhibitor in the bulk electrolyte, θ is the degree of surface coverage (θ = %IE/100), and K_{ads} is the adsorption equilibrium constant. A plot of C/θ versus C should give straight lines with slope equal to unity and the intercept is $(1/K_{ads})$. In order to get a comparative view, the variation of the adsorption equilibrium constant (K_{ads}) of the inhibitors with their molar concentrations was calculated according to (3). The experimental data give good curves fitting for the applied adsorption isotherm as the correlation coefficients (R^2) were in the range (0.943–0.999). The values obtained are given in Table 4.

These results confirm the assumption that these compounds are adsorbed on the metal surface through the protonated (N, S, and O) atoms or via the lone pair of electrons of (N, S, and O) atoms. The extent of inhibition is directly related to the performance of adsorption layer which is a sensitive function of the molecular structure. The equilibrium constant of adsorption K_{ads} obtained from the intercepts of Langmuir adsorption isotherm is related to the free energy of adsorption ΔG°_{ads} as follows:

$$K_{ads} = \frac{1}{55.5} \exp\left[\frac{-\Delta G^{\circ}_{ads}}{RT}\right], \quad (3)$$

where 55.5 is the molar concentration of water in the solution in M^{-1}.

Plot of (ΔG°_{ads}) versus T (Figure 5) gave the heat of adsorption (ΔH°_{ads}) and the standard entropy (ΔS°_{ads}) according to the thermodynamic basic equation (4) as follows:

$$\Delta G^{\circ}_{ads} = \Delta H^{\circ}_{ads} - T\Delta S^{\circ}_{ads}. \quad (4)$$

Table 5 clearly shows a good dependence of ΔG°_{ads} on T, indicating the good correlation among thermodynamic parameters. The negative value of ΔG°_{ads} ensures the spontaneity of the adsorption process and stability of the adsorbed layer on the steel surface. Generally, values of ΔG°_{ads} around $-20\,\mathrm{kJ\,mol^{-1}}$ or lower are consistent with the electrostatic interaction between the charged molecules and the charged metal (physisorption); those around $-40\,\mathrm{kJ\,mol^{-1}}$ or higher involve charge sharing or transfer from organic molecules to the metal surface to form a coordinate type of bond (chemisorption) [25]. The calculated ΔG°_{ads} values are closer to $-40\,\mathrm{kJ\,mol^{-1}}$ indicating that the adsorption mechanism of the inhibitors on steel in 1 M HCl solutions was typical of chemisorption. The unshared electron pairs in sulphur and nitrogen as well as in oxygen may interact with d-orbitals of steel to provide a protective chemisorbed film [26].

TABLE 3: Values of inhibition efficiencies (%IE) and surface coverage (θ) of inhibitors for the corrosion of carbon steel in 1 M HCl from weight loss measurements at different concentrations and at 25°C.

$C_{inh.} \times 10^{-5}$ (M)	A		B		C	
	%IE	θ	%IE	θ	%IE	θ
2	46.1	0.461	35.6	0.356	23.9	0.239
4	54.3	0.543	47.9	0.479	33.8	0.338
6	58.9	0.589	54.3	0.543	41.7	0.417
8	64.3	0.643	60.8	0.608	50.2	0.502
10	66	0.66	62.5	0.625	56.7	0.567
12	72.5	0.725	66.6	0.666	64.9	0.649

TABLE 4: Adsorption parameters for inhibitors in 1 M HCl obtained from Langmuir adsorption isotherm at different temperatures.

Temperature (°C)	A			B			C		
	$K_{ads} \times 10^{-3}$ (M^{-1})	Slope	R^2	$K_{ads} \times 10^{-3}$ (M^{-1})	Slope	R^2	$K_{ads} \times 10^{-3}$ (M^{-1})	Slope	R^2
25	42.82	1.24	0.988	30.01	1.24	0.988	13.37	1.00	0.943
35	36.87	1.13	0.985	43.23	1.19	0.985	14.87	1.02	0.957
45	57.13	1.04	0.987	58.75	1.12	0.997	16.46	1.03	0.964
55	139.41	1.05	0.999	57.13	1.00	0.999	45.99	1.08	0.995

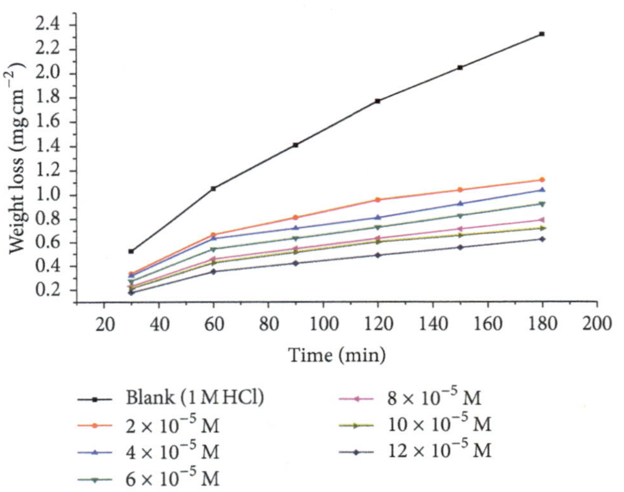

FIGURE 1: Weight loss time curves for the corrosion of carbon steel in 1 M HCl in the absence and presence of different concentrations of inhibitor (A) at 25°C.

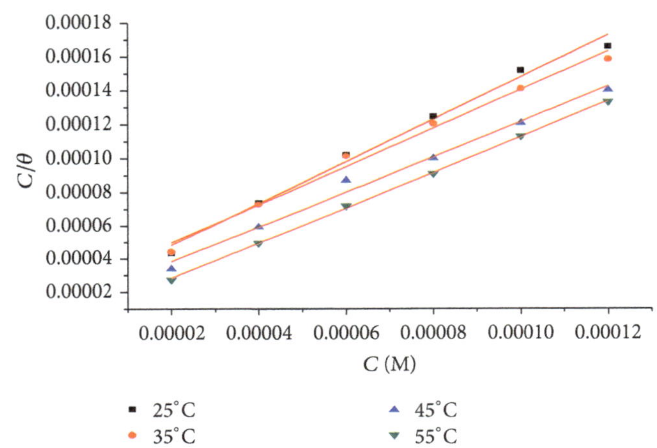

FIGURE 2: Langmuir adsorption isotherm of inhibitor (A) on carbon steel surface in 1 M HCl at different temperatures.

The values of thermodynamic parameter for the adsorption of inhibitors (Table 5) can provide valuable information about the mechanism of corrosion inhibition. While an endothermic adsorption process ($\Delta H^\circ_{ads} > 0$) is attributed unequivocally to chemisorption [27], an exothermic adsorption process ($\Delta H^\circ_{ads} < 0$) may involve either physisorption or chemisorption or mixture of both processes. The ΔS°_{ads} values in the presence of inhibitors in 1 M HCl are positive. This indicates that an increase in disorder takes places from reactants to the metal-adsorbed reaction complex [28].

3.1.2. Effect of Temperature. The effect of temperature on the corrosion rate of carbon steel in 1 M HCl and in presence of different inhibitors concentrations was studied in the temperature range of 298–328 K using weight loss measurements. Similar curves were obtained for other inhibitors (not shown). As the temperature increases, the rate of corrosion increases; nevertheless, the inhibition efficiency of the additives increases as shown in Table 6 for inhibitor (A), the most effective one. Similar tables were obtained for other inhibitors (not shown). Table 6 illustrates that the adsorption is aided by increasing the temperature. This behavior proves that the adsorption of inhibitors on carbon steel surface occurs through chemical adsorption. Ivanov [29] considers the increase of %IE with increasing the temperature as the change in the nature of the adsorption mode; the inhibitor is being physically adsorbed at lower temperatures, while chemisorption is favored as temperature increases. Similar observations were reported by other authors [30–32].

TABLE 5: Thermodynamic parameters for the adsorption of inhibitors on carbon steel surface in 1 M HCl at different temperatures.

Inhibitor	Temperature (°C)	$K_{ads.} \times 10^{-3}$ (M^{-1})	$-\Delta G^{\circ}_{ads.}$ (kJ mol^{-1})	$\Delta H^{\circ}_{ads.}$ (kJ mol^{-1})	$\Delta S^{\circ}_{ads.}$ (kJ mol^{-1})
A	25	42.8	36.4	2.8	0.13
	35	36.9	37.2		0.12
	45	57.1	39.6		0.12
	55	139.4	43.3		0.13
B	25	30	35.5	1.7	0.12
	35	43.2	37.6		0.12
	45	58.8	39.7		0.12
	55	57.1	40.8		0.12
C	25	13.4	33.5	3	0.12
	35	14.9	34.9		0.11
	45	16.5	36.3		0.11
	55	46	40.2		0.12

TABLE 6: Values of inhibition efficiencies %IE, coverage factor (θ), and corrosion rate (C.R.) of inhibitor (A) for the corrosion of carbon steel in 1 M HCl from weight loss measurements at different concentrations at temperature range of 298–328 K.

Concentration $\times 10^{-5}$ (M)	298 K			308 K			318 K			328 K		
	θ	%IE	C.R.	θ	%IE	C.R.	θ	%IE	C.R.	θ	%IE	C.R.
2	0.461	46.1	0.008	0.454	45.4	0.014	0.589	58.9	0.03	0.732	73.2	0.03
4	0.543	54.3	0.007	0.549	54.9	0.012	0.675	67.5	0.02	0.813	81.3	0.02
6	0.589	58.9	0.006	0.593	59.3	0.011	0.69.1	69.1	0.02	0.835	83.5	0.02
8	0.643	64.3	0.005	0.67	66.66	0.009	0.801	80.1	0.01	0.882	88.2	0.01
10	0.66	66	0.005	0.712	71.2	0.008	0.83	83	0.01	0.889	88.9	0.01
12	0.725	72.5	0.004	0.761	76.1	0.006	0.86	86	0.01	0.905	90.5	0.01

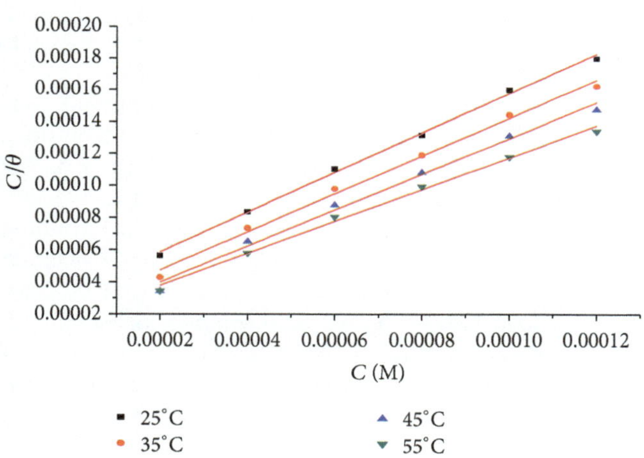

FIGURE 3: Langmuir adsorption isotherm of inhibitor (B) on carbon steel surface in 1 M HCl at different temperatures.

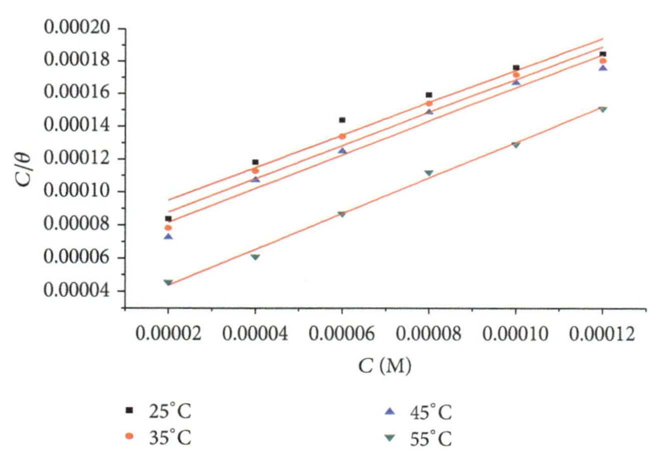

FIGURE 4: Langmuir adsorption isotherm of inhibitor (C) on carbon steel surface in 1 M HCl at different temperatures.

3.1.3. Kinetic-Thermodynamic Corrosion Parameter. The activation parameters for the corrosion process were calculated from Arrhenius-type plot according to (5) as follows:

$$k_{corr.} = A\exp\left(\frac{E_a^*}{RT}\right), \quad (5)$$

where E_a^* is the apparent activation corrosion energy, R is the universal gas constant, T is the absolute temperature, and A is the Arrhenius preexponential constant. Values of apparent activation energy of corrosion (E_a^*) for carbon steel in 1 M HCl shown in Table 7 without and with various concentrations of compound (A) determined from the slope of $\log(k_{corr.})$ versus $1/T$ plots are shown in Figure 6. Inspection of the data shows that the activation energy is lower in the presence of inhibitors than in its absence. This was attributed to the slow rate of inhibitor adsorption with a resultant closer

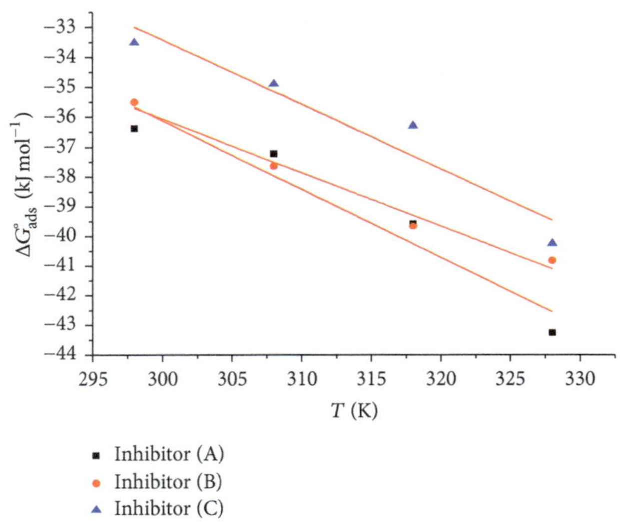

FIGURE 5: Variation of ΔG°_{ads} versus T for the adsorption of inhibitors on carbon steel surface in 1 M HCl at different temperatures.

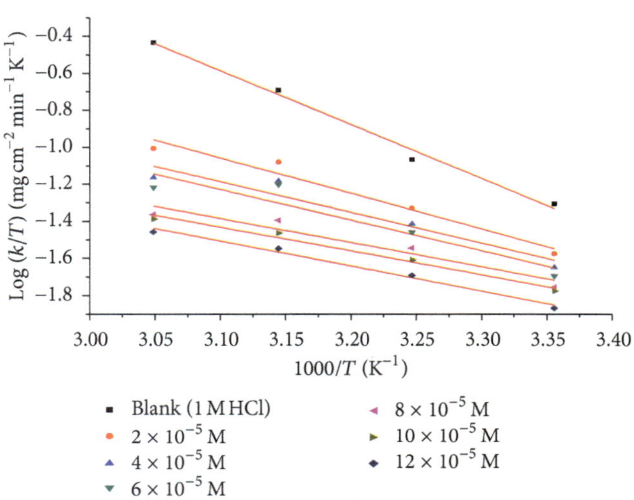

FIGURE 6: log(k) versus (1/T) curves for Arrhenius plots for carbon steel corrosion rates ($k_{corr.}$) after 120 minutes of immersion in 1 M HCl in the absence and presence of various concentrations of inhibitor (A).

approach to equilibrium during the experiments at higher temperatures according to Hour and Holliday [33]. But Riggs and Hurd [34] explained that the decrease in the activation energy of corrosion at higher levels of inhibition arises from a shift of the net corrosion reaction from the uncovered part of the metal surface to the covered one. Schmid and Huang [35] found that organic molecules inhibit both the anodic and cathodic partial reactions on the electrode surface and a parallel reaction takes place on the covered area, but the reaction rate on the covered area is substantially less than that on uncovered area which is similar to the present study. The alternative formulation of transition state equation is shown in (6) as follows:

$$k_{corr} = \left(\frac{RT}{Nh}\right) \exp\left(\frac{\Delta S^*}{R}\right) \exp\left(-\frac{\Delta H^*}{RT}\right), \quad (6)$$

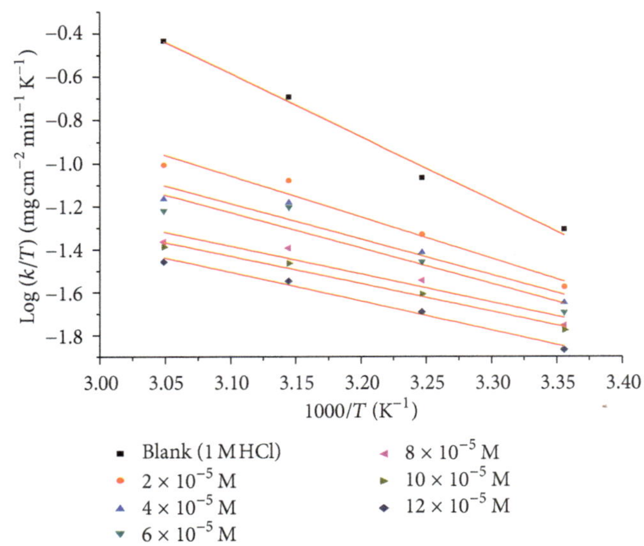

FIGURE 7: log(k/T) versus (1/T) curves for Arrhenius plots for carbon steel corrosion rates ($k_{corr.}$) after 120 minutes of immersion in 1 M HCl in the absence and presence of various concentrations of inhibitor (A).

where $k_{corr.}$ is the rate of metal dissolution, h is Planck's constant, N is Avogadro's number, ΔS^* is the entropy of activation, and ΔH^* is the enthalpy of activation.

Figure 7 shows a plot of (log k/T) against (1/T) in the case of inhibitor (A) in 1 M HCl. Similar behavior is observed in the case of inhibitors B and C (not shown). Straight lines are obtained with slopes equal to ($\Delta H^*/2.303R$) and intercepts are [$\log(R/Nh + \Delta S^*/2.303R)$] calculated in Table 7.

The decrease in E_a^* with the increase in inhibitor concentration (Table 7) is typical of chemisorption. The positive signs of the enthalpies (ΔH^*) reflect the endothermic nature of the steel dissolution process. Value of entropies (ΔS^*) implies that the activated complex at the rate determining step represents an association rather than a dissociation step, meaning that a decrease in disordering takes place form reactants toward the activated complex formation [30, 36]. However, the value of (ΔS^*) decreases gradually with increasing the inhibitor concentrations in all the acid media.

3.2. Electrochemical Measurements

3.2.1. Electrochemical Frequency Modulation Technique (EFM). EFM is a nondestructive corrosion measurement technique that can directly and quickly determine the corrosion current values without prior knowledge of Tafel slopes and with only a small polarizing signal. These advantages of EFM technique make it an ideal candidate for online corrosion monitoring [37]. The great strength of the EFM is the causality factors which serve as an internal check on the validity of EFM measurement. The causality factors CF-2 and CF-3 are calculated from the frequency spectrum of the current responses.

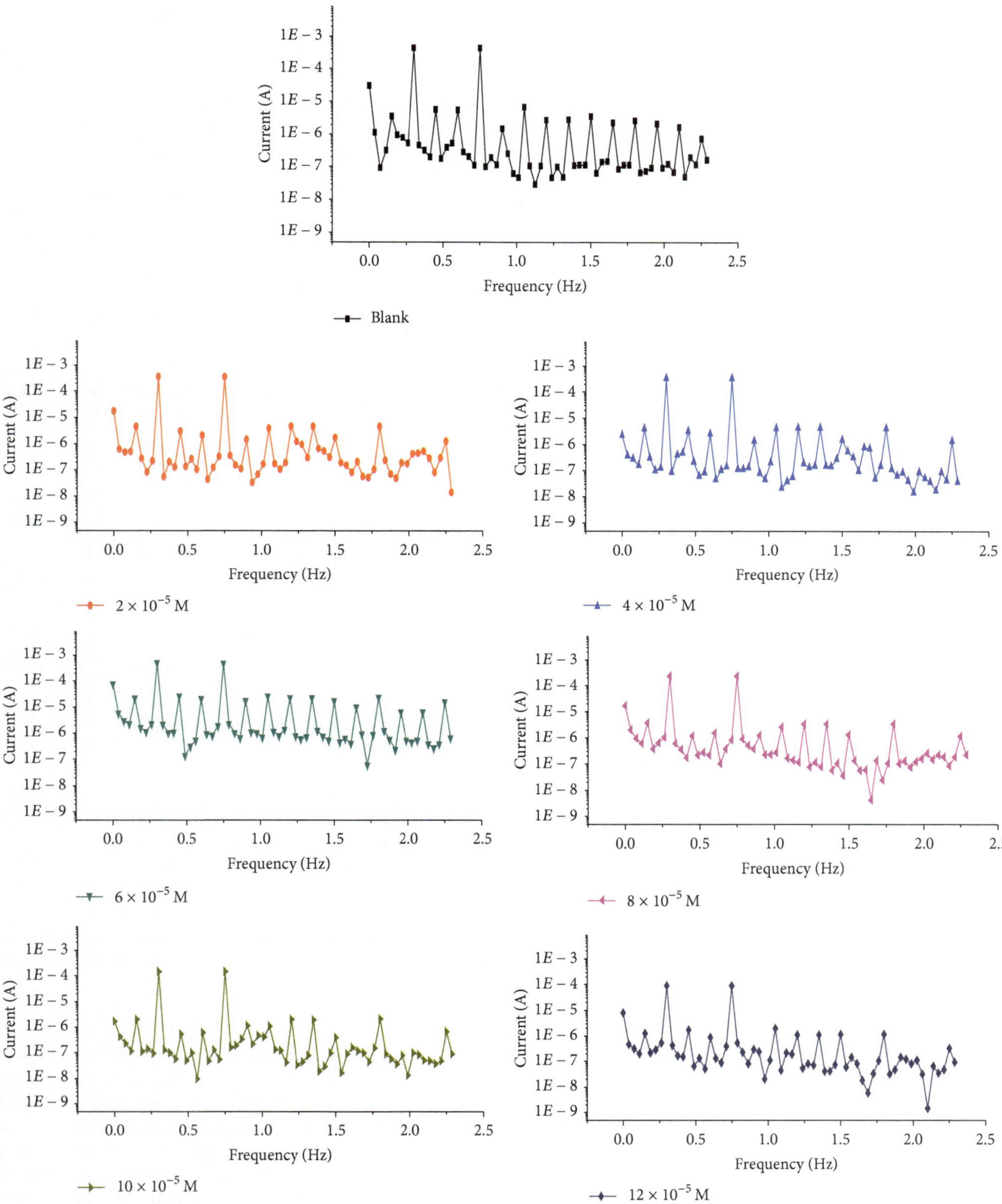

FIGURE 8: EFM spectra for carbon steel in 1 M HCl in the absence and presence of different concentrations of compound (A).

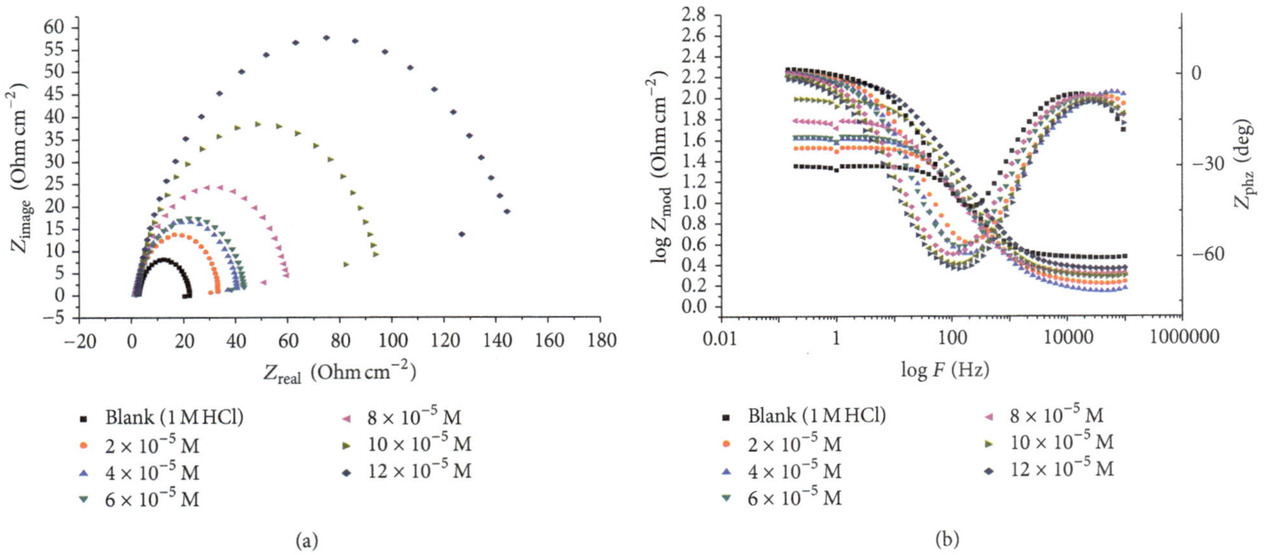

FIGURE 9: (a) The Nyquist plots for the corrosion of carbon steel in 1 M HCl in the absence and presence of different concentrations of inhibitor (A) at 25°C; (b) the Bode plots for the corrosion of carbon steel in 1 M HCl in the absence and presence of different concentrations of inhibitor (A) at 25°C.

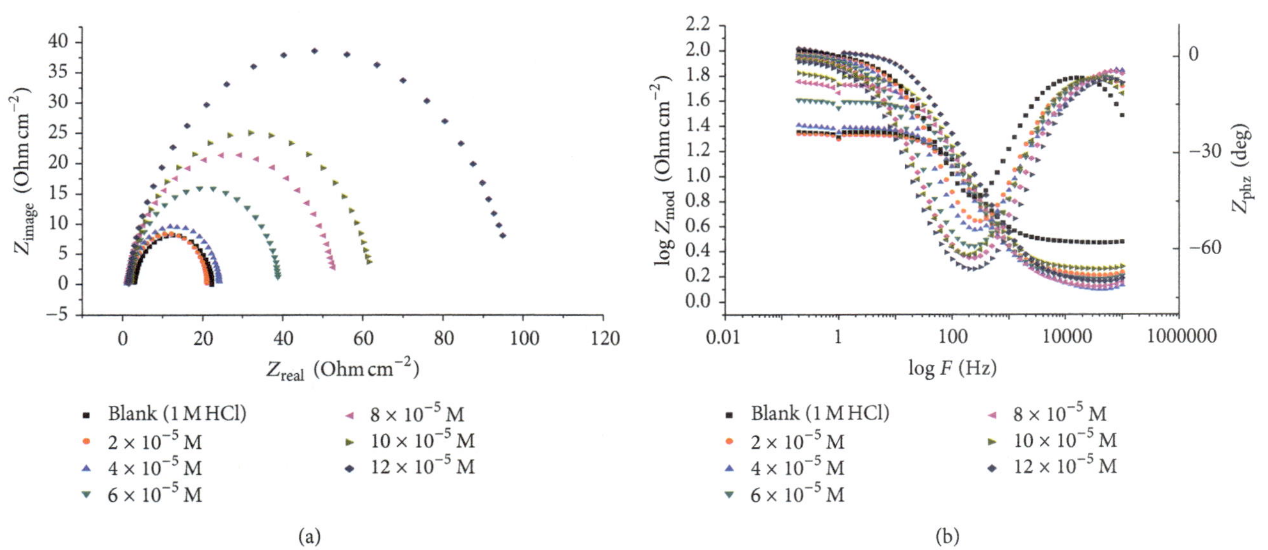

FIGURE 10: (a) The Nyquist plots for the corrosion of carbon steel in 1 M HCl in the absence and presence of different concentrations of inhibitor (B) at 25°C; (b) the Bode plots for the corrosion of carbon steel in 1 M HCl in the absence and presence of different concentrations of inhibitor (B) at 25°C.

Figure 8 shows the EFM intermodulation spectra (current versus frequency) of carbon steel in HCl solution containing different concentrations of compound (A). Similar curves were obtained for compounds (B and C) (not shown). The harmonic and intermodulation peaks are clearly visible and are much larger than the background noise. The two large peaks, with amplitude of about 200 μA, are the response to the 40 and 100 mHz (2 and 5 Hz) excitation frequencies. It is important to note that between the peaks there is nearly no current response (<100 nA). The experimental EFM data were treated using two different models: complete diffusion control of the cathodic reaction and the "activation" model. For the latter, a set of three nonlinear equations had been solved, assuming that the corrosion potential does not change due to the polarization of the working electrode [38]. The larger peaks were used to calculate the corrosion current density (i_{corr}), the Tafel slopes (β_c and β_a), and the causality factors (CF-2 and CF-3). These electrochemical parameters were listed in Table 8. The data presented in Table 8 obviously show that the addition of any one of tested compounds at a given concentration to the acidic solution decreases the corrosion current density, indicating that these compounds inhibit the corrosion of carbon steel in 1 M HCl through adsorption. The causality factors obtained under different

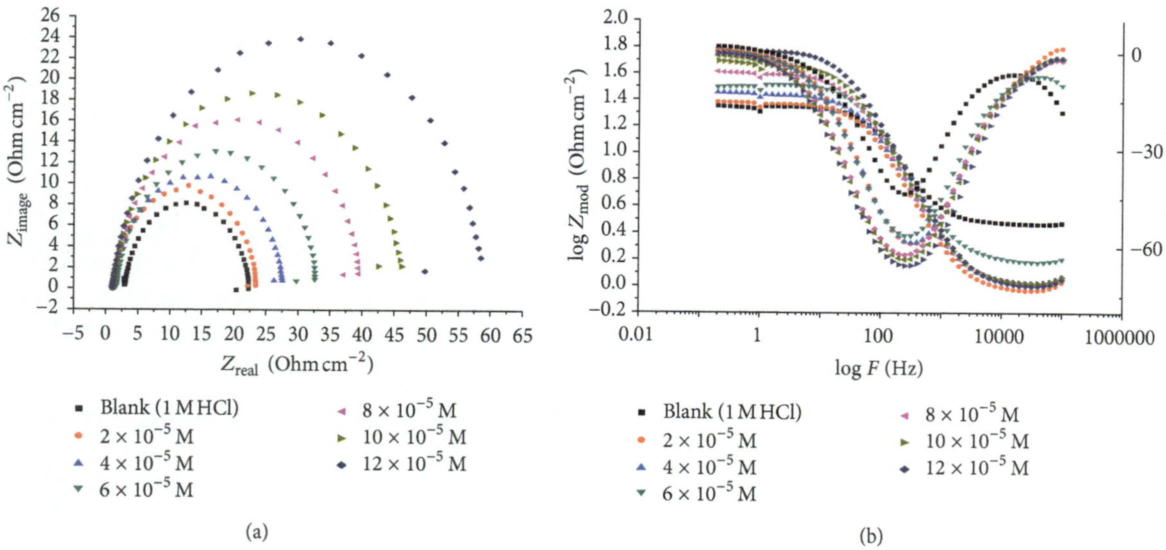

FIGURE 11: (a) The Nyquist plots for the corrosion of carbon steel in 1 M HCl in the absence and presence of different concentrations of inhibitor (C) at 25°C; (b) the Bode plots for the corrosion of carbon steel in 1 M HCl in the absence and presence of different concentrations of inhibitor (C) at 25°C.

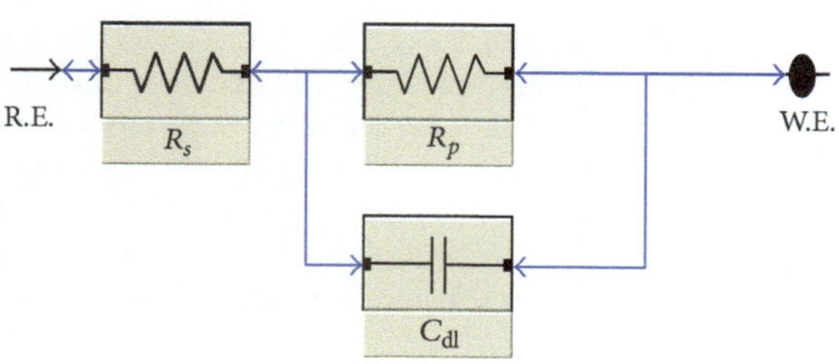

FIGURE 12: Equivalent circuit model used to fit experimental EIS.

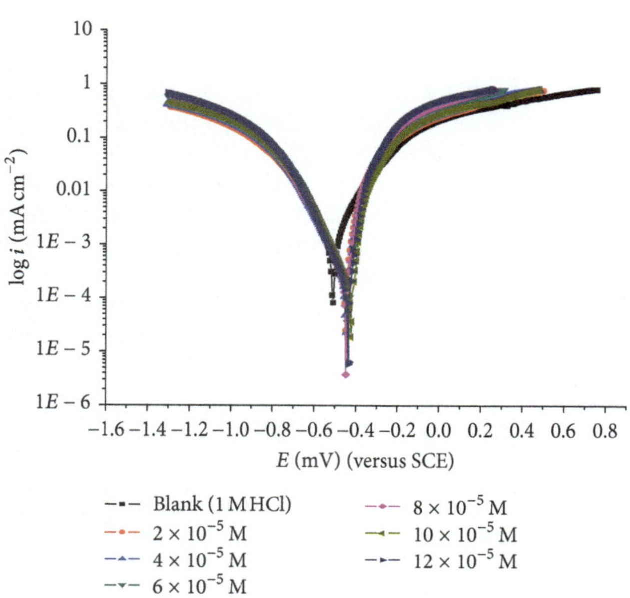

FIGURE 13: Potentiodynamic polarization curves for the dissolution of carbon steel in 1 M HCl in the absence and presence of different concentrations of compound (A) at 25°C.

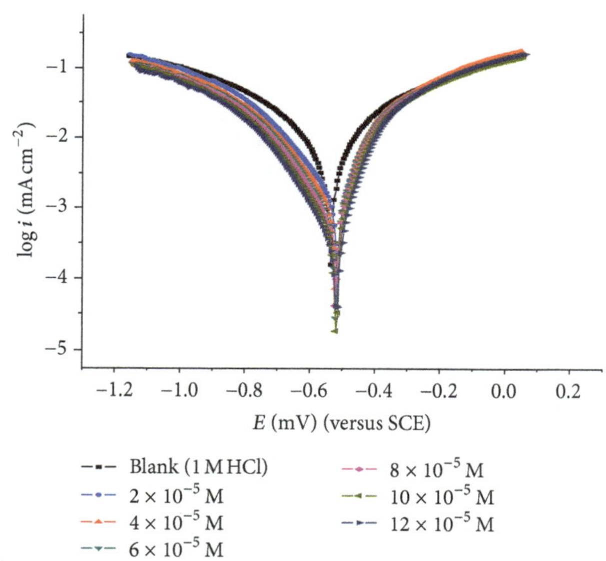

FIGURE 14: Potentiodynamic polarization curves for the dissolution of carbon steel in 1 M HCl in the absence and presence of different concentrations of compound (B) at 25°C.

experimental conditions are approximately equal to the theoretical values (2 and 3) indicating that the measured data are verified and of good quality. The inhibition efficiencies $\%IE_{EFM}$ increase by increasing the inhibitor concentrations and were calculated as follows from (7):

$$\%IE_{EFM} = \left[1 - \left(\frac{i_{corr}}{i^{o}_{corr}}\right)\right] \times 100, \qquad (7)$$

where i^{o}_{corr} and i_{corr} are corrosion current densities in the absence and presence of inhibitor, respectively. The inhibition

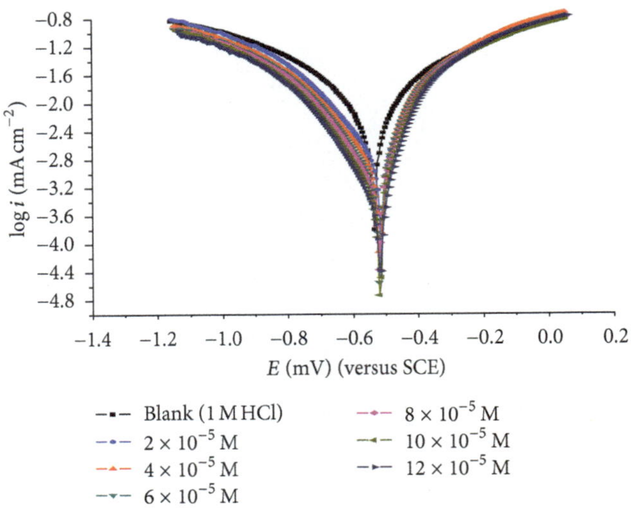

FIGURE 15: Potentiodynamic polarization curves for the dissolution of carbon steel in 1 M HCl in the absence and presence of different concentrations of compound (C) at 25°C.

sufficiency obtained from this method is in the the following order:

$$(A) > (B) > (C). \quad (8)$$

3.2.2. Electrochemical Impedance Spectroscopy (EIS). Impedance is a measure of the ability of a circle to resist the flow of electrical current. Electrochemical impedance is usually measured by applying an AC potential to an electrochemical cell and measuring the current through the cell. Surface properties, electrode kinetics, and mechanistic information can be obtained from impedance diagrams [39–43]. Figure 9 shows the Nyquist (a) and Bode (b) plots obtained at open-circuit potential both in the absence and presence of increasing concentrations of investigated compounds at 25°C. The increase in the size of the capacitive loop with the addition of investigated compounds shows that a barrier gradually forms on the carbon steel surface. The increase in the capacitive loop size (Figure 9(a)) is enhanced, at a fixed inhibitor concentration, following the order: (A) > (B) > (C), confirming the highest inhibitive influence of compound (A). Bode plots (Figure 9(b)) show that the total impedance increases with increasing inhibitor concentration ($\log Z$ versus $\log f$). But $\log f$ versus phase also Bode plot shows the continuous increase in the phase angle shift, obviously correlating with the increase of inhibitor adsorbed on carbon steel surface. The Nyquist plots do not yield perfect semicircles as expected from the theory of EIS. The deviation from ideal semicircle was generally attributed to the frequency dispersion [44] as well as to the inhomogeneities of the surface.

EIS spectra of the investigated compounds were analyzed using the equivalent circuit, Figure 12, which represents a single charge transfer reaction and fits well with our experimental results. The constant phase element, CPE, is introduced in the circuit instead of a pure double layer capacitor to give a more accurate fit [45]. The double layer capacitance, C_{dL}, for a circuit including a CPE parameter (Y_0 and n) was calculated from [46]

$$C_{dL} = \frac{Y_0 \omega^{n-1}}{\sin[n(\pi/2)]}, \quad (9)$$

where Y_0 is the magnitude of the CPE, $\omega = 2\pi f_{max}$, f_{max} is the frequency at which the imaginary component of the impedance is maximal, and the factor n is an adjustable parameter that usually lies between 0.50 and 1.0. From Nyquist plots (Figures 9, 10, and 11), it was clear that the curves approximated by a single capacitive semicircles and showing that the corrosion process was mainly charged-transfer controlled [47, 48]. The general shape of the curves is very similar for all samples (in presence or absence of inhibitors at different immersion times) indicating that there is no change in the corrosion mechanism [49]. From the impedance data (Table 9), we concluded that the value of R_{ct} increases with increasing the concentration of the inhibitors and this indicates an increase in $\%IE_{EIS}$, which is in concord with the EFM results obtained. In fact the presence of inhibitors enhances the value of R_{ct} in acidic solution. Values of double layer capacitance are also brought down to the maximum extent in the presence of inhibitor, and the decrease in the values of CPE follows the order similar to that obtained for i_{corr} in this study. The decrease in CPE/C_{dL} results from a decrease in local dielectric constant and/or an increase in the thickness of the double layer, suggesting that organic derivatives inhibit the iron corrosion by adsorption at metal/acid [50, 51]. The inhibition efficiency was calculated from the charge transfer resistance data from (10) [52]:

$$\%IE_{EIS} = \left[1 - \left(\frac{R^{\circ}_{ct}}{R_{ct}}\right)\right] \times 100, \quad (10)$$

where R°_{ct} and R_{ct} are the charge-transfer resistance values without and with inhibitor, respectively.

3.3. Potentiodynamic Polarization Measurements. Figures 13, 14, and 15 show the potentiodynamic polarization curves of carbon steel dissolution in 1 M HCl in the absence and presence of different concentrations of inhibitor (A, B, and C) at 25°C. The numerical values of the variation of the corrosion current density ($i_{corr.}$), the corrosion potential ($E_{corr.}$), Tafel slopes (β_a and β_c), the degree of surface coverage (θ), and the inhibition efficiency were recorded in Table 9. The inhibition efficiency and degree of surface coverage were calculated using

$$\%IE_p = \theta \times 100 = \left[\frac{(i^{o}_{corr} - i_{corr})}{i^{o}_{corr}}\right] \times 100, \quad (11)$$

where i^{o}_{corr} and i_{corr} are the uninhibited and inhibited corrosion current densities, respectively. The results of Table 9 indicate the following.

(1) The cathodic and anodic curves obtained exhibit Tafel-type behavior. Addition of thiophene derivatives increased both the cathodic and anodic overvoltages and inhibited both the hydrogen evolution

TABLE 7: Activation parameters for carbon steel corrosion in the absence and presence of various concentrations of inhibitor (A) in 1 M HCl.

$C_{inh.} \times 10^{-5}$ (M)	Activation parameters		
	E_a^* (kJ mol^{-1})	ΔH^* (kJ mol^{-1})	ΔS^* (J mol^{-1}K^{-1})
Blank	58.50	24274.22	5574.8599
2	39.39	15975.93	1.5162
4	34.38	13798.83	0.1738
6	34.30	13764.91	0.1532
8	27.54	10830.32	0.0086
10	27.20	10682.49	0.0068
12	25.41	8712.32	0.0048

and the anodic dissolution processes that are mixed-type inhibitors.

(2) The corrosion current density ($i_{corr.}$) decreases with increasing the concentration of the thiophene derivatives which indicates that these compounds act as inhibitors, and the degree of inhibition depends on the concentration and type of inhibitors present.

(3) The slopes of anodic and cathodic Tafel lines (β_a and β_c) were slightly changed by increasing the concentration of the tested compounds. This indicates that these inhibitors act as mixed-type inhibitors. Tafel lines are parallel, which indicates that there is no change of the mechanism of inhibition in the presence and absence of inhibitors.

(4) The orders of inhibition efficiency of all inhibitors at different concentrations as given by polarization measurements are listed in Table 9. The results are in good agreement with those obtained from weight loss measurements.

4. Mechanism

The inhibition mechanism involves the adsorption of the inhibitor on the metal surface immersed in aqueous HCl solution. Four types of adsorption [53] may take place involving organic molecules at the metal-solution interface:

(1) electrostatic attraction between the charged molecules and the charged metal;

(2) interaction of unshared electron pairs in the molecule with the metal;

(3) interaction of π-electrons with the metal;

(4) combination of all the above.

The fact that the adsorption of inhibitors on C-steel surface proceeded via chemical adsorption indicates that the mechanism is consistent with the transferred electron from the inhibitor's molecule to the empty orbital of Fe in C-steel. In acid solution, the protonation of the inhibitors may occur easily, so it is difficult for the protonated compound to approach the positively charged carbon steel surface (H_3O^+/metal interface) due to the electrostatic repulsion. Since chloride ions have a smaller degree of hydration; thus, they could bring excess negative charges in the vicinity of the interface and favour more adsorption of the positively charged inhibitor molecules and the protonated compounds adsorbed through electrostatic interactions between the positively charged molecules and the negatively charged metal surface. We hereby propose that the mechanism involves the protonation of the nitrogen to form NH_3^+ by HCl solution. Therefore, the cationic form of the inhibitors immersed in the acidic medium can compete with hydrogen proton (H^+) for the electrons on the metallic surface. Since the size of the inhibitor's cation is much larger than that of hydrogen molecule (due to the presence of the heterocyclic structures in the inhibitors), after the release of H_2, the inhibitor returns to its neutral form with the electronegative group having a free electron pair that can facilitate its adsorption on C-steel surface. The adsorbed inhibitor can form complex with iron surface and then protect the metal against further corrosion attack by blocking its active sites by functional groups as C–S, C–O, N–H, O–H, and C=O and heterocyclic rings in their molecular structures. These functional groups have been reported in other literature as effective corrosion inhibitors [54, 55].

The results show that compound (A) exhibits better performance because it has 7 centers of adsorption (5N, 1S, and 1O), so that it cover a wide area of the metal surface and protect it, also the presence of $-CH_3$ group which acts as donating group increases the electron density on the active centers, that leads to increase coverage surface, thereby giving higher inhibition efficiency.

Compound (B) has also 7 centers of adsorption (4N, 1S, and 2O from ester group). Due to the existence of ester group which in the presence of acidic media converts to carboxylic group, which attack C-steel surface but with increase temperature the reversible reaction takes place [19] and the formed organic acid convert to ester again.

Table 8: Electrochemical kinetic parameters obtained from EFM technique for carbon steel in 1 M HCl in the absence and presence of different concentrations of compounds.

Compound	Concentration × 10^{-5} (M)	i_{corr} (μA cm^{-2})	β_c (mV dec^{-1})	β_a (mV dec^{-1})	$k_{corr.}$ (mpy)	CF-2	CF-3	θ	%IE
Blank	0	913.3	149	132	420.8	1.90	3.34	—	—
A	2	535.4	103	97	246.7	1.86	2.88	0.414	41.4
	4	522.7	106	98	240.8	1.85	2.20	0.428	42.8
	6	408.0	59	50	188.0	2.01	2.19	0.553	55.3
	8	338.7	97	92	156.0	1.37	2.37	0.629	62.9
	10	222.7	100	97	102.6	1.88	2.60	0.756	75.6
	12	138.9	110	97	64.0	1.88	2.42	0.848	84.8
B	2	529.9	92	106	242.1	1.88	3.08	0.420	42.0
	4	375.1	103	92	171.4	2.25	1.73	0.589	58.9
	6	322.8	43	37	147.5	1.95	2.29	0.647	64.7
	8	314.9	101	93	143.9	2.10	2.88	0.655	65.5
	10	302.0	39	35	138.0	1.94	2.87	0.669	66.9
	12	204.9	102	93	93.6	1.98	2.45	0.776	77.6
C	2	503.1	101	90	229.9	1.86	2.88	0.449	44.9
	4	433.5	102	94	198.1	1.85	2.20	0.525	52.5
	6	341.1	98	88	155.9	2.01	2.19	0.627	62.7
	8	337.1	101	95	154.0	1.37	2.37	0.631	63.1
	10	307.5	54.0	78	140.5	1.88	2.60	0.663	66.3
	12	288.8	102.6	93	131.9	1.88	2.42	0.684	68.4

Table 9: Corrosion potential (E_{corr}), corrosion current density (i_{corr}), Tafel slopes (β_c, β_a), degree of surface coverage (θ), and inhibition efficiency (%IE_p) of carbon steel in 1 M HCl at 25°C for compounds A, B, and C.

Inhibitor	Concentration × 10^{-5} (M)	$-E_{corr.}$ (mV)	$i_{corr.}$ (μA cm^{-2})	β_c (mV dec^{-1})	β_a (mV dec^{-1})	$k_{corr.}$ (mpy)	θ	%IE_p
1 M HCl	Blank	511	2170	206.7	205.3	992.3	—	—
A	2	449	695	199	111	317.6	0.680	68.0
	4	447	543	195	118	248.1	0.750	75.0
	6	438	458	183	99	209.1	0.789	78.9
	8	447	409	168	85	186.9	0.812	81.2
	10	422	278	157	79	127.0	0.872	87.2
	12	436	181	145	79	82.8	0.917	91.7
B	2	507	1900	257	182	866.8	0.124	12.4
	4	504	1130	188	136	516.3	0.479	47.9
	6	497	800	189	154	365.6	0.631	63.1
	8	486	453	188	133	207.1	0.791	79.1
	10	482	310	140	105	141.5	0.857	85.7
	12	472	205	162	108	93.9	0.906	90.6
C	2	522	2050	211	188	936.6	0.055	5.5
	4	526	1780	165	122	815.3	0.180	18.0
	6	516	1200	167	125	549.2	0.447	44.7
	8	474	455	171	119	196.9	0.790	79.0
	10	491	387	130	85	176.8	0.822	82.2
	12	478	310	140	97	141.7	0.857	85.7

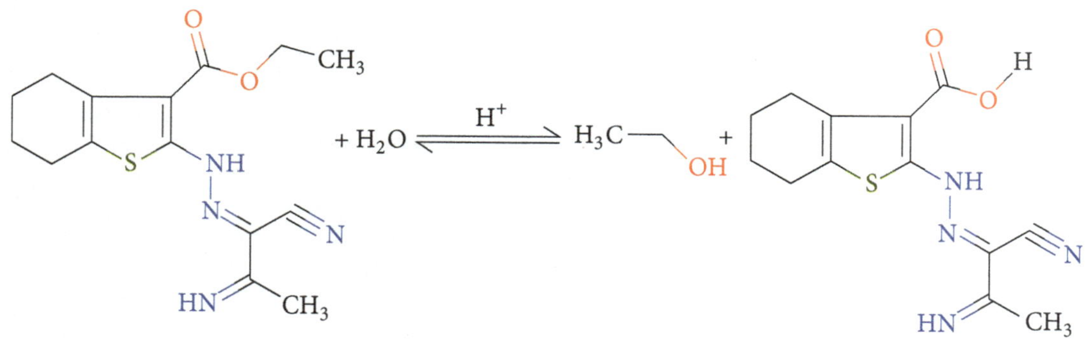

(12)

Compound (C) is less active than (A) and (B) because of its small molecular surface area (294.83), and t has 6 adsorption centers (5N and 1S). The presence of cyanide group (–CN) which acts as electron withdrawing group decreases the electron density on the active centers, which led to enhance the surface coverage, thereby giving lower inhibition efficiency.

5. Conclusions

From the results of the study the following may be concluded.

(I) All the investigated compounds are good corrosion inhibitors for carbon steel in 1 M HCl solution. The effectiveness of these inhibitors depends on their structures. The variation in inhibitive efficiency depends on the type and the nature of the substituent present in the inhibitor molecule.

(II) Reasonably good agreement was observed between the values obtained by the weight loss and electrochemical measurements were in good agreement. The order of %IE of these investigated compounds is as follows: (A) > (B) > (C).

(III) Results obtained from potentiodynamic polarization indicated that the investigated derivatives are mixed-type inhibitors.

(IV) The thermodynamic parameters revealed that the inhibition of corrosion by investigated compounds is due to the formation of a chemisorbed film on the metal surface.

(V) The adsorption of all investigated compounds onto steel surface follows the Langmuir adsorption isotherm model.

Conflict of Interests

The authors declare that there is no conflict of interests regarding the publication of this paper.

Acknowledgments

The authors are thankful to the Head of Chemistry Department, Mansoura University, and the Head of East Delta Electricity Production Company, Damietta, for their encouragement and necessary laboratory facilities.

References

[1] M. Elayyachy, M. Elkodadi, A. Aouniti et al., "New bipyrazole derivatives as corrosion inhibitors for steel in hydrochloric acid solutions," *Materials Chemistry and Physics*, vol. 93, no. 2-3, pp. 281–285, 2005.

[2] A. Popova, M. Christov, S. Raicheva, and E. Sokolova, "Adsorption and inhibitive properties of benzimidazole derivatives in acid mild steel corrosion," *Corrosion Science*, vol. 46, no. 6, pp. 1333–1350, 2004.

[3] B. Mernari, H. Elatarri, M. Traisnel, F. Bentiss, and M. Lagrenee, "Inhibiting effects of 3,5-bis(n-pyridyl)-4-amino-1,2,4-triazoles on the corrosion for mild steel in 1 M HCl medium," *Corrosion Science*, vol. 40, no. 2-3, pp. 391–399, 1998.

[4] X. Li, S. Deng, H. Fu, T. Li, and G. Mu, "Inhibition of the corrosion of cold rolled steel in hydrochloric acid solution by Tween-40," *Anti-Corrosion Methods and Materials* , vol. 56, no. 4, pp. 232–238, 2009.

[5] Y. Sürme and A. A. Gürten, "Role of polyethylene glycol tert-octylphenyl ether on corrosion behaviour of mild steel in acidic solution," *Corrosion Engineering, Science and Technology*, vol. 44, no. 4, pp. 304–3011, 2009.

[6] G. Y. Elewady, "Pyrimidine derivatives as corrosion inhibitors for carbon-steel in 2M hydrochloric acid solution," *International Journal of Electrochemical Science*, vol. 3, pp. 1149–1161, 2008.

[7] I. Zaafarany and M. Abdallah, "Ethoxylated fatty amide as corrosion inhibitors for carbon steel in hydrochloric acid solution," *International Journal of Electrochemical Science*, vol. 5, no. 1, pp. 18–28, 2010.

[8] A. Yurt, A. Balaban, S. U. Kandemir, and G. Bereket, "Investigation on some Schiff bases as HCl corrosion inhibitors for carbon steel," *Materials Chemistry and Physics*, vol. 85, no. 2-3, pp. 420–426, 2004.

[9] S. A. Ali, H. A. Al-Muallema, S. U. Rahman, and M. T. Saeed, "Bis-isoxazolidines: a new class of corrosion inhibitors of mild steel in acidic media," *Corrosion Science*, vol. 50, no. 11, pp. 3070–3077, 2008.

[10] H. Ju, Z. P. Kai, and Y. Li, "Aminic nitrogen-bearing polydentate Schiff base compounds as corrosion inhibitors for iron in acidic media: a quantum chemical calculation," *Corrosion Science*, vol. 50, no. 3, pp. 865–871, 2008.

[11] D. I. Gopi, K. Govindaraju, and L. Kavhith, "Investigation of triazole derived Schiff bases as corrosion inhibitors for mild steel in hydrochloric acid medium," *Journal of Applied Electrochemistry* , vol. 40, no. 7, pp. 1349–1356, 2010.

[12] M. A. Migahed, A. M. Abdul-Raheim, A. M. Atta, and W. Brostow, "Synthesis and evaluation of a new water soluble corrosion inhibitor from recycled poly(ethylene terphethalate)," *Materials Chemistry and Physics*, vol. 121, no. 1-2, pp. 208–214, 2010.

[13] A. Singh and M. Quraish, "Inhibiting effects of 5-substituted isatin-based Mannich bases on the corrosion of mild steel in hydrochloric acid solution," *Journal of Applied Electrochemistry*, vol. 40, no. 7, pp. 1293–1306, 2010.

[14] I. B. Obot, N. O. Obi-Egbedi, and N. W. Odozi, "Acenaphtho [1,2-b] quinoxaline as a novel corrosion inhibitor for mild steel in 0.5 M H_2SO_4," *Corrosion Science*, vol. 52, no. 3, pp. 923–926, 2010.

[15] I. B. Obot and N. O. Obi-Egbedi, "Indeno-1-one [2,3-b]quinoxaline as an effective inhibitor for the corrosion of mild steel in 0.5 M H_2SO_4 solution," *Materials Chemistry and Physics*, vol. 122, no. 2-3, pp. 325–328, 2010.

[16] H. Ashassi-Sorkhabi and M. Es'haghi, "Corrosion inhibition of mild steel in hydrochloric acid by betanin as a green inhibitor," *Journal of Solid State Electrochemistry*, vol. 13, no. 8, pp. 1297–1301, 2009.

[17] F. M. Mahgoub, B. A. Abdel-Nabey, and Y. A. El-Samadisy, "Adopting a multipurpose inhibitor to control corrosion of ferrous alloys in cooling water systems," *Materials Chemistry and Physics*, vol. 120, no. 1, pp. 104–108, 2010.

[18] B. I. Zerg, M. Sfair, M. Tale et al., "Effect of some tripodal bipyrazolic compounds on C38 steel corrosion in hydrochloric acid solution," *Journal of Applied Electrochemistry*, vol. 40, no. 9, pp. 1575–1582, 2010.

[19] H. F. Eldien, *Synthesis of some new thiophenes with expected biological activity [M.S. thesis]*, Mansoura University, Mansoura, Egypt, 2012.

[20] E. E. Oguzie, "Corrosion inhibition of mild steel in hydrochloric acid solution by methylene blue dye," *Materials Letters*, vol. 59, no. 8-9, pp. 1076–1079, 2005.

[21] K. F. Khaled, "Application of electrochemical frequency modulation for monitoring corrosion and corrosion inhibition of iron by some indole derivatives in molar hydrochloric acid," *Materials Chemistry and Physics*, vol. 112, no. 1, pp. 290–300, 2008.

[22] K. F. Khaled, "Evaluation of electrochemical frequency modulation as a new technique for monitoring corrosion and corrosion inhibition of carbon steel in perchloric acid using hydrazine carbodithioic acid derivatives," *Journal of Applied Electrochemistry*, vol. 39, no. 3, pp. 429–438, 2009.

[23] R. W. Bosch, J. Hubrecht, W. F. Bogaerts, and B. C. Syrett, "Electrochemical frequency modulation: a new electrochemical technique for online corrosion monitoring," *Corrosion*, vol. 57, no. 1, pp. 60–70, 2001.

[24] S. S. Abdel-Rehim, K. F. Khaled, and N. S. Abd-Elshafi, "Electrochemical frequency modulation as a new technique for monitoring corrosion inhibition of iron in acid media by new thiourea derivative," *Electrochimica Acta*, vol. 51, no. 16, pp. 3269–3277, 2006.

[25] A. Yurt, G. Bereket, A. Kivrak, A. Balaban, and B. Erk, "Effect of Schiff bases containing pyridyl group as corrosion inhibitors for low carbon steel in 0.1 M HCl," *Journal of Applied Electrochemistry*, vol. 35, no. 10, pp. 1025–1032, 2005.

[26] F. Bentiss, M. Traisnel, and M. Lagrenee, "The substituted 1,3,4-oxadiazoles: a new class of corrosion inhibitors of mild steel in acidic media," *Corrosion Science*, vol. 42, no. 1, pp. 127–146, 2000.

[27] W. Durnie, R. de Marco, A. Jefferson, and B. Kinsella, "Development of a structure-activity relationship for oil field corrosion inhibitors," *Journal of the Electrochemical Society*, vol. 146, no. 5, pp. 1751–1756, 1999.

[28] G. Banerjee and S. N. Malhotra, "Contribution to adsorption of aromatic amines on mild steel surface from HCl solutions by impedance, UV, and Raman spectroscopy," *Corrosion*, vol. 48, no. 1, pp. 10–15, 1992.

[29] E. S. Ivanov, *Inhibitors for Metal Corrosion in Acid Media*, Metallurgy, Moscow, Russia, 1986.

[30] F. Bentiss, M. Lebrini, and M. Lagrenee, "Thermodynamic characterization of metal dissolution and inhibitor adsorption processes in mild steel/2,5-bis(n-thienyl)-1,3,4-thiadiazoles/hydrochloric acid system," *Corrosion Science*, vol. 47, no. 12, pp. 2915–2931, 2005.

[31] M. Lebrini, F. Bentiss, H. Vezin, and M. Lagrenee, "The inhibition of mild steel corrosion in acidic solutions by 2,5-bis(4-pyridyl)-1,3,4-thiadiazole: structure-activity correlation," *Corrosion Science*, vol. 48, no. 5, pp. 1279–1291, 2006.

[32] M. Bouklah, B. Hammouti, M. Lagrenee, and F. Bentiss, "Thermodynamic properties of 2,5-bis(4-methoxyphenyl)-1,3,4-oxadiazole as a corrosion inhibitor for mild steel in normal sulfuric acid medium," *Corrosion Science*, vol. 48, no. 9, pp. 2831–2842, 2006.

[33] T. P. Hour and R. D. Holliday, "The inhibition by quinolines and thioureas of the acid dissolution of mild steel," *Journal of Applied Chemistry*, vol. 3, no. 11, pp. 502–513, 1953.

[34] L. O. Riggs Jr. and T. J. Hurd, "Temperature coefficient of corrosion inhibition," *Corrosion*, vol. 23, no. 8, pp. 252–260, 1967.

[35] G. M. Schmid and H. J. Huang, "Spectro-electrochemical studies of the inhibition effect of 4, 7-diphenyl -1, 10-phenanthroline on the corrosion of 304 stainless steel," *Corrosion Science*, vol. 20, no. 8-9, pp. 1041–1057, 1980.

[36] J. Marsh, *Advanced Organic Chemistry*, Wieley Eastern, New Delhi, India, 3rd edition, 1988.

[37] E. Kus and F. Mansfeld, "An evaluation of the electrochemical frequency modulation (EFM) technique," *Corrosion Science*, vol. 48, no. 4, pp. 965–979, 2006.

[38] G. A. Caigman, S. K. Metcalf, and E. M. Holt, "Thiophene substituted dihydropyridines," *Journal of Chemical Crystallography*, vol. 30, no. 6, pp. 415–422, 2000.

[39] D. C. Silverman and J. E. Carrico, "Electrochemical impedance technique—a practical tool for corrosion prediction," *Corrosion*, vol. 44, no. 5, pp. 280–287, 1988.

[40] W. J. Lorenz, "Determination of corrosion rates by electrochemical DC and AC methods," *Corrosion Science*, vol. 21, no. 9-10, pp. 647–672, 1981.

[41] D. D. Macdonald and M. C. Mckubre, "Impedance measurements in Electrochemical systems," in *Modern Aspects of Electrochemistry*, J. O. M. Bockris, B. E. Conway, and R. E. White, Eds., vol. 14, p. 61, Plenum Press, New York, NY, USA, 1982.

[42] F. Mansfeld, "Recording and analysis of AC impedance data for corrosion studies," *Corrosion*, vol. 37, no. 5, pp. 301–307, 1981.

[43] C. Gabrielli, Identification of Electrochemical processes by Frequency Response Analysis, Solarton Instrumentation Group, 1980.

[44] M. El Achouri, S. Kertit, H. M. Gouttaya et al., "Corrosion inhibition of iron in 1 M HCl by some gemini surfactants in the series of alkanediyl-α,ω-bis-(dimethyl tetradecyl ammonium bromide)," *Progress in Organic Coatings*, vol. 43, no. 4, pp. 267–273, 2001.

[45] J. R. Macdonald and W. B. Johanson, *Theory in Impedance Spectroscopy*, John Wiley & Sons, New York, NY, USA, 1987.

[46] S. F. Mertens, C. Xhoffer, B. C. de Cooman, and E. Temmerman, "Short-term deterioration of polymer-coated 55% Al-Zn. Part 1. Behavior of thin polymer films," *Corrosion*, vol. 53, no. 5, pp. 381–387, 1997.

[47] G. Trabanelli, C. Montecelli, V. Grassi, and A. Frignani, "Electrochemical study on inhibitors of rebar corrosion in carbonated concrete," *Cement and Concrete Research*, vol. 35, pp. 1804–1813, 2005.

[48] A. J. Trowsdate, B. Noble, S. J. Haris, I. S. R. Gibbins, G. E. Thomson, and G. C. Wood, "The influence of silicon carbide reinforcement on the pitting behaviour of aluminium," *Corrosion Science*, vol. 38, no. 2, pp. 177–191, 1996.

[49] F. M. Reis, H. G. de Melo, and I. Costa, "EIS investigation on Al 5052 alloy surface preparation for self-assembling monolayer," *Electrochimica Acta*, vol. 51, no. 8-9, pp. 1780–1788, 2006.

[50] M. Lagrenée, B. Mernari, M. Bouanis, M. Traisnel, and F. Bentiss, "Study of the mechanism and inhibiting efficiency of 3,5-bis(4-methylthiophenyl)-4H-1,2,4-triazole on mild steel corrosion in acidic media," *Corrosion Science*, vol. 44, no. 3, pp. 573–588, 2002.

[51] E. McCafferty and N. Hackerman, "Double layer capacitance of iron and corrosion inhibition with polymethylene diamines," *Journal of The Electrochemical Society*, vol. 119, pp. 146–154, 1972.

[52] H. Ma, S. Chen, L. Niu, S. Zhao, S. Li, and D. Li, "Inhibition of copper corrosion by several Schiff bases in aerated halide solutions," *Journal of Applied Electrochemistry*, vol. 32, no. 1, pp. 65–72, 2002.

[53] D. P. Schweinsberg, G. A. George, A. K. Nanayakkara, and D. A. Steinert, "The protective action of epoxy resins and curing agents-inhibitive effects on the aqueous acid corrosion of iron and steel," *Corrosion Science*, vol. 28, no. 1, pp. 33–42, 1988.

[54] M. Shyamala and P. K. Kasthuri, "The inhibitory action of the extracts of *Adathoda vasica, Eclipta alba, and Centella asiatica* on the corrosion of mild steel in hydrochloric acid medium: a comparative study," *International Journal of Corrosion*, vol. 2012, Article ID 852827, 13 pages, 2012.

[55] I. E. Uwah, P. C. Okafor, and V. E. Ebiekpe, "Inhibitive action of ethanol extracts from Nauclea latifolia on the corrosion of mild steel in H_2SO_4 solutions and their adsorption characteristics," *Arabian Journal of Chemistry*, vol. 6, no. 3, pp. 285–293, 2013.

Permissions

The contributors of this book come from diverse backgrounds, making this book a truly international effort. This book will bring forth new frontiers with its revolutionizing research information and detailed analysis of the nascent developments around the world.

We would like to thank all the contributing authors for lending their expertise to make the book truly unique. They have played a crucial role in the development of this book. Without their invaluable contributions this book wouldn't have been possible. They have made vital efforts to compile up to date information on the varied aspects of this subject to make this book a valuable addition to the collection of many professionals and students.

This book was conceptualized with the vision of imparting up-to-date information and advanced data in this field. To ensure the same, a matchless editorial board was set up. Every individual on the board went through rigorous rounds of assessment to prove their worth. After which they invested a large part of their time researching and compiling the most relevant data for our readers. Conferences and sessions were held from time to time between the editorial board and the contributing authors to present the data in the most comprehensible form. The editorial team has worked tirelessly to provide valuable and valid information to help people across the globe.

Every chapter published in this book has been scrutinized by our experts. Their significance has been extensively debated. The topics covered herein carry significant findings which will fuel the growth of the discipline. They may even be implemented as practical applications or may be referred to as a beginning point for another development. Chapters in this book were first published by Hindawi Publishing Corporation; hereby published with permission under the Creative Commons Attribution License or equivalent.

The editorial board has been involved in producing this book since its inception. They have spent rigorous hours researching and exploring the diverse topics which have resulted in the successful publishing of this book. They have passed on their knowledge of decades through this book. To expedite this challenging task, the publisher supported the team at every step. A small team of assistant editors was also appointed to further simplify the editing procedure and attain best results for the readers.

Our editorial team has been hand-picked from every corner of the world. Their multi-ethnicity adds dynamic inputs to the discussions which result in innovative outcomes. These outcomes are then further discussed with the researchers and contributors who give their valuable feedback and opinion regarding the same. The feedback is then collaborated with the researches and they are edited in a comprehensive manner to aid the understanding of the subject.

Apart from the editorial board, the designing team has also invested a significant amount of their time in understanding the subject and creating the most relevant covers. They scrutinized every image to scout for the most suitable representation of the subject and create an appropriate cover for the book.

The publishing team has been involved in this book since its early stages. They were actively engaged in every process, be it collecting the data, connecting with the contributors or procuring relevant information. The team has been an ardent support to the editorial, designing and production team. Their endless efforts to recruit the best for this project, has resulted in the accomplishment of this book. They are a veteran in the field of academics and their pool of knowledge is as vast as their experience in printing. Their expertise and guidance has proved useful at every step. Their uncompromising quality standards have made this book an exceptional effort. Their encouragement from time to time has been an inspiration for everyone.

The publisher and the editorial board hope that this book will prove to be a valuable piece of knowledge for researchers, students, practitioners and scholars across the globe.

List of Contributors

S. Mousavi Nasab, M. Aboutalebi and S. H. Seyedein
Centre of Excellence for Advanced Materials & Processing (CEAMP), Department of Metallurgy and Materials Engineering, Iran University of Science & Technology, Tehran 16846-13114, Iran

A. Molavi Kakhki
Department of Materials Science & Engineering, Sharif University of Technology, Tehran 11365-11155, Iran

J. Vahdati Khaki
Department of Materials Science and Engineering, Ferdowsi University of Mashhad, Mashhad 91779-48974, Iran

Chih-Chun Hsieh and Weite Wu
Department of Materials Science and Engineering, National Chung Hsing University, 250 Kuo-Kuang Road, Taichung 402, Taiwan

Saeed Ghali, Mamdouh Eissa and Hoda El-Faramawy
Steel Technology Department, Central Metallurgical Research & Development Institute (CMRDI), Helwan 11421, Egypt

Z. He, T. Xia, W. Xiong, Q. Shen and B. Li
Key Laboratory for Ferrous Metallurgy and Resources Utilization of Ministry of Education, Wuhan University of Science and Technology, Wuhan 430081, China

M. Yadav, Sumit Kumar and Debasis Behera
Department of Applied Chemistry, Indian School of Mines, Dhanbad 826004, India

A. S. Bolokang and S. Bhero
Department of Engineering Metallurgy, University of Johannesburg, P.O. Box 17011, Doornfontein 2028, South Africa

M. J. Phasha
Council for Scientific and Industrial Research (CSIR), Materials Science and Manufacturing, Meiring Naude Road, Brummeria, P.O. Box 395, Pretoria 0001, South Africa

D. E. Motaung
DST/CSIR Nanotechnology Innovation Centre, National Centre for Nano-Structured Materials, Council for Scientific Industrial Research, P.O. Box 395, Pretoria 0001, South Africa

Yici Wang and Guoping Luo
College of Metallurgical and Ecological Engineering, University of Science and Technology Beijing, Beijing 100083, China
College of Material and Metallurgy, Inner Mongolia University of Science and Technology, Baotou 014010, China

Wenwu Yu, Yan Ban and Qi Jiang
College of Material and Metallurgy, Inner Mongolia University of Science and Technology, Baotou 014010, China

Kuiyu Cheng
Superalloy Division, Institute of Metal Research, Chinese Academy of Sciences, 72 Wenhua Road, Shenyang 110016, China
High Temperature Materials Group, KIMS, 531 Changwondaero, Changwon 641-010, Republic of Korea

ChangYong Jo
High Temperature Materials Group, KIMS, 531 Changwondaero, Changwon 641-010, Republic of Korea

Doo Hyun Kim
Materials Test and Characterization Group, KIMS, 531 Changwondaero, Changwon 641-010, Republic of Korea

Tao Jin and Zhuangqi Hu
Superalloy Division, Institute of Metal Research, Chinese Academy of Sciences, 72 Wenhua Road, Shenyang 110016, China

R. R. M. de Sousa and A. de S. Brandim
1Grupo de Pesquisa em Materiais e Metrologia, Departamento de Mec^anica, Instituto Federal IF-PI, 64000040 Teresina, PI, Brazil

F. O. de Araújo
Departamento de Ci^encias Exatas e Naturais, Universidade Federal Rural do Semi- ´ Arido, 59625900 Mossor´o, RN, Brazil

J. A. P. da Costa
Departamento de F´ısica, UERN, Campus Central, 59610210 Mossor´o, RN, Brazil

R. A. de Brito and C. Alves Jr.
LabPlasma, Departamento de F´ısica, UFRN, Campus Universit´ario, 59072970 Natal, RN, Brazil

Sergio Villanueva Bravo
Department of Materials Science and Engineering, Kyushu University, Fukuoka 819-0395, Japan
Autonomous San Luis Potosi University, S. L. P., Mexico

Kaoru Yamamoto
Kurume National College of Technology, Fukuoka 830-8555, Japan

Hirofumi Miyahara
Department of Materials Science and Engineering, Kyushu University, Fukuoka 819-0395, Japan

Keisaku Ogi
Oita National College of Technology, 1666 Maki, Oita 870-0152, Japan

P. H. Kien
Department of Physics, Thai Nguyen University of Education, Luong Ngoc Quyen, Thai Nguyen, Vietnam

H. V. Hue
Faculty of General Science, Ho Chi Minh City University of Food Industry, 140 Le Trong Tan, Ho Chi Minh, Vietnam

P. K. Hung
Department of Computational Physics, Ha Noi University of Science and Technology, 1 Dai Co Viet, Ha Noi, Vietnam

V. Rajagopal Reddy, B. Prasanna Lakshmi and R. Padma
Department of Physics, Sri Venkateswara University, Tirupati 517 502, India

Abdlmanam S. A. Elmaryami and Badrul Omar
Mechanical Engineering Department, Universiti Tun Hussein Onn Malaysia, Batu Pahat, Johor, 86400 Parit Raja, Malaysia

Samavedam Santhi
Department of Metallurgical and Materials Engineering, MGIT, Hyderabad 500075, India

S. B. Sakri
Special Fabrication Division, DRDL, Hyderabad 500058, India

Dharwada Hanumantha Rao
Mechanical Engineering Department, M.V.S.R. Engineering College, Hyderabad 501510, India

Srinivasan Sundarrajan
National Institute of Technology, Trichy 620015, India

T. Satish Kumar, S. Rajesh Kumar and T. L. Prakash
Refractory Metals Division, Centre for Materials for Electronics Technology (C-MET), IDA Phase-III, HCL (PO), Cherlapally, Hyderabad 500051, India

M. Lakshmipathi Rao
Department of Physics, Osmania University, Hyderabad 500007, India

A. Mustapha, E. A. Charles and D. Hardie
School of Chemical Engineering and Advanced Materials, Newcastle University, Newcastle upon Tyne NE1 7RU, UK

C. Wang
Environmental Corrosion Center, Institute of Metal Research, Chinese Academy of Sciences, Shenyang 110016, China
Shenyang National Laboratory for Materials Science, Institute of Metal Research, Chinese Academy of Sciences, Shenyang 110016, China

Y. B. Xu
Shenyang National Laboratory for Materials Science, Institute of Metal Research, Chinese Academy of Sciences, Shenyang 110016, China

E. H. Han
Environmental Corrosion Center, Institute of Metal Research, Chinese Academy of Sciences, Shenyang 110016, China

T. Poornima
Department of Science and Humanities, PESIT, Bangalore 560085, India

Jagannath Nayak
Department of Metallurgical and Materials Engineering, National Institute of Technology Karnataka, Surathkal, Srinivasnagar, Karnataka 75025, India

A. Nityananda Shetty
Department of Chemistry, National Institute of Technology Karnataka, Surathkal, Srinivasnagar, Karnataka 575025, India

Jenő Takács
Department of Engineering Science, University of Oxford, Parks Road, Oxford OX1 3PJ, UK

György Kovács
Department of Materials, Physics of Eotvos Lorand University, P.O. Box 32, Budapest 1518, Hungary

Lajos K. Varga
Research Institute for Solid State Physics and Optics, P.O. Box 49, Budapest 1525, Hungary

X. D. Zhang, S. Z. Hao, Y. Qin, X. N. Li and C. Dong
Key Laboratory of Materials Modification, Department of Materials Engineering, Dalian University of Technology, Dalian 116024, China

T. Grosdidier, B. Bolle and N. Allain-Bonasso
Laboratoire d'Etude des Microstructures et de M´ecanique des Mat´eriaux (LEM3), CNRS UMR 7239, Universit´e Paul Verlaine-Metz, Ile du Saulcy, 57045 Metz, France

J. X. Zou
Shanghai Engineering Research Center of Mg Materials and Applications, School of Materials Science and Engineering, Shanghai Jiao Tong University, Shanghai 200240, China

B. Gao
School of Materials Science and Engineering, Shenyang University, Shenyang 200240, China

Ajay Kumar
Department of Mechanical Engineering, Noida Institute of Engineering and Technology, Greater Noida 201306, India

Hari Singh
Department of Mechanical Engineering, National Institute of Technology, Kurukshetra, Haryana 136119, India

Sachin Maheshwari
Division of MPAE, Netaji Subhas Institute of Technology, New Delhi 110078, India

O. I. Sekunowo, G. I. Lawal and S. O. Adeosun
Department of Metallurgical and Materials Engineering, Faculty of Engineering, University of Lagos, Lagos 101017, Nigeria

N. T. Kumbhar
Materials Science Division, Bhabha Atomic Research Centre, Trombay, Mumbai 400 085, India

K. Bhanumurthy
Scientific Information Resource Division, Bhabha Atomic Research Centre, Trombay, Mumbai 400 085, India

Jong-Taek Yeom, Jeoung Han Kim, Jae-Keun Hong and Nho-Kwang Park
Special Alloys Group, Korea Institute of Materials Science, 531 Changwondaero, Changwon 641-831, Republic of Korea

Chong Soo Lee
Department of Materials Science and Engineering, POSTECH, San 31, Hyojadong, Namgu, Pohang 790-784, Republic of Korea

Bolarinwa Komolafe and Mamoun Medraj
Department of Mechanical Engineering, Concordia University, 1455 de Maisonneuve Boulevard West, Montreal, QC, Canada H3G 1M8

A. S. Fouda
Department of Chemistry, Faculty of Science, Mansoura University, El-Mansoura, Egypt

A. A. Attia
Department of Chemistry, Faculty of Science, Zagazig University, Zagazig, Egypt

A. A. Negm
East Delta Electricity Production Company, Damietta, Egypt